Statistical Matching

WILEY SERIES IN SURVEY METHODOLOGY
Established in part by WALTER A. SHEWHART AND SAMUEL S. WILKS

A complete list of the titles in this series appears at the end of this volume.

Statistical Matching

Theory and Practice

Marcello D'Orazio, Marco Di Zio and Mauro Scanu
ISTAT – Istituto Nazionale di Statistica, Rome, Italy

John Wiley & Sons, Ltd

Telephone (+44) 1243 779777

Email (for orders and customer service enquiries): cs-books@wiley.co.uk
Visit our Home Page on www.wiley.com

Other Wiley Editorial Offices

John Wiley & Sons Inc., 111 River Street, Hoboken, NJ 07030, USA

Jossey-Bass, 989 Market Street, San Francisco, CA 94103-1741, USA

Wiley-VCH Verlag GmbH, Boschstr. 12, D-69469 Weinheim, Germany

John Wiley & Sons Australia Ltd, 42 McDougall Street, Milton, Queensland 4064, Australia

John Wiley & Sons (Asia) Pte Ltd, 2 Clementi Loop #02-01, Jin Xing Distripark, Singapore 129809

John Wiley & Sons Canada Ltd, 22 Worcester Road, Etobicoke, Ontario, Canada M9W 1L1

Wiley also publishes its books in a variety of electronic formats. Some content that appears in print may not be available in electronic books.

Library of Congress Cataloging-in-Publication Data

D'Orazio, Marcello.
Statistical matching : theory and practice / Marcello D'Orazio, Marco Di Zio, and Mauro Scanu.
p. cm.
Includes bibliographical references and index.
ISBN-13: 978-0-470-02353-2 (acid-free paper)
ISBN-10: 0-470-02353-8 (acid-free paper)
1. Statistical matching. I. Di Zio, Marco. II. Scanu, Mauro. III. Title.
QA276.6.D67 2006
519.5′2–dc22

2006040184

British Library Cataloguing in Publication Data

A catalogue record for this book is available from the British Library

ISBN-13: 978-0-470-02353-2 (HB)
ISBN-10: 0-470-02353-8 (HB)

Typeset in 10/12pt Times by Laserwords Private Limited, Chennai, India
Printed and bound in Great Britain by TJ International, Padstow, Cornwall
This book is printed on acid-free paper responsibly manufactured from sustainable forestry in which at least two trees are planted for each one used for paper production.

Contents

Preface

Statistical matching is a relatively new area of research which has been receiving increasing attention in response to the flood of data which are now available. It has the practical objective of drawing information piecewise from different independent sample surveys.

The origins of statistical matching can be traced back to the mid-1960s, when a comprehensive data set with information on socio-demographic variables, income and tax returns by family was created by matching the 1966 Tax File and the 1967 Survey of Economic Opportunities; see Okner (1972). Interest in procedures for producing information from distinct sample surveys rose in the following years, although not without controversy. Is it possible to draw joint information on two variables never jointly observed but distinctly available in two independent sample surveys? Are standard statistical techniques able to solve this problem? As a matter of fact, there are two opposite aspects: the practical aspect that aims to produce a large amount of information rapidly and at low cost, and the theoretical aspect that needs to assess whether this production process is justifiable. This book is positioned at the boundary of these two aspects.

Chapters 1–4 are the methodological core of the book. Details of the mathematical-statistical framework of the statistical matching problem are given, together with examples. One of the objectives of this book is to give a complete, formalized treatment of the statistical matching procedures which have been defined or applied hitherto. More precisely, the data sets will always be samples generated by appropriate models or populations (archives and other nonstatistical sources will not be considered). When dealing with sample surveys, the different statistical matching approaches can be justified according to different paradigms. Most (but not all) of the book will rely on a likelihood based inference. The nonparametric case will also be addressed in some detail throughout the book. Other approaches, based on the Bayesian paradigm or on model assisted approaches for finite populations, will be also highlighted. By comparing and contrasting the various statistical matching procedures we hope to produce a synthesis that justifies their use.

Chapters 5–7 are more related to the practical aspects of statistically matching two files. An experience of the construction of a social accounting matrix (Coli *et al.*, 2005) is described in detail, in order to illustrate the peculiarities of the different phases of statistical matching, and the effect of the use of statistical matching techniques without a preliminary analysis of all the aspects.

Finally, sophisticated methods for statistical matching inevitably require the use of computers. The Appendix details some algorithms written in the R language. (the codes are also available on the following webpage: http://www.wiley.com/go/matching).

This book is intended for researchers in the national statistical institutes, and for applied statisticians who face (perhaps for the first time) the problem of statistical matching and could benefit from a structured summary of results in the relevant literature. Readers should possess a background that includes maximum likelihood methods as well as basic concepts in regression analysis and the analysis of contingency tables (some reminders are given in the Appendix). At the same time, we hope the book will also be of interest to methodological researchers. There are many aspects of statistical matching still in need of further exploration.

We are indebted to all those who encouraged us to work on this problem. We particularly thank Pier Luigi Conti, Francesca Tartamella and Barbara Vantaggi for their helpful suggestions and for careful reading of some parts of this book.

The views expressed in this book are those of the authors and do not necessarily reflect the policy of ISTAT.

Marcello, Marco, Mauro
Roma

1

The Statistical Matching Problem

1.1 Introduction

Nowadays, decision making requires as much rich and timely information as possible. This can be obtained by carrying out appropriate surveys. However, there are constraints that make this approach difficult or inappropriate.

(i) It takes an appreciable amount of time to plan and execute a new survey. Timeliness, one of the most important requirements for statistical information, risks being compromised.

(ii) A new survey demands funds. The total cost of a survey is an inevitable constraint.

(iii) The need for information may require the analysis of a large number of variables. In other words, the survey should be characterized by a very long questionnaire. It is well established that the longer the questionnaire, the lower the quality of the responses and the higher the frequency of missing responses.

(iv) Additional surveys increase the response burden, affecting data quality, especially in terms of total nonresponse.

A practical solution is to exploit as much as possible all the information already available in different data sources, i.e. to carryout a statistical integration of information already collected. This book deals with one of these data integration procedures: *statistical matching*. Statistical matching (also called data fusion

Statistical Matching: Theory and Practice M. D'Orazio, M. Di Zio and M. Scanu

or synthetical matching) aims to integrate two (or more) data sets characterized by the fact that:

(a) the different data sets contain information on (i) a set of common variables and (ii) variables that are not jointly observed;

(b) the units observed in the data sets are different (disjoint sets of units).

Remark 1.1 Sometimes there is terminological confusion about different procedures that aim to integrate two or more data sources. For instance, Paass (1985) uses the term 'record linkage' to describe the state of the art of statistical matching procedures. Nowadays record linkage refers to an integration procedure that is substantially different from the statistical matching problem in terms of both (a) and (b). First of all, the sets of units of the two (or more) files are at least partially overlapping, contradicting requirement (b). Secondly, the common variables can sometimes be misreported, or subject to change (statistical matching procedures have not hitherto dealt with the problem of the quality of the data collected). The lack of stability of the common variables makes it difficult to link those records in the files that refer to the same units. Hence, record linkage procedures are mostly based on appropriate discriminant analysis procedures in order to distinguish between those records that are actually a match and those that refer to distinct units; see Winkler (1995) and references therein.

A different set of procedures is also called statistical matching. This is characterized by the fact that the two files are completely overlapping, in the sense that each unit observed in one file is also observed in the other file, contradicting requirement (b). However, the common variables are unable to identify the units. These procedures are well established in the literature (see DeGroot *et al.*, 1971; DeGroot and Goel 1976; Goel and Ramalingam 1989) and will not be considered in the rest of this book.

A natural question arises: what is meant by integration? As a matter of fact, integration of two or more sources means the possibility of having joint information on the not jointly observed variables of the different sources. There are two apparently distinct ways to pursue this aim.

- Micro approach – The objective in this case is the construction of a *synthetic* file which is *complete*. The file is complete in the sense that all the variables of interest, although collected in different sources, are contained in it. It is synthetic because it is not a product of direct observation of a set of units in the population of interest, but is obtained by exploiting information in the source files in some appropriate way. We remark that the synthetic nature of data is useful in overcoming the problem of confidentiality in the public use of micro files.

- Macro approach – The source files are used in order to have a direct estimation of the joint distribution function (or of some of its key characteristics, such as the correlation) of the variables of interest which have not been observed in common.

Actually, statistical matching has mostly been analysed and applied following the micro approach. There are a number of reasons for this fact. Sometimes it is a necessary input of some procedures, such as the application of microsimulation models. In other cases, a synthetic complete data set is preferred simply because it is much easier to analyse than two or more incomplete data sets. Finally, joint information on variables never jointly observed in a unique data set may be of interest to multiple subjects (universities, research centres): the complete synthetic data set becomes the source which satisfies the information needs of these subjects.

On the other hand, when the need is just for a contingency table of variables not jointly observed or a set of correlation coefficients, the macro approach can be used more efficiently without resorting to synthetic files. It will be emphasized throughout this book that the two approaches are not distinct. The micro approach is always a byproduct of an estimation of the joint distribution of all the variables of interest. Sometimes this relation is explicitly stated, while in other cases it is implicitly assumed.

Before analysing statistical matching procedures in detail, it is necessary to define the notation and the statistical/mathematical framework for the statistical matching problem; see Sections 1.2 and 1.3. These details will open up a set of different issues that correspond to the different chapters and sections of this book. The outline of the book is given in Section 1.5.

1.2 The Statistical Framework

Throughout the book, we will analyse the problem of statistically matching two independent sample surveys, say A and B. We will also assume that these two samples consist of records independently generated from appropriate models. The case of samples drawn from finite populations will be treated separately in Chapter 5.

Let $(\mathbf{X}, \mathbf{Y}, \mathbf{Z})$ be a random variable with density $f(\mathbf{x}, \mathbf{y}, \mathbf{z})$, $\mathbf{x} \in \mathcal{X}$, $\mathbf{y} \in \mathcal{Y}$, $\mathbf{z} \in \mathcal{Z}$, and $\mathcal{F} = \{f\}$ be a suitable family of densities.[1] Without loss of generality, let $\mathbf{X} = (X_1, \ldots, X_P)'$, $\mathbf{Y} = \left(Y_1, \ldots, Y_Q\right)'$ and $\mathbf{Z} = (Z_1, \ldots, Z_R)'$ be vectors of random variables (r.v.s) of dimension P, Q and R, respectively. Assume that A and B are two samples consisting of n_A and n_B independent and identically distributed (i.i.d.) observations generated from $f(\mathbf{x}, \mathbf{y}, \mathbf{z})$. Furthermore, let the units in A have $\mathbf{Z}$ missing, and the units in B have $\mathbf{Y}$ missing. Let

$$\left(\mathbf{x}_a^A, \mathbf{y}_a^A\right) = \left(x_{a1}^A, \ldots, x_{aP}^A, y_{a1}^A, \ldots, y_{aQ}^A\right),$$

[1]We will use the term 'density' for both absolutely continuous and discrete variables, in the former case with respect to the Lebesgue measure, and in the latter case with respect to the counting measure. Hence, in the discrete case $f(\mathbf{x}, \mathbf{y}, \mathbf{z})$ should be interpreted as the probability that $\mathbf{X}$ assumes category $\mathbf{x}$, $\mathbf{Y}$ category $\mathbf{y}$ and $\mathbf{Z}$ category $\mathbf{z}$.

$a = 1, \ldots, n_A$, be the observed values of the units in sample A, and

$$\left(\mathbf{x}_b^B, \mathbf{z}_b^B\right) = \left(x_{b1}^B, \ldots, x_{bP}^B, z_{b1}^B, \ldots, z_{bR}^B\right),$$

$b = 1, \ldots, n_B$, be the observed values of the units in sample B (for the sake of simplicity, we will omit the superscripts A and B and identify the observed values in the two samples by the sample counters a and b, unless otherwise specified). When the objective is to gain information on the joint distribution of $(\mathbf{X}, \mathbf{Y}, \mathbf{Z})$ from the observed samples A and B, we are dealing with the statistical matching problem.

Table 1.1 shows typical statistical matching samples A and B. These samples can be considered as a unique sample $A \cup B$ of $n_A + n_B$ i.i.d. observations from $f(\mathbf{x}, \mathbf{y}, \mathbf{z})$ characterized by:

- the presence of missing data, and hence of a missing data generation mechanism;
- the absence of joint information on $\mathbf{X}$, $\mathbf{Y}$, and $\mathbf{Z}$.

The first point has been the focus of a very large statistical literature (see also Appendix A). The possible characterizations of the missing data generation mechanisms for the statistical matching problem are treated in Section 1.3. It will be seen that standard inferential procedures for partially observed samples are also appropriate for the statistical matching problem.

The second issue is actually the essence of the statistical matching problem. Its treatment is the focus throughout this book.

Remark 1.2 The previous framework for the statistical matching problem has frequently been used (at least implicitly) in practice. However, real statistical matching applications may not fit such a framework. One of the strongest assumptions is that $A \cup B$ is a unique data set of i.i.d. records from $f(\mathbf{x}, \mathbf{y}, \mathbf{z})$. When, for instance, the two samples are drawn at different times, this assumption may no longer hold.

Without loss of generality, let A be the most up-to-date sample of size n_A still from $f(\mathbf{x}, \mathbf{y}, \mathbf{z})$ (which is the joint distribution of interest), with $\mathbf{Z}$ missing. Let B be a sample independent of A whose n_B sample units are i.i.d. from the distribution $g(\mathbf{X}, \mathbf{Y}, \mathbf{Z})$, with g distinct from f. It is questionable whether these samples can be statistically matched. Matching can actually be performed when, although the two distributions f and g differ, the conditional distribution of $\mathbf{Z}$ given $\mathbf{X}$ is the same on both occasions. In this case, appropriate statistical matching procedures have been defined which assign different roles to the two samples A and B: B should lend information on $\mathbf{Z}$ to the A file. In the following it will be made clear whenever this alternative framework is under consideration.

Table 1.1 Sample data $A \cup B$ for the statistical matching problem. The shaded cells correspond to the unobserved variables in samples A and B, respectively

Sample	Y_1	…	Y_q	…	Y_Q	X_1	…	X_p	…	X_P	Z_1	…	Z_r	…	Z_R
	y_{11}^A	…	y_{1q}^A	…	y_{1Q}^A	x_{11}^A	…	x_{1p}^A	…	x_{1P}^A					
	…					…									
A	y_{a1}^A	…	y_{aq}^A	…	y_{aQ}^A	x_{a1}^A	…	x_{ap}^A	…	x_{aP}^A					
	…					…									
	$y_{n_A 1}^A$	…	$y_{n_A q}^A$	…	$y_{n_A Q}^A$	$x_{n_A 1}^A$	…	$x_{n_A p}^A$	…	$x_{n_A P}^A$					
						x_{11}^B	…	x_{1p}^B	…	x_{1P}^B	z_{11}^B	…	z_{1r}^B	…	z_{1R}^B
						…					…				
B						x_{b1}^B	…	x_{bp}^B	…	x_{bP}^B	z_{b1}^B	…	z_{br}^B	…	z_{bR}^B
						…					…				
						$x_{n_B 1}^B$	…	$x_{n_B p}^B$	…	$x_{n_B P}^B$	$z_{n_B 1}^B$	…	$z_{n_B r}^B$	…	$z_{n_B R}^B$

1.3 The Missing Data Mechanism in the Statistical Matching Problem

Before going into the details of the statistical matching procedures, let us describe the overall sample $A \cup B$. As already described in Section 1.2, it is a sample of $n_A + n_B$ units from $f(\mathbf{x}, \mathbf{y}, \mathbf{z})$ with $\mathbf{Z}$ missing in A and $\mathbf{Y}$ missing in B. Hence, the statistical matching problem can be regarded as a problem of analysis of a partially observed data set. Generally speaking, when missing items are present, it is necessary to take into account a set of additional r.v.s $\mathbf{R} = \left(\mathbf{R}_x, \mathbf{R}_y, \mathbf{R}_z\right)$, where $\mathbf{R}_x$, $\mathbf{R}_y$ and $\mathbf{R}_z$ are respectively random vectors of dimension P, Q and R:

$$\mathbf{R}_x = \left(R_{X_1}, \ldots, R_{X_P}\right)',$$

$$\mathbf{R}_y = \left(R_{Y_1}, \ldots, R_{Y_Q}\right)',$$

$$\mathbf{R}_z = \left(R_{Z_1}, \ldots, R_{Z_R}\right)'.$$

The indicator r.v. R_{X_j} shows when X_j has been observed ($R_{X_j} = 1$) or not ($R_{X_j} = 0$), $j = 1, \ldots, P$. Similar definitions hold for the random vectors $\mathbf{R}_y$ and $\mathbf{R}_z$. Appropriate inferences when missing items are present should consider a model that takes into account the variables of interest $(\mathbf{X}, \mathbf{Y}, \mathbf{Z})$ and the missing data mechanism $\mathbf{R}$. Particularly important is the relationship among these variables, defined by the conditional distribution of $\mathbf{R}$ given the variables of interest: $h(\mathbf{r}_x, \mathbf{r}_y, \mathbf{r}_z|\mathbf{x}, \mathbf{y}, \mathbf{z})$. Rubin (1976) defines three different missing data models, which are generally assumed by the analyst: missing completely at random (MCAR), missing at random (MAR), and missing not at random (MNAR); see Appendix A. Indeed, the statistical matching problem has a particular property: missingness is induced by the sampling design. When A and B are jointly considered as a unique data set of $n_A + n_B$ independent units generated from the same distribution $f(\mathbf{x}, \mathbf{y}, \mathbf{z})$, with $\mathbf{Z}$ missing in A and $\mathbf{Y}$ missing in B, i.e. for the statistical matching problem, the missing data mechanism is MCAR. A missing data mechanism is MCAR when $\mathbf{R}$ is independent of either the observed and the unobserved r.v.s $\mathbf{X}$, $\mathbf{Y}$ and $\mathbf{Z}$. Consequently,

$$h(\mathbf{r}_x, \mathbf{r}_y, \mathbf{r}_z|\mathbf{x}, \mathbf{y}, \mathbf{z}) = h(\mathbf{r}_x, \mathbf{r}_y, \mathbf{r}_z). \tag{1.1}$$

In order to show this assertion, it is enough to consider that $\mathbf{R}$ is independent of $(\mathbf{X}, \mathbf{Y}, \mathbf{Z})$, i.e. equation (1.1), or, equivalently for the symmetry of the concept of independence between r.v.s, that the conditional distribution of $(\mathbf{X}, \mathbf{Y}, \mathbf{Z})$ given $\mathbf{R}$, say $\phi(\mathbf{x}, \mathbf{y}, \mathbf{z}|\mathbf{r}_x, \mathbf{r}_y, \mathbf{r}_z)$ does not depend on $\mathbf{R}$:

$$\phi(\mathbf{x}, \mathbf{y}, \mathbf{z}|\mathbf{r}_x, \mathbf{r}_y, \mathbf{r}_z) = \phi(\mathbf{x}, \mathbf{y}, \mathbf{z}),$$

for every $\mathbf{x} \in \mathcal{X}$, $\mathbf{y} \in \mathcal{Y}$, $\mathbf{z} \in \mathcal{Z}$.

As a matter of fact, the statistical matching problem is characterized by just two patterns of $\mathbf{R}$:

- $\mathbf{R} = (\mathbf{1}_P, \mathbf{1}_Q, \mathbf{0}_R)$ for the units in A and
- $\mathbf{R} = \left(\mathbf{1}_P, \mathbf{0}_Q, \mathbf{1}_R\right)$ for the units in B,

where $\mathbf{1}_j$ and $\mathbf{0}_j$ are two j-dimensional vectors of ones and zeros, respectively. Due to the i.i.d. assumption of the generation of the $n_A + n_B$ values for $(\mathbf{X}, \mathbf{Y}, \mathbf{Z})$, we have that

$$\phi\left(\mathbf{x}, \mathbf{y}, \mathbf{z}|\mathbf{1}_P, \mathbf{1}_Q, \mathbf{0}_R\right) = \phi\left(\mathbf{x}, \mathbf{y}, \mathbf{z}|\mathbf{1}_P, \mathbf{0}_Q, \mathbf{1}_R\right) = f\left(\mathbf{x}, \mathbf{y}, \mathbf{z}\right) \tag{1.2}$$

for every $\mathbf{x} \in \mathcal{X}$, $\mathbf{y} \in \mathcal{Y}$, $\mathbf{z} \in \mathcal{Z}$, where $\phi\left(\mathbf{x}, \mathbf{y}, \mathbf{z}|\mathbf{1}_P, \mathbf{1}_Q, \mathbf{0}_R\right)$ is the distribution which generates the records in sample A and $\phi\left(\mathbf{x}, \mathbf{y}, \mathbf{z}|\mathbf{1}_P, \mathbf{0}_Q, \mathbf{1}_R\right)$ is the distribution which generates the records in sample B. In other words, the missing data mechanism is independent of both observed and missing values of the variables under study, which is the definition of the MCAR mechanism. This fact allows the possibility of making inference on the overall joint distribution of $(\mathbf{X}, \mathbf{Y}, \mathbf{Z})$ without considering (i.e. *ignoring*) the random indicators $\mathbf{R}$. Additionally, inferences can be based on the observed sampling distribution. This is obtained by marginalizing the overall distribution $f(\mathbf{x}, \mathbf{y}, \mathbf{z})$ with respect to the unobserved variables. As a consequence, the observed sampling distribution for the $n_A + n_B$ units is easily computed:

$$\prod_{a=1}^{n_A} f_{\mathbf{XY}}\left(\mathbf{x}_a, \mathbf{y}_a\right) \prod_{b=1}^{n_B} f_{\mathbf{XZ}}\left(\mathbf{x}_b, \mathbf{z}_b\right). \tag{1.3}$$

The observed sampling distribution (1.3) is the reference distribution for this book, as it is for most papers on statistical matching; see, for instance, Rässler (2002, pg. 78). The following remark underlines which alternatives can be considered, what missing data generation mechanism refers to them, and their feasibility.

Remark 1.3 Remark 1.2 states that A and B cannot always be considered as generated from an identical distribution. In this case, equation (1.1) no longer holds and the missing data mechanism in $A \cup B$ cannot be assumed MCAR. In the notation of Remark 1.2, the distributions of $(\mathbf{X}, \mathbf{Y}, \mathbf{Z})$ given the patterns of missing data are:

$$\phi\left(\mathbf{x}, \mathbf{y}, \mathbf{z}|\mathbf{1}_P, \mathbf{1}_Q, \mathbf{0}_R\right) = f\left(\mathbf{x}, \mathbf{y}, \mathbf{z}\right),$$
$$\phi\left(\mathbf{x}, \mathbf{y}, \mathbf{z}|\mathbf{1}_P, \mathbf{0}_Q, \mathbf{1}_R\right) = g\left(\mathbf{x}, \mathbf{y}, \mathbf{z}\right),$$

for every $\mathbf{x} \in \mathcal{X}$, $\mathbf{y} \in \mathcal{Y}$, $\mathbf{z} \in \mathcal{Z}$. This situation can be formalized via the so-called pattern mixture models (Little, 1993): if the two samples are analysed as a unique sample of $n_A + n_B$ units, the corresponding generating model is a mixture of the two distributions f and g. Little warns that this approach usually leads to underidentified models, and shows which restrictions that tie unidentified parameters with the identified ones should be used. In general, as already underlined in Remark 1.2, the interest is not in the mixture of the two distributions, but only in the most up-to-date one, $f(\mathbf{x}, \mathbf{y}, \mathbf{z})$ (an exception will be illustrated in Remark 6.1). For this reason, these models will not be considered any more. The framework illustrated in Remark 1.2 will just consider B as a donor of information on $\mathbf{Z}$, when possible.

1.4 Accuracy of a Statistical Matching Procedure

Sections 1.2 and 1.3 have described the input of the statistical matching problem: a partially observed data set with the absence of joint information on the variables of interest and some basic assumptions on the data generating model. This section deals with the output. As declared in Section 1.1, the statistical matching problem may be addressed using either the micro or macro approach. These approaches can be adopted by using many different statistical procedures, i.e. different transformations of the available (observed) data. Are there any guidelines as to the choice of procedure? In other words, how is it possible to assess the accuracy of a statistical matching procedure?

It must be remarked that it is not easy to draw definitive conclusions. Papers that deal explicitly with this problem are few in number, among them Barr and Turner (1990); see also D'Orazio *et al.* (2002) and references therein. A number of different issues should be taken into account.

(a) What assumptions can be reasonably considered for the joint model $(\mathbf{X}, \mathbf{Y}, \mathbf{Z})$?

(b) What estimator for $f(\mathbf{x}, \mathbf{y}, \mathbf{z})$ is preferable, if any, under the model assumed in (a)?

(c) What method of generating appropriate values for the missing variables can be used under the model chosen in (a) and according to the estimator chosen in (b)?

As a matter of fact, (a) is a very general question related to the data generation process, (b) is related to the macro approach, and (c) to the micro approach. They are interrelated in the sense that an apparently reasonable answer to a question is not reasonable if the previous questions are unanswered. Actually, there is yet another question that should be considered when a synthetic file is distributed and inferential methods are applied to it.

(d) What inferential procedure can be used on the synthetic data set?

The combination of (a) and (b) for the macro approach, (a), (b) and (c) for the micro approach, and (a), (b), (c), and (d) for the analysis of the result of the micro approach gives an overall sketch of the accuracy of the corresponding statistical matching result. A general measure that amalgamates all these aspects has not been yet defined. It can only be assessed via appropriate Monte Carlo experiments in a simulated framework.

Let us investigate each of the accuracy issues (a)–(d) in more detail.

1.4.1 Model assumptions

Table 1.1 shows that the statistical matching problem is characterized by a very annoying situation: there is no observation where all the variables of interest are jointly recorded. A consequence is that, of all the possible statistical models for $(\mathbf{X}, \mathbf{Y}, \mathbf{Z})$, only a few are actually identifiable for $A \cup B$. In other words, $A \cup B$ does not contain enough information for the estimation of parameters such as the correlation matrix or the contingency table of $(\mathbf{Y}, \mathbf{Z})$. Furthermore, for the same reason, it is not possible to test on $A \cup B$ which model is appropriate for $(\mathbf{X}, \mathbf{Y}, \mathbf{Z})$. There are different possibilities.

- Further information (e.g. previous experience or an *ad hoc* survey) justifies the use of an identifiable model for $A \cup B$.
- Further information (e.g. previous experience or an *ad hoc* survey) is used together with $A \cup B$ in order to make other models also identifiable.
- No assumptions are made on the $(\mathbf{X}, \mathbf{Y}, \mathbf{Z})$ model. This problem is studied as a problem characterized by uncertainty on some of the model properties.

The first two assumptions are able to produce a unique point estimate of the parameters. For the third choice, which is a conservative one, a set rather than a point estimate of the inestimable parameters, such as the correlation matrix of $(\mathbf{Y}, \mathbf{Z})$, will be the output. The features of this set of estimates describe uncertainty for that parameter.

The first two choices are assumptions that should be well justified by additional sources of information. If these assumptions are wrong, no matter what sophisticated inferential machinery is used, the results of the macro and, hence, of the micro approaches will reflect the assumption and not the real underlying model. Also in these cases, evaluation of uncertainty is a precious source of information. In fact, reliability of conclusions based on one of the first two choices can be based on the evaluation of their uncertainty when no assumptions are considered. For instance, if a correlation coefficient for the never jointly observed variables $\mathbf{Y}$ and $\mathbf{Z}$ is estimated under a specific identifiable model for $A \cup B$ or with the help of further auxiliary information, an indication of the reliability of these estimates is given by the width of the uncertainty set: the smaller it is, the higher is the reliability of the estimates with respect to model misspecification.

1.4.2 Accuracy of the estimator

Let us assume that a model for $(\mathbf{X}, \mathbf{Y}, \mathbf{Z})$ has been firmly established. When the approach is macro, accuracy of a statistical matching procedure means accuracy of the estimator of the distribution function $f(\mathbf{x}, \mathbf{y}, \mathbf{z})$. In this case, appropriate measures such as the mean square error (MSE) or, accounting for its components, the bias and variance are well known in both the parametric and nonparametric case.

In a parametric framework, minimization of the MSE of each parameter estimator can (almost) be obtained, at least for large data sets and under minimal regularity conditions, when maximum likelihood (ML) estimators are used. More precisely, the consistency property of ML estimators is claimed in most of the results of this book. It must be emphasized that the ML approach given the overall set $A \cup B$

has an additional property in this case: every parameter estimate is coherent with the other estimates. Sometimes a partially observed data set may suggest distinct estimators for each parameter of the joint distribution that are not coherent. It will be seen that this issue is fundamental in statistical matching, given that it deals with the partially observed data set of Table 1.1.

In a nonparametric framework, consistency of the results is also one of the most important aspects to consider. Consistency of estimators is a very important characterization for the statistical matching problem. In fact, it ensures that, for large samples, estimates are very close to the true but unknown distribution $f(\mathbf{x}, \mathbf{y}, \mathbf{z})$. In the next subsection it will be seen that this aspect is relevant also to the micro approach.

1.4.3 Representativeness of the synthetic file

This aspect is the most commonly investigated issue for assessing the accuracy of a statistical matching procedure. Generally speaking, four large categories of accuracy evaluation procedures can be defined (Rässler, 2002), from the most difficult goal to the simplest:

(i) Synthetic records should coincide with the true (but unobserved) values.

(ii) The joint distribution of all variables is reflected in the statistically matched file.

(iii) The correlation structure of the variables is preserved.

(iv) The marginal and joint distributions of the variables in the source files are preserved in the matched file.

The first point is the most ambitious and difficult requirement to fulfil. It can be achieved when logical or mathematical rules determining a single value for-each single unit are available. However, when using statistical rules, it is not as important to reproduce the exact value as it is the joint distribution $f(\mathbf{x}, \mathbf{y}, \mathbf{z})$, which contains all the relevant statistical information.

The third and fourth points do not ensure that the final synthetic data set is appropriate for any kind of inferences for $(\mathbf{X}, \mathbf{Y}, \mathbf{Z})$, contradicting one of the main characteristics that a synthetic data set should possess. For instance, the fourth point ensures only reasonable inferences for the distributions of $(\mathbf{X}, \mathbf{Y})$ and $(\mathbf{X}, \mathbf{Z})$.

When the second goal is fulfilled, the synthetic data set can be considered as a sample generated from the joint distribution of $(\mathbf{X}, \mathbf{Y}, \mathbf{Z})$. Hence, the synthetic data set is *representative* of $f(\mathbf{x}, \mathbf{y}, \mathbf{z})$, and can be used as a general purpose sample in order to infer its characteristics.

Any discrepancy between the real data generating model and the underlying model of the synthetic complete data set is called *matching noise*; see Paass (1985).

Focusing on the second point, under identifiable models or with the help of additional information (Section 1.4.1), the relevant question is whether the data synthetically generated via the estimated distribution $f(\mathbf{x}, \mathbf{y}, \mathbf{z})$ are affected by the matching noise or not. It is not always a simple matter. As claimed in Section 1.4.2, when the available data sets are large and the macro approach is used with a consistent estimator of $f(\mathbf{x}, \mathbf{y}, \mathbf{z})$, it is possible to define micro approaches with a reduced matching noise. Note that a good estimate of $f(\mathbf{x}, \mathbf{y}, \mathbf{z})$ is a necessary but not a sufficient condition to ensure that the matching noise is as low as possible. In fact, the generation of the synthetic data should be also done appropriately.

1.4.4 Accuracy of estimators applied on the synthetic data set

This is a critical issue for the micro approach. If the synthetic data set can be considered as a sample generated according to $f(\mathbf{x}, \mathbf{y}, \mathbf{z})$ (or approximately so), it is appropriate to use estimators that would be applied in complete data cases. Hence, the objective of reducing the matching noise (Section 1.4.3) is fundamental.

In fact, estimators preserve their inferential properties (e.g. unbiasedness, consistency) with respect to the model that has generated the synthetic data. When the matching noise is large, these results are a misleading indication as to the true model $f(\mathbf{x}, \mathbf{y}, \mathbf{z})$.

As a matter of fact, this last problem resembles that of Section 1.4.1. In Section 1.4.1 there was a model misspecification problem. Now the problem is that the data generating model of the synthetic data set differs from the data generating model of the observed data set. In both cases the result is similar: inferences are related to models that differ from the target one.

1.5 Outline of the Book

This book aims to explore the statistical matching problem and its possible solutions. This task will be addressed by considering features of its input (Sections 1.2 and 1.3) and, more importantly, of its output (Section 1.4).

One of the key issues is model assumption. As remarked in Section 1.4.1, a first set of techniques refer to the case where the overall model family $\mathcal{F}$ is identifiable for $A \cup B$. A natural identifiable model is one that assumes the independence of $\mathbf{Y}$ and $\mathbf{Z}$ given $\mathbf{X}$. This assumption is usually called the *conditional independence assumption* (CIA). Chapter 2 is devoted to the description and analysis of the different statistical matching approaches under the CIA.

The set of identifiable models for $A \cup B$ is rather narrow, and may be inappropriate for the phenomena under study. In order to overcome this problem, further auxiliary information beyond just $A \cup B$ is needed. This *auxiliary information* may be either in parametric form, i.e. knowledge of the values of some of the parameters of the model for $(\mathbf{X}, \mathbf{Y}, \mathbf{Z})$, or as an additional data sample C. The use of auxiliary information in the statistical matching process is described in Chapter 3.

Both Chapters 2 and 3 will consider the following aspects:

(i) macro and micro approaches;

(ii) parametric and nonparametric definition of the set of possible distribution functions $\mathcal{F}$;

(iii) the possibility of departures from the i.i.d. case (as in Remark 1.2).

As claimed in Section 1.4.1, a very important issue deals with the situation where no model assumptions are hypothesized. In this case, it is possible to study the uncertainty associated to the parameters due to lack of sample information. Given the importance of this topic, it is described in considerable detail in Chapter 4.

The framework developed in Section 1.2 is not the most appropriate for samples drawn from finite populations according to complex survey designs, unless ignorability of the sample design is claimed; see, for example, Gelman *et al.* (2004, Chapter 7). Despite the amount of data sets of this kind, only few methodological results for statistical matching are available. A general review of these approaches and the link with the corresponding results under the framework of Section 1.2 is given in Chapter 5.

Generally speaking, statistical integration of different sources is strictly connected to the integration of the data production processes. Actually, statistical integration of sources would be particularly successful when applied to sources already standardized in terms of definitions and concepts. Unfortunately, this is not always true. Some considerations on the preliminary operations needed for statistically matching two samples are reported in Chapter 6.

Finally, Chapter 7 presents some statistical matching applications. A particular statistical matching application is described in some detail in order to make clear all the tasks that should be considered when matching two real data sets. Furthermore, this example allows the comparison of the results of different statistical matching procedures.

All the original codes used for simulations and experiments, developed in the R environment (R Development Core Team, 2004), are reported in Appendix E in order to enable the reader to make practical use of the techniques discussed in the book. The same codes can also be downloaded on the site http://www.wiley.com/go/matching.

2

The Conditional Independence Assumption

In this chapter, a specific model for $(\mathbf{X}, \mathbf{Y}, \mathbf{Z})$ is assumed: the independence of $\mathbf{Y}$ and $\mathbf{Z}$ given $\mathbf{X}$. This assumption is usually referred to as the conditional independence assumption or CIA.

This model has had a very important role in statistical matching: it was assumed, explicitly or implicitly, in all the early statistical matching applications. The reason is simple: this model is identifiable for $A \cup B$ (i.e. for Table 1.1), and directly estimable. In fact, when the CIA holds, the structure of the density function for $(\mathbf{X}, \mathbf{Y}, \mathbf{Z})$ is the following:

$$f(\mathbf{x}, \mathbf{y}, \mathbf{z}) = f_{\mathbf{Y}|\mathbf{X}}(\mathbf{y}|\mathbf{x})\, f_{\mathbf{Z}|\mathbf{X}}(\mathbf{z}|\mathbf{x})\, f_{\mathbf{X}}(\mathbf{x}), \qquad \forall\, \mathbf{x} \in \mathcal{X}, \mathbf{y} \in \mathcal{Y}, \mathbf{z} \in \mathcal{Z}, \tag{2.1}$$

where $f_{\mathbf{Y}|\mathbf{X}}$ is the conditional density of $\mathbf{Y}$ given $\mathbf{X}$, $f_{\mathbf{Z}|\mathbf{X}}$ is the conditional density of $\mathbf{Z}$ given $\mathbf{X}$, and $f_{\mathbf{X}}$ is the marginal density of $\mathbf{X}$. In order to estimate (2.1), it is enough to gain information on the marginal distribution of $\mathbf{X}$ and on the pairwise relationship between respectively $\mathbf{X}$ and $\mathbf{Y}$, and $\mathbf{X}$ and $\mathbf{Z}$. This information is actually available in the distinct samples A and B.

Remark 2.1 The CIA is an assumption that cannot be tested from the data set $A \cup B$. It can be a wrong assumption and, hence, misleading. In the rest of this chapter, we will rely on the CIA, i.e. we firmly believe that this model holds true for the data at hand. The effects of an incorrect model assumption have already been anticipated (Section 1.4.1).

As usual, it is possible to use the available observed information for the statistical matching problem (the overall sample $A \cup B$ of Table 1.1) in many different ways. At first sight, the most natural ones are those that aim at the direct estimation of the joint distribution (2.1) or of any important characteristic of the joint

Statistical Matching: Theory and Practice M. D'Orazio, M. Di Zio and M. Scanu

distribution (e.g. a correlation coefficient), i.e. a *macro approach*. However, papers on statistical matching in the CIA context have also given special consideration to the reconstruction of a synthetic data set, i.e. a *micro approach*.

We will describe both the alternatives, respectively when $\mathcal{F}$ is a parametric set of distributions (Sections 2.1 and 2.2) and in a nonparametric framework (Sections 2.3 and 2.4). Mixed procedures, i.e. two-step procedures which are partly parametric and partly nonparametric, are treated in Section 2.5. A Bayesian approach is discussed in Section 2.7.

Finally, identifiable models for $A \cup B$ other than the CIA are shown in Section 2.8.

2.1 The Macro Approach in a Parametric Setting

Let $\mathcal{F}$ be a parametric family of distributions, i.e. each density $f(\mathbf{x}, \mathbf{y}, \mathbf{z}; \boldsymbol{\theta}) \in \mathcal{F}$ is defined by a finite-dimensional vector parameter $\boldsymbol{\theta} \in \Theta \subseteq \mathbb{R}^T$, for some integer T. Under the CIA, $\mathcal{F}$ may be decomposed into three different sets of distribution functions according to equation (2.1): $f_{\mathbf{X}}(\cdot; \boldsymbol{\theta}_{\mathbf{X}}) \in \mathcal{F}_{\mathbf{X}}$ for the marginal distribution of $\mathbf{X}$; $f_{\mathbf{Y}|\mathbf{X}}\left(\cdot; \boldsymbol{\theta}_{\mathbf{Y}|\mathbf{X}}\right) \in \mathcal{F}_{\mathbf{Y}|\mathbf{X}}$ for the conditional distribution of $\mathbf{Y}$ given $\mathbf{X}$; and $f_{\mathbf{Z}|\mathbf{X}}\left(\cdot; \boldsymbol{\theta}_{\mathbf{Z}|\mathbf{X}}\right) \in \mathcal{F}_{\mathbf{Z}|\mathbf{X}}$ for the conditional distribution of $\mathbf{Z}$ given $\mathbf{X}$. Given the decomposition in (2.1), the distribution of $(\mathbf{X}, \mathbf{Y}, \mathbf{Z})$ is perfectly identified by the parameter vectors $\boldsymbol{\theta}_{\mathbf{X}}$, $\boldsymbol{\theta}_{\mathbf{Y}|\mathbf{X}}$ and $\boldsymbol{\theta}_{\mathbf{Z}|\mathbf{X}}$:

$$f(\mathbf{x}, \mathbf{y}, \mathbf{z}; \boldsymbol{\theta}) = f_{\mathbf{X}}(\mathbf{x}; \boldsymbol{\theta}_{\mathbf{X}})\, f_{\mathbf{Y}|\mathbf{X}}\left(\mathbf{y}|\mathbf{x}; \boldsymbol{\theta}_{\mathbf{Y}|\mathbf{X}}\right) f_{\mathbf{Z}|\mathbf{X}}\left(\mathbf{z}|\mathbf{x}; \boldsymbol{\theta}_{\mathbf{Z}|\mathbf{X}}\right), \qquad (2.2)$$

$\boldsymbol{\theta}_{\mathbf{X}} \in \Theta_{\mathbf{X}}$, $\boldsymbol{\theta}_{\mathbf{Y}|\mathbf{X}} \in \Theta_{\mathbf{Y}|\mathbf{X}}$, $\boldsymbol{\theta}_{\mathbf{Z}|\mathbf{X}} \in \Theta_{\mathbf{Z}|\mathbf{X}}$. In this framework, the macro approach consists in estimating the parameters $(\boldsymbol{\theta}_{\mathbf{X}}, \boldsymbol{\theta}_{\mathbf{Y}|\mathbf{X}}, \boldsymbol{\theta}_{\mathbf{Z}|\mathbf{X}})$.

By (1.3) and equation (2.2), the observed likelihood function of the overall sample $A \cup B$ is:

$$\begin{aligned} L(\boldsymbol{\theta}|A \cup B) &= \prod_{a=1}^{n_A} f_{\mathbf{XY}}(\mathbf{x}_a, \mathbf{y}_a; \boldsymbol{\theta}) \prod_{b=1}^{n_B} f_{\mathbf{XZ}}(\mathbf{x}_b, \mathbf{z}_b; \boldsymbol{\theta}) \\ &= \prod_{a=1}^{n_A} f_{\mathbf{Y}|\mathbf{X}}\left(\mathbf{y}_a|\mathbf{x}_a; \boldsymbol{\theta}_{\mathbf{Y}|\mathbf{X}}\right) \prod_{b=1}^{n_B} f_{\mathbf{Z}|\mathbf{X}}\left(\mathbf{z}_b|\mathbf{x}_b; \boldsymbol{\theta}_{\mathbf{Z}|\mathbf{X}}\right) \\ &\quad \times \prod_{a=1}^{n_A} f_{\mathbf{X}}(\mathbf{x}_a; \boldsymbol{\theta}_{\mathbf{X}}) \prod_{b=1}^{n_B} f_{\mathbf{X}}(\mathbf{x}_b; \boldsymbol{\theta}_{\mathbf{X}}). \end{aligned} \qquad (2.3)$$

Although the data set $A \cup B$ is affected by missing items (Table 1.1), the maximum likelihood estimates of the parameters $\boldsymbol{\theta}_{\mathbf{X}}$, $\boldsymbol{\theta}_{\mathbf{Y}|\mathbf{X}}$ and $\boldsymbol{\theta}_{\mathbf{Z}|\mathbf{X}}$ can be derived directly on the appropriate subsets of complete data without the use of iterative procedures; see Rubin (1974) and Section A.1.2. More precisely, the ML estimator of $\boldsymbol{\theta}_{\mathbf{X}}$ is computed on the overall sample $A \cup B$, while the ML estimators for $\boldsymbol{\theta}_{\mathbf{Y}|\mathbf{X}}$ and $\boldsymbol{\theta}_{\mathbf{Z}|\mathbf{X}}$ are computed respectively on the subsets A and B.

2.1.1 Univariate normal distributions case

For illustrative purposes, the simple case of three univariate normal distributions is considered. The generalization to the multivariate case is given in Section 2.1.2.

Let (X, Y, Z) be a trivariate normal distribution with parameters

$$\boldsymbol{\theta} = (\boldsymbol{\mu}, \boldsymbol{\Sigma}) = \left[\begin{pmatrix} \mu_X \\ \mu_Y \\ \mu_Z \end{pmatrix}, \begin{pmatrix} \sigma_X^2 & \sigma_{XY} & \sigma_{XZ} \\ \sigma_{XY} & \sigma_Y^2 & \sigma_{YZ} \\ \sigma_{XZ} & \sigma_{YZ} & \sigma_Z^2 \end{pmatrix} \right]$$

where $\boldsymbol{\mu}$ is the mean vector and $\boldsymbol{\Sigma}$ is the covariance matrix. Hence, the joint (X, Y, Z) distribution is:

$$f(x, y, z|\boldsymbol{\mu}, \boldsymbol{\Sigma}) = \frac{1}{\sqrt{(2\pi)^3|\boldsymbol{\Sigma}|}} e^{-\frac{1}{2}\left((x,y,z)-\boldsymbol{\mu}'\right)\boldsymbol{\Sigma}^{-1}\left((x,y,z)'-\boldsymbol{\mu}\right)},$$

with $(x, y, z) \in \mathbb{R}^3$. Under the CIA, the joint distribution of (X, Y, Z) can be equivalently expressed through the factorization (2.2). By the properties of the multinormal distribution (see Anderson, 1984) we have the following:

(a) The marginal distribution of X is normal with parameters

$$\boldsymbol{\theta}_X = (\mu_X, \sigma_X^2).$$

(b) The conditional distribution of Y given X is also normal with mean given by the linear regression of Y on X and variance given by the residual variance of Y with respect to the regression on X. Hence, the conditional distribution of Y given X can equivalently be defined by the parameters

$$\boldsymbol{\theta}_{Y|X} = \left(\mu_{Y|X}, \sigma_{Y|X}^2\right).$$

These conditional parameters can be expressed in terms of those in $\boldsymbol{\theta}$ through the following equations:

$$\begin{aligned} \mu_{Y|X} &= \alpha_Y + \beta_{YX} X, \\ \alpha_Y &= \mu_Y - \beta_{YX}\mu_X, \\ \beta_{YX} &= \frac{\sigma_{XY}}{\sigma_X^2}, \\ \sigma_{Y|X}^2 &= \sigma_Y^2 - \frac{\sigma_{XY}^2}{\sigma_X^2} = \sigma_Y^2 - \beta_{YX}^2\sigma_X^2. \end{aligned}$$

Note that the conditional distribution of Y given X can also be defined through the regression model:

$$Y = \mu_{Y|X} + \epsilon_{Y|X} = \alpha_Y + \beta_{YX} X + \epsilon_{Y|X} \tag{2.4}$$

where $\epsilon_{Y|X}$ is normally distributed with zero mean and variance $\sigma_{Y|X}^2$.

(c) The same holds for the conditional distribution of Z given X, which is still normal with parameters

$$\boldsymbol{\theta}_{Z|X} = \left(\mu_{Z|X}, \sigma^2_{Z|X}\right),$$

where

$$\begin{aligned} \mu_{Z|X} &= \alpha_Z + \beta_{ZX} X, \\ \alpha_Z &= \mu_Z - \beta_{ZX}\mu_X, \\ \beta_{ZX} &= \frac{\sigma_{XZ}}{\sigma^2_X}, \\ \sigma^2_{Z|X} &= \sigma^2_Z - \frac{\sigma^2_{XZ}}{\sigma^2_X} = \sigma^2_Z - \beta^2_{ZX}\sigma^2_X. \end{aligned} \tag{2.5}$$

The corresponding regression equation of Z given X is:

$$Z = \mu_{Z|X} + \epsilon_{Z|X} = \alpha_Z + \beta_{ZX} X + \epsilon_{Z|X}, \tag{2.6}$$

where $\epsilon_{Z|X}$ follows a normal distribution with zero mean and variance $\sigma^2_{Z|X}$.

Remark 2.2 The parameters $\boldsymbol{\theta}_X, \boldsymbol{\theta}_{Y|X}$ and $\boldsymbol{\theta}_{Z|X}$ are obtained from the subset of $\boldsymbol{\theta}$ defined by the parameters $\left(\boldsymbol{\mu}, \sigma^2_X, \sigma^2_Y, \sigma^2_Z, \sigma_{XY}, \sigma_{XZ}\right)$. In fact, the only parameter which is not used in the density decomposition (2.2) is σ_{YZ}, which is perfectly determined by the other parameters under the CIA:

$$\sigma_{YZ} = \frac{\sigma_{XY}\sigma_{XZ}}{\sigma^2_X}.$$

Analogously, under the CIA the partial correlation coefficient is $\rho_{YZ|X} = 0$ and the bivariate (Y, Z) correlation coefficient is $\rho_{YZ} = \rho_{XY}\rho_{XZ}$.

Remark 2.3 The CIA implies that Z is useless as a Y regressor, given that its partial regression coefficient is null. Equivalently, Y is useless as a regressor for Z.

Let $\{x_a, y_a\}$, $a = 1, \ldots, n_A$, and $\{x_b, z_b\}$, $b = 1, \ldots, n_B$, be $n_A + n_B$ independent observations from (X, Y, Z). The problem of ML estimation of the parameters of normal distributions when the data set is only partially observed has a long history. One of the first articles (Wilks, 1932) deals with the problem of bivariate normal partially observed data. Extensions are given in Matthai (1951) and Edgett (1956). The statistical matching framework (although not yet denoted in this way) for a trivariate normal data set can be found in Lord (1955) and Anderson (1957). Particularly interesting is the paper by Anderson, which can be considered as a precursor of Rubin (1974) (see also Section A.1.2) as far as the normal distribution is concerned. Anderson notes that the ML estimates of the parameters in $\boldsymbol{\theta}$ can be split into three different ML problems on complete data sets, respectively for $\boldsymbol{\theta}_X$, $\boldsymbol{\theta}_{Y|X}$ and $\boldsymbol{\theta}_{Z|X}$, following the decomposition in (2.3).

(i) For the marginal distribution of X, the ML estimate of $\boldsymbol{\theta}_X$ is given by the usual ML estimates on the overall data set $A \cup B$:

$$\hat{\mu}_X = \bar{x}_{A\cup B} = \frac{1}{n_A + n_B}\left[\sum_{a=1}^{n_A} x_a + \sum_{b=1}^{n_B} x_b\right],$$

$$\hat{\sigma}_X^2 = s_{X;A\cup B}^2 = \frac{1}{n_A + n_B}\left[\sum_{a=1}^{n_A} (x_a - \hat{\mu}_X)^2 + \sum_{b=1}^{n_B} (x_b - \hat{\mu}_X)^2\right].$$

(ii) For the distribution of Y given X, the ML estimate of $\boldsymbol{\theta}_{Y|X}$ should follow these steps. First, it is possible to estimate the parameters of the regression equation (2.4):

$$\hat{\beta}_{YX} = \frac{s_{XY;A}}{s_{X;A}^2} = \frac{\sum_{a=1}^{n_A} (x_a - \bar{x}_A)(y_a - \bar{y}_A)}{\sum_{a=1}^{n_A} (x_a - \bar{x}_A)^2},$$

$$\hat{\alpha}_Y = \bar{y}_A - \hat{\beta}_{YX}\bar{x}_A,$$

$$\hat{\sigma}_{Y|X}^2 = s_{Y;A}^2 - \hat{\beta}_{YX}^2 s_{X;A}^2$$

$$= \frac{1}{n_A}\sum_{a=1}^{n_A} (y_a - \bar{y}_A)^2 - \hat{\beta}_{YX}^2 \frac{1}{n_A}\sum_{a=1}^{n_A} (x_a - \bar{x}_A)^2$$

where $\bar{x}_A$ and $\bar{y}_A$ are the sample means of X and Y respectively in A, and s denotes the sample variance or covariance, according to the subscripts.

The previous ML estimates, together with those described in step (i), allow the computation of the ML estimates of the following marginal parameters, useful for $\boldsymbol{\theta}$:

$$\hat{\mu}_Y = \hat{\alpha}_Y + \hat{\beta}_{YX}\hat{\mu}_X = \bar{y}_A + \hat{\beta}_{YX}\left(\bar{x}_{A\cup B} - \bar{x}_A\right),$$

$$\hat{\sigma}_Y^2 = \hat{\sigma}_{Y|X}^2 + \hat{\beta}_{YX}^2\hat{\sigma}_X^2 = s_{Y;A}^2 + \hat{\beta}_{YX}^2\left(\hat{\sigma}_X^2 - s_{X;A}^2\right)$$

$$\hat{\sigma}_{XY} = \hat{\beta}_{YX}\hat{\sigma}_X^2.$$

(iii) The same arguments hold for the distribution of Z given X. The ML estimate of $\boldsymbol{\theta}_{Z|X}$ is given, in obvious notation, by the following parameter estimates:

$$\hat{\beta}_{ZX} = \frac{s_{XZ;B}}{s_{X;B}^2} = \frac{\sum_{b=1}^{n_B} (x_b - \bar{x}_B)(z_b - \bar{z}_B)}{\sum_{b=1}^{n_B} (x_b - \bar{x}_B)^2},$$

$$\hat{\alpha}_Z = \bar{z}_B - \hat{\beta}_{ZX}\bar{x}_B,$$

$$\hat{\sigma}_{Z|X}^2 = s_{Z;B}^2 - \hat{\beta}_{ZX}^2 s_{X;B}^2$$

$$= \frac{1}{n_B}\sum_{b=1}^{n_B} (z_b - \bar{z}_B)^2 - \hat{\beta}_{ZX}^2 \frac{1}{n_B}\sum_{b=1}^{n_B} (x_b - \bar{x}_B)^2.$$

The marginal parameters of Z are computed accordingly:

$$\begin{aligned} \hat{\mu}_Z &= \hat{\alpha}_Z + \hat{\beta}_{ZX}\hat{\mu}_X = \bar{z}_B + \hat{\beta}_{ZX}\left(\bar{x}_{A\cup B} - \bar{x}_B\right), \\ \hat{\sigma}_Z^2 &= \hat{\sigma}_{Z|X}^2 + \hat{\beta}_{ZX}^2\hat{\sigma}_X^2 = s_{Z;B}^2 + \hat{\beta}_{ZX}^2\left(\hat{\sigma}_X^2 - s_{X;B}^2\right) \\ \hat{\sigma}_{XZ} &= \hat{\beta}_{ZX}\hat{\sigma}_X^2. \end{aligned}$$

Remark 2.4 Since Wilks (1932), researchers have been interested in the gain in efficiency from using the ML estimates as compared to their observed counterparts (i.e. the usual estimators computed on the relevant complete part of data set, e.g. $\bar{z}_B$, $s_{Z;B}^2$, $s_{XZ;B}$). As Anderson (1957) notes, the ML estimate of μ_Y is the so-called regression estimate in double sampling. The same also holds for the other marginal parameters of respectively Z and Y. Lord (1955) proves that the gain in efficiency in using $\hat{\mu}_Y$ instead of $\bar{y}_A$ is

$$\frac{\mathrm{Var}(\hat{\mu}_Y)}{\mathrm{Var}(\bar{y}_A)} = 1 - \frac{n_B}{n_A + n_B}\rho_{XY}^2,$$

where ρ_{XY} is the correlation coefficient between X and Y. Hence, whenever ρ_{XY}^2 is large and the proportion of cases in B is large, using $\hat{\mu}_Y$ is expected to lead to a great improvement.

Remark 2.5 At first sight, it may seem that the previous estimators do not take into account all the available information. For instance, the estimator of the regression parameter $\beta_{YX} = \sigma_{XY}/\sigma_X^2$ is computed by means of $\bar{s}_{X;A}^2$ instead of the ML estimator of the variance of X: $\hat{\sigma}_X^2$. This fact is well discussed in Rubin (1974), and can easily be understood from the likelihood in (2.3). Each parameter $\boldsymbol{\theta}_X$, $\boldsymbol{\theta}_{Y|X}$ and $\boldsymbol{\theta}_{Z|X}$ defines a factor in (2.3). When each factor is maximized, the overall likelihood function itself is maximized. Each factor is defined on a complete data subset, and the ML estimates can be expressed in closed form.

It has also been argued in Moriarity and Scheuren, (2001) that the use of $\hat{\sigma}_X^2$ in the computation of β_{YX} leads to unpleasant results. In particular, the associated estimated covariance matrix for the pair (X, Y) would be:

$$\begin{pmatrix} s_{X;A\cup B}^2 & s_{XY;A} \\ s_{XY;A} & s_{Y;A}^2 \end{pmatrix}, \tag{2.7}$$

which may not be positive semidefinite. This follows from the fact that the Cauchy–Schwarz inequality (i.e. the square of the covariance of a couple of variables should not be greater than the product of the variances of the two variables) does not generally hold for the matrix (2.7).

Remark 2.6 Despite the claims of the previous remarks, different authors have considered alternatives to the ML estimator.

(i) Rässler (2002) uses least squares estimators of the regression parameters. Note that the main difference consists in substituting the denominator of the ML estimators (sample size) with the difference between sample size and degrees of freedom. For large samples, this difference is very slight.

(ii) Moriarity and Scheuren, (2001) estimate $\boldsymbol{\theta}$ with its sample observed counterpart (e.g. estimate means with the average of the observed values, and variances with the sample variances of the observed data).

2.1.2 The multinormal case

The previous arguments can easily be extended to the general case of multivariate **X**, **Y** and **Z**.

Let $(\mathbf{X}, \mathbf{Y}, \mathbf{Z})$ be respectively P, Q and R-dimensional r.v.s jointly distributed as a multinormal with parameters

$$\boldsymbol{\theta} = (\boldsymbol{\mu}, \boldsymbol{\Sigma}) = \left[\begin{pmatrix} \boldsymbol{\mu}_{\mathbf{X}} \\ \boldsymbol{\mu}_{\mathbf{Y}} \\ \boldsymbol{\mu}_{\mathbf{Z}} \end{pmatrix}, \begin{pmatrix} \boldsymbol{\Sigma}_{\mathbf{XX}} & \boldsymbol{\Sigma}_{\mathbf{XY}} & \boldsymbol{\Sigma}_{\mathbf{XZ}} \\ \boldsymbol{\Sigma}_{\mathbf{YX}} & \boldsymbol{\Sigma}_{\mathbf{YY}} & \boldsymbol{\Sigma}_{\mathbf{YZ}} \\ \boldsymbol{\Sigma}_{\mathbf{ZX}} & \boldsymbol{\Sigma}_{\mathbf{ZY}} & \boldsymbol{\Sigma}_{\mathbf{ZZ}} \end{pmatrix} \right].$$

Hence, the joint distribution is

$$f(\mathbf{x}, \mathbf{y}, \mathbf{z} | \boldsymbol{\mu}, \boldsymbol{\Sigma}) = \frac{1}{\sqrt{(2\pi)^{P+Q+R} |\boldsymbol{\Sigma}|}} e^{-\frac{1}{2}((\mathbf{x},\mathbf{y},\mathbf{z}) - \boldsymbol{\mu}')\boldsymbol{\Sigma}^{-1}((\mathbf{x},\mathbf{y},\mathbf{z})' - \boldsymbol{\mu})},$$

with $(\mathbf{x}, \mathbf{y}, \mathbf{z}) \in \mathbb{R}^{P+Q+R}$.

The CIA imposes the restriction that

$$\boldsymbol{\Sigma}_{\mathbf{YZ}} = \boldsymbol{\Sigma}_{\mathbf{YX}} \boldsymbol{\Sigma}_{\mathbf{XX}}^{-1} \boldsymbol{\Sigma}_{\mathbf{XZ}},$$

or, in other words, that the covariance matrix of **Y** and **Z** given **X**, $\boldsymbol{\Sigma}_{\mathbf{YZ}|\mathbf{X}}$, is null. Under the CIA, the decomposition (2.2) of the joint distribution $(\mathbf{X}, \mathbf{Y}, \mathbf{Z})$ is the following.

(a) The marginal distribution for **X** is multinormal with parameters

$$\boldsymbol{\theta}_{\mathbf{X}} = (\boldsymbol{\mu}_{\mathbf{X}}, \boldsymbol{\Sigma}_{\mathbf{XX}}).$$

(b) The distribution of **Y** given **X** is multinormal with parameters

$$\boldsymbol{\theta}_{\mathbf{Y}|\mathbf{X}} = \left(\boldsymbol{\mu}_{\mathbf{Y}|\mathbf{X}}, \boldsymbol{\Sigma}_{\mathbf{YY}|\mathbf{X}}\right).$$

These parameters can be expressed in terms of those in $\boldsymbol{\theta}$ through the following equations:

$$\boldsymbol{\mu}_{\mathbf{Y}|\mathbf{X}} = \boldsymbol{\alpha}_{\mathbf{Y}} + \boldsymbol{\beta}_{\mathbf{YX}} \mathbf{X}, \tag{2.8}$$

$$\boldsymbol{\alpha}_{\mathbf{Y}} = \boldsymbol{\mu}_{\mathbf{Y}} - \boldsymbol{\beta}_{\mathbf{YX}} \boldsymbol{\mu}_{\mathbf{X}}, \tag{2.9}$$

$$\boldsymbol{\beta}_{\mathbf{YX}} = \boldsymbol{\Sigma}_{\mathbf{YX}} \boldsymbol{\Sigma}_{\mathbf{XX}}^{-1}, \tag{2.10}$$

$$\boldsymbol{\Sigma}_{\mathbf{YY}|\mathbf{X}} = \boldsymbol{\Sigma}_{\mathbf{YY}} - \boldsymbol{\Sigma}_{\mathbf{YX}} \boldsymbol{\Sigma}_{\mathbf{XX}}^{-1} \boldsymbol{\Sigma}_{\mathbf{XY}}. \tag{2.11}$$

Equivalently, the regression equation of $\mathbf{Y}$ on $\mathbf{X}$ is

$$\mathbf{Y} = \boldsymbol{\alpha}_{\mathbf{Y}} + \boldsymbol{\beta}_{\mathbf{YX}}\mathbf{X} + \boldsymbol{\epsilon}_{\mathbf{Y}|\mathbf{X}}, \tag{2.12}$$

where $\boldsymbol{\epsilon}_{\mathbf{Y}|\mathbf{X}}$ is a multinormal Q-dimensional r.v. with null mean vector and covariance matrix (the residual variance of the regression) equal to

$$\boldsymbol{\Sigma}_{\mathbf{YY}|\mathbf{X}} = \boldsymbol{\Sigma}_{\mathbf{YY}} - \boldsymbol{\Sigma}_{\mathbf{YX}}\boldsymbol{\Sigma}_{\mathbf{XX}}^{-1}\boldsymbol{\Sigma}_{\mathbf{XY}}.$$

(c) The same holds for the conditional distribution of $\mathbf{Z}$ given $\mathbf{X}$, which is distributed as a multinormal with parameters

$$\boldsymbol{\theta}_{\mathbf{Z}|\mathbf{X}} = \left(\boldsymbol{\mu}_{\mathbf{Z}|\mathbf{X}}, \boldsymbol{\Sigma}_{\mathbf{ZZ}|\mathbf{X}}\right),$$

where

$$\boldsymbol{\mu}_{\mathbf{Z}|\mathbf{X}} = \boldsymbol{\alpha}_{\mathbf{Z}} + \boldsymbol{\beta}_{\mathbf{ZX}}\mathbf{X}, \tag{2.13}$$

$$\boldsymbol{\alpha}_{\mathbf{Z}} = \boldsymbol{\mu}_{\mathbf{Z}} - \boldsymbol{\beta}_{\mathbf{ZX}}\boldsymbol{\mu}_{\mathbf{X}}, \tag{2.14}$$

$$\boldsymbol{\beta}_{\mathbf{ZX}} = \boldsymbol{\Sigma}_{\mathbf{ZX}}\boldsymbol{\Sigma}_{\mathbf{XX}}^{-1}, \tag{2.15}$$

$$\boldsymbol{\Sigma}_{\mathbf{ZZ}|\mathbf{X}} = \boldsymbol{\Sigma}_{\mathbf{ZZ}} - \boldsymbol{\Sigma}_{\mathbf{ZX}}\boldsymbol{\Sigma}_{\mathbf{XX}}^{-1}\boldsymbol{\Sigma}_{\mathbf{XZ}}. \tag{2.16}$$

The corresponding regression equation of $\mathbf{Z}$ given $\mathbf{X}$ is

$$\mathbf{Z} = \boldsymbol{\alpha}_{\mathbf{Z}} + \boldsymbol{\beta}_{\mathbf{ZX}}\mathbf{X} + \boldsymbol{\epsilon}_{\mathbf{Z}|\mathbf{X}}, \tag{2.17}$$

where $\boldsymbol{\epsilon}_{\mathbf{Z}|\mathbf{X}}$ is a multinormal R-dimensional r.v. with null mean vector and covariance matrix equal to

$$\boldsymbol{\Sigma}_{\mathbf{ZZ}|\mathbf{X}} = \boldsymbol{\Sigma}_{\mathbf{ZZ}} - \boldsymbol{\Sigma}_{\mathbf{ZX}}\boldsymbol{\Sigma}_{\mathbf{XX}}^{-1}\boldsymbol{\Sigma}_{\mathbf{XZ}}.$$

The ML estimators of $\boldsymbol{\theta}_{\mathbf{X}}$, $\boldsymbol{\theta}_{\mathbf{Y}|\mathbf{X}}$ and $\boldsymbol{\theta}_{\mathbf{Z}|\mathbf{X}}$ can be computed through the following steps.

(i) The ML estimator of $\boldsymbol{\theta}_{\mathbf{X}}$ is obtained from the overall sample $A \cup B$:

$$\hat{\boldsymbol{\mu}}_{\mathbf{X}} = \frac{1}{n_A + n_B}\left(\sum_{a=1}^{n_A}\mathbf{x}_a + \sum_{b=1}^{n_B}\mathbf{x}_b\right), \tag{2.18}$$

$$\hat{\boldsymbol{\Sigma}}_{\mathbf{XX}} = \frac{1}{n_A + n_B}\sum_{a=1}^{n_A}\left(\mathbf{x}_a - \hat{\boldsymbol{\mu}}_{\mathbf{X}}\right)\left(\mathbf{x}_a - \hat{\boldsymbol{\mu}}_{\mathbf{X}}\right)' + \frac{1}{n_A + n_B}\sum_{b=1}^{n_B}\left(\mathbf{x}_b - \hat{\boldsymbol{\mu}}_{\mathbf{X}}\right)\left(\mathbf{x}_b - \hat{\boldsymbol{\mu}}_{\mathbf{X}}\right)', \tag{2.19}$$

where $\mathbf{x}_a$ and $\mathbf{x}_b$ are column vectors representing respectively the ath and bth records (observations) of the data sets A and B.

(ii) The ML estimator of $\boldsymbol{\theta}_{\mathbf{Y}|\mathbf{X}}$, i.e. the parameters of the regression equation (2.12), is computed on A:

$$\hat{\boldsymbol{\beta}}_{\mathbf{YX}} = S_{\mathbf{YX};A} S^{-1}_{\mathbf{XX};A}, \tag{2.20}$$

$$\hat{\boldsymbol{\alpha}}_{\mathbf{Y}} = \bar{\mathbf{y}}_A - \hat{\boldsymbol{\beta}}_{\mathbf{YX}} \bar{\mathbf{x}}_A, \tag{2.21}$$

$$\hat{\boldsymbol{\Sigma}}_{\mathbf{YY}|\mathbf{X}} = S_{\mathbf{YY};A} - \hat{\boldsymbol{\beta}}_{\mathbf{YX}} S_{\mathbf{XY};A}, \tag{2.22}$$

where $\bar{\mathbf{x}}_A$ and $\bar{\mathbf{y}}_A$ are the sample vector means of $\mathbf{X}$ and $\mathbf{Y}$ respectively in A, and S denotes the sample covariance matrices, according to the subscripts. Note that the estimated regression function of $\mathbf{Y}$ on $\mathbf{X}$ is

$$\bar{\mathbf{y}}_A + S_{\mathbf{YX};A} S^{-1}_{\mathbf{XX};A} \left(\mathbf{X} - \bar{\mathbf{x}}_A\right).$$

(iii) The same arguments hold for the distribution of $\mathbf{Z}$ given $\mathbf{X}$. The ML estimator of $\boldsymbol{\theta}_{\mathbf{Z}|\mathbf{X}}$ is, in obvious notation:

$$\hat{\boldsymbol{\beta}}_{\mathbf{ZX}} = S_{\mathbf{ZX};B} S^{-1}_{\mathbf{XX};B}, \tag{2.23}$$

$$\hat{\boldsymbol{\alpha}}_{\mathbf{Z}} = \bar{\mathbf{z}}_B - \hat{\boldsymbol{\beta}}_{\mathbf{ZX}} \bar{\mathbf{x}}_B, \tag{2.24}$$

$$\hat{\boldsymbol{\Sigma}}_{\mathbf{ZZ}|\mathbf{X}} = S_{\mathbf{ZZ};B} - \hat{\boldsymbol{\beta}}_{\mathbf{ZX}} S_{\mathbf{XZ};B}. \tag{2.25}$$

The estimated regression function of $\mathbf{Z}$ on $\mathbf{X}$ is:

$$\bar{\mathbf{z}}_B + S_{\mathbf{ZX};B} S^{-1}_{\mathbf{XX};B} \left(\mathbf{X} - \bar{\mathbf{x}}_B\right).$$

The ML estimator of $\boldsymbol{\theta}$ is obtained through the previous steps (i)–(iii). In particular, through equations (2.9)–(2.11), the following ML estimators can be computed:

$$\hat{\boldsymbol{\mu}}_{\mathbf{Y}} = \hat{\boldsymbol{\alpha}}_{\mathbf{Y}} + \hat{\boldsymbol{\beta}}_{\mathbf{YX}} \hat{\boldsymbol{\mu}}_{\mathbf{X}}, \tag{2.26}$$

$$\hat{\boldsymbol{\Sigma}}_{\mathbf{YX}} = \hat{\boldsymbol{\beta}}_{\mathbf{YX}} \hat{\boldsymbol{\Sigma}}_{\mathbf{XX}}, \tag{2.27}$$

$$\hat{\boldsymbol{\Sigma}}_{\mathbf{YY}} = \hat{\boldsymbol{\Sigma}}_{\mathbf{YY}|\mathbf{X}} + \hat{\boldsymbol{\Sigma}}_{\mathbf{YX}} \hat{\boldsymbol{\Sigma}}^{-1}_{\mathbf{XX}} \hat{\boldsymbol{\Sigma}}_{\mathbf{XY}}. \tag{2.28}$$

In a similar fashion, through equations (2.14)–(2.16), the following ML estimators are obtained:

$$\hat{\boldsymbol{\mu}}_{\mathbf{Z}} = \hat{\boldsymbol{\alpha}}_{\mathbf{Z}} + \hat{\boldsymbol{\beta}}_{\mathbf{ZX}} \hat{\boldsymbol{\mu}}_{\mathbf{X}}, \tag{2.29}$$

$$\hat{\boldsymbol{\Sigma}}_{\mathbf{ZX}} = \hat{\boldsymbol{\beta}}_{\mathbf{ZX}} \hat{\boldsymbol{\Sigma}}_{\mathbf{XX}}, \tag{2.30}$$

$$\hat{\boldsymbol{\Sigma}}_{\mathbf{ZZ}} = \hat{\boldsymbol{\Sigma}}_{\mathbf{ZZ}|\mathbf{X}} + \hat{\boldsymbol{\Sigma}}_{\mathbf{ZX}} \hat{\boldsymbol{\Sigma}}^{-1}_{\mathbf{XX}} \hat{\boldsymbol{\Sigma}}_{\mathbf{XZ}}. \tag{2.31}$$

Example 2.1 Let A be a sample of size $n_A = 6$ (Table 2.1) from a bivariate normal r.v. (X, Y) with mean

$$(\mu_X, \mu_Y) = (10, 20)$$

Table 2.1 List of the units and of their associated variables in A

a	X	Y
1	9.23	40.61
2	8.52	27.64
3	−7.52	8.41
4	16.20	39.03
5	5.84	5.92
6	23.85	19.61

and covariance matrix

$$\boldsymbol{\Sigma}_{XY} = \begin{pmatrix} 148 & 55 \\ 55 & 153 \end{pmatrix}.$$

Let B be a further sample of size $n_B = 10$ (Table 2.2) generated from a bivariate normal r.v. (X, Z) with mean

$$(\mu_X, \mu_Z) = (10, 30)$$

and covariance matrix

$$\boldsymbol{\Sigma}_{XZ} = \begin{pmatrix} 148 & -32 \\ -32 & 10 \end{pmatrix}.$$

Under the CIA, ML estimates of $(\boldsymbol{\theta}_X, \boldsymbol{\theta}_{Y|X}, \boldsymbol{\theta}_{Z|X})$ should first be computed, and then an ML estimate for (μ_X, μ_Y, μ_Z) and $\boldsymbol{\Sigma}$ can be obtained according to the steps previously described.

Table 2.2 List of the units and of their associated variables in B

b	X	Z
1	4.27	30.49
2	1.18	31.98
3	7.70	30.94
4	22.38	26.30
5	29.92	25.11
6	5.25	31.78
7	10.95	28.29
8	13.43	30.53
9	18.73	29.93
10	13.32	27.55

(i) The ML estimate $\hat{\boldsymbol{\theta}}_{\mathbf{X}} = (\hat{\mu}_X, \hat{\sigma}_X^2)$ is given by equations (2.18) and (2.19):

$$\hat{\mu}_X = 11.45, \quad \hat{\sigma}_X^2 = 82.01.$$

(ii) The ML estimate $\hat{\boldsymbol{\theta}}_{Y|X} = \left(\hat{\mu}_{Y|X}, \hat{\sigma}_{Y|X}^2\right)$ is given by equations (2.20), (2.21) and (2.22):

$$\hat{\beta}_{YX} = 0.65, \quad \hat{\alpha}_Y = 17.46, \quad \hat{\sigma}_{Y|X}^2 = 144.76.$$

(iii) The ML estimate $\hat{\boldsymbol{\theta}}_{Z|X} = \left(\hat{\mu}_{Z|X}, \hat{\sigma}_{Z|X}^2\right)$ is given by equations (2.23), (2.24) and (2.25):

$$\hat{\beta}_{ZX} = -0.23, \quad \hat{\alpha}_Z = 32.21, \quad \hat{\sigma}_{Z|X}^2 = 1.22.$$

Finally, the ML estimates of the marginal parameters are computed through equations (2.26)–(2.31) and the relation $\sigma_{YZ} = \sigma_{YX}\sigma_{ZX}/\sigma_X^2$ due to the CIA:

$$(\hat{\mu}_X, \hat{\mu}_Y, \hat{\mu}_Z) = (11.45, 24.90, 29.58)$$

and

$$\hat{\boldsymbol{\Sigma}} = \begin{pmatrix} 82.01 & 53.31 & -18.86 \\ 53.31 & 179.41 & -12.26 \\ -18.86 & -12.26 & 5.56 \end{pmatrix}.$$

2.1.3 The multinomial case

Let us assume that (X, Y, Z) has a categorical distribution with $I \times J \times K$ categories $\Delta = \{(i, j, k) : i = 1, \ldots, I; j = 1, \ldots, J; k = 1, \ldots, K\}$, and parameter vector $\boldsymbol{\theta} = \left\{\theta_{ijk}\right\}$, $(i, j, k) \in \Delta$:

$$\theta_{ijk} = f\left(i, j, k; \boldsymbol{\theta}\right), \qquad \theta_{ijk} \geq 0, \ \ \forall \, (i, j, k) \in \Delta, \ \sum_{ijk} \theta_{ijk} = 1.$$

Under the CIA, the parameter vector $\boldsymbol{\theta}$ reduces to:

$$\boldsymbol{\theta}_X = \{\theta_{i..}\}; \boldsymbol{\theta}_{Y|X} = \left\{\theta_{j|i} = \frac{\theta_{ij.}}{\theta_{i..}}\right\}; \boldsymbol{\theta}_{Z|X} = \left\{\theta_{k|i} = \frac{\theta_{i.k}}{\theta_{i..}}\right\},$$

where the 'dot' symbol denotes marginalization of the corresponding variable. The parameters of the joint distribution are computed by

$$\theta_{ijk} = \theta_{i..}\theta_{j|i}\theta_{k|i} = \frac{\theta_{ij.}\theta_{i.k}}{\theta_{i..}}, \tag{2.32}$$

$(i, j, k) \in \Delta$. The parameters on the joint distribution of Y and Z are computed easily by

$$\theta_{.jk} = \sum_{i=1}^{I} \theta_{j|i}\theta_{k|i}\theta_{i..}, \qquad j = 1, \ldots, J; k = 1, \ldots, K.$$

When **X**, **Y** and **Z** are multivariate, it is possible to resort to appropriate loglinear models (Appendix B) for each of the following r.v.s: **X**, **Y**|**X** and **Z**|**X**. This approach simplifies the joint relationship of the r.v.s in each vector **X**, **Y**|**X** and **Z**|**X**. In the following sections, we will not consider this last case. In fact, we will assume saturated loglinear models for **X**, **Y**|**X** and **Z**|**X**. Then **X**, **Y** and **Z** can be considered as univariate r.v.s X, Y and Z with I given by the product of the number of categories of the P variables in **X**, J given by the product of the number of categories of the Q variables in **Y**, and K given by the product of the number of categories of the R variables in **Z**.

Let $n^A_{ij.}$ and $n^B_{i.k}$, $(i, j, k) \in \Delta$, be the observed marginal tables from A and B respectively. From the likelihood function (2.3), the ML estimators of $\boldsymbol{\theta}_X$, $\boldsymbol{\theta}_{Y|X}$ and $\boldsymbol{\theta}_{Z|X}$ are given by:

$$\hat{\theta}_{i..} = \frac{n^A_{i..} + n^B_{i..}}{n_A + n_B}, \qquad i = 1, \dots, I; \tag{2.33}$$

$$\hat{\theta}_{j|i} = \frac{n^A_{ij.}}{n^A_{i..}}, \qquad i = 1, \dots, I;\ j = 1, \dots, J; \tag{2.34}$$

$$\hat{\theta}_{k|i} = \frac{n^B_{i.k}}{n^B_{i..}}, \qquad i = 1, \dots, I;\ k = 1, \dots, K. \tag{2.35}$$

Note that the CIA is a particular loglinear model for (X, Y, Z) with ML estimator in closed form.

Example 2.2 Let A and B be two samples of size $n_A = 140$ and $n_B = 220$, and let Tables 2.3 and 2.4 be the corresponding contingency tables.

Maximum likelihood estimates of the parameters θ_{ijk} can be computed following equation (2.32). Thus, the estimates of $\theta_{i..}$, $\theta_{j|i}$ and $\theta_{k|i}$ are needed for the

Table 2.3 (X, Y) contingency table computed on sample A

	$Y = 1$	$Y = 2$	Total
$X = 1$	40	40	80
$X = 2$	30	30	60
Total	70	70	140

Table 2.4 (X, Z) contingency table computed on sample B

	$Z = 1$	$Z = 2$	$Z = 3$	Total
$X = 1$	40	20	40	100
$X = 2$	72	24	24	120
Total	112	64	64	220

Table 2.5 Maximum likelihood estimates of $\theta_{i..}$, $i = 1, 2$, given sample A as in Table 2.3 and sample B as in Table 2.4

$X = 1$	$X = 2$
0.50	0.50

Table 2.6 Maximum likelihood estimates of $\theta_{j|i}$, $i = 1, 2$, $j = 1, 2$, given sample A as in Table 2.3

	$Y = 1$	$Y = 2$
$X = 1$	0.50	0.50
$X = 2$	0.50	0.50

Table 2.7 Maximum likelihood estimates of $\theta_{k|i}$, $i = 1, 2$, $k = 1, 2, 3$, given sample B as in Table 2.4

	$Z = 1$	$Z = 2$	$Z = 3$
$X = 1$	0.40	0.20	0.40
$X = 2$	0.60	0.20	0.20

Table 2.8 Maximum likelihood estimates of θ_{ijk}, $j = 1, 2$, $k = 1, 2, 3$, given sample A as in Table 2.3 and sample B as in Table 2.4

	$X = 1$			$X = 2$		
	$Z = 1$	$Z = 2$	$Z = 3$	$Z = 1$	$Z = 2$	$Z = 3$
$Y = 1$	0.10	0.05	0.10	0.15	0.05	0.05
$Y = 2$	0.10	0.05	0.10	0.15	0.05	0.05

estimation of the joint distribution. Tables 2.5, 2.6, and 2.7 show the estimates $\hat{\boldsymbol{\theta}}_X$, $\hat{\boldsymbol{\theta}}_{Y|X}$ and $\hat{\boldsymbol{\theta}}_{Z|X}$. The final estimates for the joint parameters θ_{ijk} are shown in Table 2.8.

2.2 The Micro (Predictive) Approach in the Parametric Framework

The predictive approach aims to construct a synthetic complete data set for $(\mathbf{X}, \mathbf{Y}, \mathbf{Z})$, by filling in missing values in A and B. In other words, missing $\mathbf{Z}$ in

A and missing $\mathbf{Y}$ in B are predicted. Once a parametric model has been estimated, a synthetic data set of completed records may be obtained substituting the missing items in the overall file $A \cup B$ by a suitable value from the distribution of the corresponding variables given the observed variables. Actually, this approach can be considered as a single imputation method that makes use of an explicit parametric model. There are essentially two broad categories of micro approaches in the parametric framework: conditional mean matching (Section 2.2.1) and draws based on a predictive distribution (Section 2.2.2).

Remark 2.7 In this section we still consider $A \cup B$ as a unique partially observed sample of i.i.d. records from $f(\mathbf{x}, \mathbf{y}, \mathbf{z})$. Hence, A or B should be used for the estimation of $f(\mathbf{x}, \mathbf{y}, \mathbf{z})$. In this case, either A or B or both can be imputed.

Actually, the same mechanisms, with minor changes, can be applied under the framework of Remark 1.2. In this case, B is used for the estimation of the appropriate parameters of $\mathbf{Z}$ given $\mathbf{X}$, and only A is imputed.

2.2.1 Conditional mean matching

One of the most important predictive approaches substitutes each missing item with the expectation of the missing variable given the observed ones. This can be done in a straightforward way when the variables in $\mathbf{Y}$ and $\mathbf{Z}$ are continuous, i.e.

$$\tilde{\mathbf{z}}_a = E(\mathbf{Z}|\mathbf{X} = \mathbf{x}_a) = \int_{\mathcal{Z}} \mathbf{z} f_{\mathbf{Z}|\mathbf{X}}(\mathbf{z}|\mathbf{x}_a; \boldsymbol{\theta}_{\mathbf{Z}|\mathbf{X}}) d\mathbf{z}, \qquad a = 1, \ldots, n_A, \tag{2.36}$$

$$\tilde{\mathbf{y}}_b = E(\mathbf{Y}|\mathbf{X} = \mathbf{x}_b) = \int_{\mathcal{Y}} \mathbf{y} f_{\mathbf{Y}|\mathbf{X}}(\mathbf{y}|\mathbf{x}_b; \boldsymbol{\theta}_{\mathbf{Y}|\mathbf{X}}) d\mathbf{y}, \qquad b = 1, \ldots, n_B. \tag{2.37}$$

The unknown parameters $\boldsymbol{\theta}_{\mathbf{Z}|\mathbf{X}}$ and $\boldsymbol{\theta}_{\mathbf{Y}|\mathbf{X}}$ can be substituted by the corresponding ML estimates described in Section 2.1, when the variables are multinormal. Hence, the imputed values are the values defined by the estimated regression functions of $\mathbf{Z}$ on $\mathbf{X}$ and of $\mathbf{Y}$ on $\mathbf{X}$ respectively.

The substitution of the expected value of a variable for each missing item seems appealing, given that it is the best point estimate with respect to a quadratic loss function. However, it should not be considered as a good matching method. In fact, it is evident that the synthetic data set will be affected by at least two drawbacks: (i) the predicted value may be not a really observed (i.e. live) value; (ii) the synthetic distribution of the predicted values of $\mathbf{Y}$ ($\mathbf{Z}$) is concentrated on the expected value of $\mathbf{Y}$ ($\mathbf{Z}$) given $\mathbf{X}$ (further comments are postponed to Section 2.2.3). Nevertheless, these values can still be useful, as illustrated in Section 2.5.

Example 2.3 When the continuous variables are normal, the conditional mean imputation approach is the regression imputation, as in Little and Rubin (2002, p. 62). Let us consider the situation outlined in Section 2.1.1. The predictive approach would consider the following predicted values:

$$\tilde{z}_a^A = \hat{\alpha}_Z + \hat{\beta}_{ZX} x_a^A, \qquad a = 1, \ldots, n_A, \tag{2.38}$$

$$\tilde{y}_b^B = \hat{\alpha}_Y + \hat{\beta}_{YX} x_b^B, \qquad b = 1, \ldots, n_B, \tag{2.39}$$

where the ML estimates of the regression parameters are computed in Section 2.1.1. Note that some of the values $\tilde{z}_a^A$, $a = 1, \ldots, n_A$, and $\tilde{y}_b^B$, $b = 1, \ldots, n_B$, may be never observed in a real context. Furthermore, all the imputations lie on the regression line, i.e. there is no variability around it. As an example, let A and B be the samples described respectively in Tables 2.1 and 2.2 of Example 2.1. Conditional mean matching will apply equations (2.38) and (2.39) to the observed records, i.e.:

$$\tilde{z}_a^A = 32.21 - 0.23 x_a^A, \quad a = 1, \ldots, 6,$$

$$\tilde{y}_b^B = 17.46 + 0.65 x_b^B, \quad b = 1, \ldots, 10.$$

The matched files are illustrated in Tables 2.9 and 2.10.

Table 2.9 List of the units, their associated variables, and the conditional mean imputed values in A

a	X	Y	$\tilde{Z}$
1	9.23	40.61	30.09
2	8.52	27.64	30.25
3	−7.52	8.41	33.94
4	16.20	39.03	28.48
5	5.84	5.92	30.87
6	23.85	19.61	26.72

Table 2.10 List of the units, their associated variables, and the conditional mean imputed values in B

b	X	$\tilde{Y}$	Z
1	4.27	20.24	30.49
2	1.18	18.23	31.98
3	7.70	22.46	30.94
4	22.38	32.01	26.30
5	29.92	36.91	25.11
6	5.25	20.87	31.78
7	10.95	24.58	28.29
8	13.43	26.19	30.53
9	18.73	29.63	29.93
10	13.32	26.12	27.55

This imputation method was first introduced in Buck (1960). It can be shown that it allows the sample mean of the completed data to be a consistent estimator of the mean of the imputed variable and an asymptotically normal estimator, although the usual variance estimators are not consistent estimators of the variance of the imputed variable (Little and Rubin, 2002).

These drawbacks are more evident when the variables are categorical. In this case, the variables are replaced by the indicator variable of each category,

$$I_j^Y = \begin{cases} 1 & \text{if } Y = j, \\ 0 & \text{otherwise}, \end{cases} \qquad j = 1, \ldots, J,$$

and analogously I_k^Z, $k = 1, \ldots, K$. Actually, the predicted values are:

$$\tilde{I}_{k;a}^Z = E(I_k^Z | X = x_a) = \theta_{k|x_a}, \qquad a = 1, \ldots, n_A,$$

$$\tilde{I}_{j;b}^Y = E(I_j^Y | X = x_b) = \theta_{j|x_b}, \qquad b = 1, \ldots, n_B,$$

and the parameters $\theta_{k|i}$ and $\theta_{j|i}$ may be substituted by their ML counterparts described in Section 2.1.3. The only meaning of the previous expectations is the probability of belonging to a particular cell. If at a unit level the previous information does not help much, they are exactly the counts that are used for the computation of the expected unobserved contingency tables, as explained in Example 2.4. These predicted values have generally been used in a two-step procedure, playing the role of *intermediate* values, as shown in Section 2.5.

Example 2.4 Let (X, Y, Z) be as in Section 2.1.3. The predicted values $\tilde{I}_{k;a}^Z$, $a = 1, \ldots, n_A$, and $\tilde{I}_{j;b}^Y$, $b = 1, \ldots, n_B$, are the counts used for the computation of the ML estimate of the overall (unknown) contingency table for the variables X, Y and Z, denoted by $\mathbf{n}_{XYZ} = \{n_{ijk}\}$, among the $n_A + n_B$ units in $A \cup B$, i.e. of the table compatible with the ML estimates of the parameters, as in Section 2.1.3. This is easily seen from the following:

$$\begin{aligned}\hat{n}_{ijk} &= (n_A + n_B)\hat{\theta}_{ijk} = (n_A + n_B)\hat{\theta}_{i..}\hat{\theta}_{j|i}\hat{\theta}_{k|i} = \left(n_{i..}^A + n_{i..}^B\right)\hat{\theta}_{j|i}\hat{\theta}_{k|i} \\ &= n_{i..}^A \frac{n_{ij.}^A}{n_{i..}^A}\hat{\theta}_{k|i} + n_{i..}^B \hat{\theta}_{j|i}\frac{n_{i.k}^B}{n_{i..}^B} = \sum_{a=1}^{n_A} I_{i;a}^X I_{j;a}^Y \hat{\theta}_{k|i} + \sum_{b=1}^{n_B} I_{i;b}^X \hat{\theta}_{j|i} I_{k;b}^Z \\ &= \sum_{a=1}^{n_A} I_{i;a}^X I_{j;a}^Y \tilde{I}_{k;a}^Z + \sum_{b=1}^{n_B} I_{i;b}^X \tilde{I}_{j;b}^Y I_{k;b}^Z, \qquad (i, j, k) \in \Delta.\end{aligned}$$

It must be emphasized that the missing items are not replaced by a particular value, but by a distribution. For instance, a generic unit $a \in A$ replaces the missing z_a value with the estimated distribution $\hat{\theta}_{k|i}$, $k = 1, \ldots, K$. Nevertheless, this procedure has an optimal property, i.e. the marginal observed distributions for X on the overall sample $A \cup B$, for $Y|X$ on A, and for $Z|X$ on B are preserved in $\hat{\mathbf{n}}_{XYZ}$, which is the contingency table consisting of the estimated $\hat{n}_{ijk}$. On the other hand, the marginal Y and Z distributions observed respectively on A and B are not preserved unless $n_{i..}^A = n_{i..}^B$ for all $i = 1, \ldots, I$.

2.2.2 Draws based on conditional predictive distributions

As already noted, one of the drawbacks of the conditional mean matching method is the absence of variability for the imputations relative to the same conditioning variables. Little and Rubin (2002, p. 66) show that, under the assumption that missing data follow a MAR mechanism, the data generating multivariate distributions are better preserved by imputing missing values with a random draw from a predictive distribution. In the statistical matching problem, this corresponds to drawing a random value from $f_{\mathbf{Z}|\mathbf{X}}(\mathbf{z}|\mathbf{x}_a; \hat{\boldsymbol{\theta}}_{\mathbf{Z}|\mathbf{X}})$ for every $a = 1, \ldots, n_A$, and a random value from $f_{\mathbf{Y}|\mathbf{X}}(\mathbf{y}|\mathbf{x}_b; \hat{\boldsymbol{\theta}}_{\mathbf{Y}|\mathbf{X}})$ for every $b = 1, \ldots, n_B$ (where the two densities are estimated respectively in B and A). Note that we are not considering a predictive distribution from a Bayesian point of view (this topic is deferred to Section 2.7). In fact, the distributions used for the random draw are obtained by substituting the unknown parameter values $\boldsymbol{\theta}_{\mathbf{Y}|\mathbf{X}}$ and $\boldsymbol{\theta}_{\mathbf{Z}|\mathbf{X}}$ with their ML estimates, as shown in Section 2.1.

Example 2.5 This method is particularly suitable when $\mathbf{X}$, $\mathbf{Y}$ and $\mathbf{Z}$ are multinormal. In this case, this approach is referred to as *stochastic regression imputation*. It consists in estimating the regression parameters by ML, following the results of Section 2.1.2, and imputing for each $b = 1, \ldots, n_B$ the value

$$\tilde{\mathbf{y}}_b = \hat{\boldsymbol{\alpha}}_{\mathbf{Y}} + \hat{\boldsymbol{\beta}}_{\mathbf{YX}}\mathbf{x}_b + \mathbf{e}_b, \qquad (2.40)$$

where $\mathbf{e}_b$ is a value generated randomly from a multinormal r.v. with zero mean vector and estimated residual variance $\hat{\boldsymbol{\Sigma}}_{\mathbf{YY}|\mathbf{X}}$. The same holds for the completion of the data set A.

Again, as in Example 2.3, let A (Table 2.1) and B (Table 2.2) be completed through draws based on predictive distributions. Formula (2.40) is

$$\tilde{y}_b^B = 17.46 + 0.65x_b^B + e_b, \quad b = 1, \ldots, 10,$$

where e_b is a value generated randomly from a normal r.v. with zero mean and estimated residual variance $\hat{\sigma}_{Y|X}^2 = 144.76$. Analogously, for the imputation of the Z values in B, the formula to use is

$$\tilde{z}_a^A = 32.21 - 0.23x_a^A + e_a, \quad a = 1, \ldots, 6,$$

where e_b is a value generated randomly form a normal r.v. with zero mean and estimated residual variance $\hat{\sigma}_{Z|X}^2 = 1.22$. One of the possible matched files is illustrated in Tables 2.11 (completion of A) and 2.12 (completion of B).

Table 2.11 List of the units, their associated variables, and the imputed values in A

a	X	Y	$\tilde{Z}$
1	9.23	40.61	28.98
2	8.52	27.64	29.68
3	−7.52	8.41	33.69
4	16.20	39.03	28.31
5	5.84	5.92	32.53
6	23.85	19.61	26.29

Table 2.12 List of the units, their associated variables, and the imputed values in B

b	X	$\tilde{Y}$	Z
1	4.27	20.88	30.49
2	1.18	28.27	31.98
3	7.70	26.37	30.94
4	22.38	34.84	26.30
5	29.92	27.99	25.11
6	5.25	10.72	31.78
7	10.95	22.56	28.29
8	13.43	25.54	30.53
9	18.73	34.30	29.93
10	13.32	35.81	27.55

2.2.3 Representativeness of the predicted files

When dealing with a micro approach, it is necessary to understand whether the synthetic data set created may satisfy the user's objectives, i.e. to make inference on the joint distribution of $(\mathbf{X}, \mathbf{Y}, \mathbf{Z})$, $f(\mathbf{x}, \mathbf{y}, \mathbf{z}; \boldsymbol{\theta})$. As a consequence, the synthetic data set should be *representative* of the distribution $f(\mathbf{x}, \mathbf{y}, \mathbf{z}; \boldsymbol{\theta})$ or, in other words, the synthetic sample should be considered as generated from the distribution $f(\mathbf{x}, \mathbf{y}, \mathbf{z}; \boldsymbol{\theta})$. In both the previous approaches, ML estimators of the parameters have been considered. One of the properties of the ML estimator is that, under quite general conditions, it is consistent. Consequently, at least for large data sets, $\hat{\boldsymbol{\theta}}$ can be considered approximately equal to the real and unknown parameter vector $\boldsymbol{\theta}$. As a result, the synthetic data set created in Section 2.2.2 may be considered as approximately representative of $f(\mathbf{x}, \mathbf{y}, \mathbf{z}; \boldsymbol{\theta})$.

The same does not hold for the data set created in Section 2.2.1 through conditional mean matching, as first noted by Kadane (1978). Without loss of generality, consider the completed A data set, $(x_a, y_a, \tilde{z}_a)$, $a = 1, \ldots, n_A$, when X, Y and Z are normal, as in Section 2.3. Again, assume the sample sizes are large enough to ensure that the ML estimates of the parameters are (approximately) equal to the true ones. The predicted values $\tilde{z}_a$ have been computed by means of the regression function (2.38), i.e. are generated by the r.v.

$$\tilde{Z} = \alpha_Z + \beta_{ZX} X.$$

Consequently, the completed data set should be considered as generated by the r.v. $(X, Y, \tilde{Z})$ which is normal with mean vector $\boldsymbol{\mu}$ but variance matrix

$$\boldsymbol{\Phi} = \begin{pmatrix} \sigma_X^2 & \sigma_{XY} & \sigma_{XZ} \\ \sigma_{XY} & \sigma_Y^2 & \sigma_{XY}\sigma_{XZ}/\sigma_X^2 \\ \sigma_{XZ} & \sigma_{XY}\sigma_{XZ}/\sigma_X^2 & \sigma_{XZ}^2/\sigma_X^2 \end{pmatrix}.$$

In other words, the variance of Z, σ_Z^2, is actually underestimated and equals the variance due to regression (see, for example (2.5), noting that $\sigma_{Z|X}^2$ is always nonnegative). Furthermore, $(X, Y, \tilde{Z})$ is a singular distribution, given that $\tilde{Z}$ is a linear combination of the other variables. The same holds for the completed file B, considering that the predicted values in (2.39) are generated by a linear combination of the other variables.

2.3 Nonparametric Macro Methods

The previous sections have presented some results when the family $\mathcal{F}$ of distributions of interest is parametric. Actually, when the variables are categorical or discrete, the multinomial distribution is a very flexible one. However, when the variables are continuous, there may not be enough information to restrict $\mathcal{F}$ to a parametric family of distributions (e.g. the multinormal). In such cases, nonparametric methods are preferable, in the sense that they are not affected by misleading assumptions on the parametric form of $\mathcal{F}$. As in the previous paragraphs, it is possible to consider two approaches: the macro and the micro. While the micro approach has been widely used for solving the statistical matching problem (see Section 2.4), the macro approach has not received much attention, with the remarkable exception of Paass (1985). The macro approach consists of the estimation of the joint distribution of **X**, **Y**, **Z**. This approach deserves some attention given the large amount of literature now available on the topic: a general reference for nonparametric density estimation on complete data sets is Wand and Jones (1995).

A first nonparametric approach consists in the computation of the empirical cumulative distribution function as an estimate of the joint cumulative distribution function of $(\mathbf{Y}, \mathbf{Z})$ given $\mathbf{X}$, for $\mathbf{X}$ categorical:

$$F_{\mathbf{YZ}|\mathbf{X}}(\mathbf{y}, \mathbf{z}|\mathbf{x}) = \int_{\mathbf{t} \leq \mathbf{y}} \int_{\mathbf{v} \leq \mathbf{z}} f_{\mathbf{YZ}|\mathbf{X}}(\mathbf{t}, \mathbf{v}|\mathbf{x})\, d\mathbf{t} d\mathbf{v}, \tag{2.41}$$

where $\mathbf{t} \le \mathbf{y}$ and $\mathbf{v} \le \mathbf{z}$ are componentwise inequalities. Following the CIA, it is useful to factorize (2.41):

$$F_{\mathbf{YZ}|\mathbf{X}}(\mathbf{y}, \mathbf{z}|\mathbf{x}) = F_{\mathbf{Y}|\mathbf{X}}(\mathbf{y}|\mathbf{x})\, F_{\mathbf{Z}|\mathbf{X}}(\mathbf{z}|\mathbf{x}). \tag{2.42}$$

The factors in (2.42) can be estimated consistently by:

$$\hat{F}_{\mathbf{Y}|\mathbf{X}}(\mathbf{y}|\mathbf{x}) = \frac{\sum_{a=1}^{n_A} I(\mathbf{y}_a \le \mathbf{y})\, I(\mathbf{x}_a = \mathbf{x})}{\sum_{a=1}^{n_A} I(\mathbf{x}_a = \mathbf{x})}, \tag{2.43}$$

$$\hat{F}_{\mathbf{Z}|\mathbf{X}}(\mathbf{z}|\mathbf{x}) = \frac{\sum_{b=1}^{n_B} I(\mathbf{z}_b \le \mathbf{z})\, I(\mathbf{x}_b = \mathbf{x})}{\sum_{b=1}^{n_B} I(\mathbf{x}_b = \mathbf{x})}. \tag{2.44}$$

Note that it may happen that some $\mathbf{X}$ categories are unobserved in A and/or B. In this case, the corresponding empirical cumulative distribution function cannot be estimated for that category of $\mathbf{X}$.

A different approach consists in estimating the density $f(\mathbf{x}, \mathbf{y}, \mathbf{z})$ directly. For the sake of simplicity, let X, Y and Z be univariate continuous r.v.s. The objective is the estimation of the different components of the joint density of X, Y and Z:

$$f(x, y, z) = f_X(x) f_{Y|X}(y|x) f_{Z|X}(z|x). \tag{2.45}$$

When the $n_A + n_B$ units in the overall sample $A \cup B$ are i.i.d. observations generated from a density $f \in \mathcal{F}$, it seems plausible to apply the procedure already considered in Section 2.1, i.e. estimate the different components of (2.45) from the appropriate subset of the overall sample $A \cup B$. The difference is in the use of nonparametric procedures for the estimation of the densities in (2.45), e.g. the *kernel density estimator* (see Wand and Jones (1995)).

(i) The marginal density $f_X(x)$ can be estimated from the overall sample $A \cup B$ with the kernel density estimator:

$$\hat{f}_X(x) = \frac{1}{n_A + n_B}\left[\sum_{a=1}^{n_A} \frac{1}{h} K_1\left(\frac{x - x_a}{h}\right) + \sum_{b=1}^{n_B} \frac{1}{h} K_1\left(\frac{x - x_b}{h}\right)\right],$$

$x \in \mathcal{X}$, for some kernel function $K_1(x)$ satisfying $\int_{-\infty}^{\infty} K_1(x)\mathrm{d}x = 1$.

(ii) The conditional density of Y given X can be estimated by first computing the joint density of X and Y on file A:

$$\hat{f}_{XY}(x, y) = \frac{1}{n_A} \sum_{a=1}^{n_A} \frac{1}{h_x h_y} K_2\left(\frac{x - x_a}{h_x}, \frac{y - y_a}{h_y}\right), \qquad x \in \mathcal{X},\ y \in \mathcal{Y},$$

with $\int_{-\infty}^{\infty}\int_{-\infty}^{\infty} K_2(x, y)\mathrm{d}x\mathrm{d}y = 1$. Then an estimate of $f_{Y|X}(y|x)$ is

$$\hat{f}_{Y|X}(y|x) = \frac{\hat{f}_{XY}(x, y)}{\int_{-\infty}^{\infty} \hat{f}_{XY}(x, y)\mathrm{d}y}$$

$$= \frac{\sum_{a=1}^{n_A} K_2\left(\frac{x - x_a}{h_x}, \frac{y - y_a}{h_y}\right)}{\sum_{a=1}^{n_A} \int_{-\infty}^{\infty} K_2(\frac{x - x_a}{h_x}, y)\mathrm{d}y}, \qquad x \in \mathcal{X},\ y \in \mathcal{Y}.$$

(iii) Analogously, the conditional density of Z given X can be estimated by first computing the joint density of X and Z on file B:

$$\hat{f}_{XZ}(x, y) = \frac{1}{n_B} \sum_{b=1}^{n_B} \frac{1}{h_x h_z} K_3 \left(\frac{x - x_b}{h_x}, \frac{z - z_b}{h_z} \right), \qquad x \in \mathcal{X},\ z \in \mathcal{Z},$$

with $\int_{-\infty}^{\infty} \int_{-\infty}^{\infty} K_3(x, z) \mathrm{d}x \mathrm{d}z = 1$. Then an estimate of $f_{Z|X}(z|x)$ is

$$\hat{f}_{Z|X}(z|x) = \frac{\hat{f}_{XZ}(x, z)}{\int_{-\infty}^{\infty} \hat{f}_{XZ}(x, z) \mathrm{d}z}$$

$$= \frac{\sum_{b=1}^{n_B} K_3 \left(\frac{x - x_b}{h_x}, \frac{z - z_b}{h_z} \right)}{\sum_{b=1}^{n_B} \int_{-\infty}^{\infty} K_3 (\frac{x - x_b}{h_x}, z) \mathrm{d}z}, \qquad x \in \mathcal{X},\ z \in \mathcal{Z}.$$

The choice of the kernel functions and of the bandwidth h depends on many factors; see Silverman (1986) and Wand and Jones (1995). More precisely, different values of h might be used for X and Y in K_2 as well as for X and Z in K_3.

An approach similar in spirit to those usually adopted in the statistical matching problem makes use of the nearest neighbour method instead of the kernel method. Let $\mathbf{X}$ be a P-dimensional vector of r.v.s. For any $\mathbf{x} \in \mathcal{X}$, let

$$d_1(\mathbf{x}) \leq d_2(\mathbf{x}) \leq \cdots \leq d_{n_A + n_B}(\mathbf{x})$$

be the $n_A + n_B$ ordered distances of each observed value versus $\mathbf{x}$, for instance, the Euclidean distance may be used (see Appendix C for different distance definitions). Following Silverman (1986, p. 98), the k nearest neighbour (kNN) estimate for the density $f_{\mathbf{X}}(\mathbf{x})$ is

$$\hat{f}_{\mathbf{X}}(\mathbf{x}) = \frac{k}{(n_A + n_B)\, c_P [d_k(\mathbf{x})]^P}, \qquad \mathbf{x} \in \mathcal{X},$$

where $c_P [d_k(\mathbf{x})]^P$ is the volume of the P-dimensional sphere of radius $d_k(\mathbf{x})$, and c_p is the volume of the P-dimensional sphere when the radius equals 1, i.e. $c_1 = 2$, $c_2 = \pi$, $c_3 = 4\pi/3$, etc. The previous arguments may be extended to the estimation of the bivariate densities $f_{XY}(x, y)$ and $f_{XZ}(x, z)$, useful for the computation of the conditional distributions $f_{Y|X}(y|x)$ and $f_{Z|X}(z|x)$.

Instead of the full distribution of (X, Y, Z), interest might centre on some important characteristics of that distribution. One important characteristic is the conditional expectation of Z (Y) given X. This function, also known as the nonparametric regression function, will be useful for predictive purposes, as in Section 2.4.3. For the sake of simplicity, we will focus on the nonparametric regression estimator of Z given X (similar results can be obtained for Y on X). The nonparametric regression function of Z given X is that function r such that

$$Z = r(X) + \epsilon,$$

where ϵ is an r.v. such that $E(\epsilon|X) = 0$.

One estimator is based on kernels (see Härdle 1992):

$$\hat{r}_k(x) = \frac{\sum_{b=1}^{n_B} K\left(\frac{x-x_b}{h}\right) z_b}{\sum_{b=1}^{n_B} K\left(\frac{x-x_b}{h}\right)}.$$

For the choice of the kernel function and the bandwidth h, the previous arguments remain valid.

Another nonparametric regression estimator, that is also frequently used in the context of nonparametric micro approach, is the kNN (see Section 2.4.3):

$$\hat{r}_k(x) = \frac{1}{n_B} \sum_{b=1}^{n_B} W_{kb}(x) z_b, \tag{2.46}$$

where W_{kb}, $b = 1, \ldots, n_B$, is a sequence of weights defined by ranking $|x - x_b|$ from the lowest to the largest. Let J_x denote the set of the first k ranks. Then

$$W_{kb}(x) = \begin{cases} \frac{n_B}{k} & b \in J_x, \\ 0 & \text{otherwise.} \end{cases}$$

The kNN estimator corresponds to estimating the conditional mean of Z given $X = x$ with a local average composed of the first k nearest neighbours of x, following the idea well illustrated by Eubank (1988): 'If m is believed to be smooth, then the observations at x_b near x should contain information about the value of r at x. Thus it should be possible to use something like a local average of the data near x to construct an estimator of $r(x)$.' The parameter k is called the smoothing parameter and balances the degree of fidelity to the data against the smoothness of the estimated curve. The choices $k = n$ and $k = 1$ correspond to independence in mean between Z and X (the regression function is a constant equal to the sample mean of Z), and to a line connecting the n_b couples (x_b, z_b), $b = 1, \ldots, n_B$. For a review of nonparametric regression methods, see Härdle (1992).

All the estimators discussed in this section satisfy important asymptotic properties. Of particular importance is their consistency, that is ensured under appropriate assumptions.

2.4 The Nonparametric Micro Approach

The first papers on statistical matching, among them Okner (1972), were focused on the definition of a complete synthetic data set through the fusion of the data sets A and B, without the assumption of any particular parametric family of distributions for the variables of interest. This objective might be pursued in two alternative ways:

(i) random draws (as in Section 2.2.2), having estimated the distribution of $(\mathbf{X}, \mathbf{Y}, \mathbf{Z})$ in a nonparametric framework;

(ii) conditional mean matching (as in Section 2.2.1), having estimated the nonparametric regression function of the variables to be imputed given the observed variables.

Although the two alternative approaches can be applied by means of a variety of nonparametric estimation procedures, such as those illustrated in Section 2.3, it is usual practice to apply a particular set of nonparametric imputation procedures, usually denoted as *hot deck imputation procedures*. These procedures are characterized by the fact that they fill missing values with observed (live) ones.

The imputation procedures in the hot deck family seem attractive because they do not need any specification of a family of distributions (i.e. they are nonparametric) and they do not need any estimate of the distribution function or of any of its characteristics. Nevertheless, it will be seen that hot deck methods implicitly assume a particular estimate of either a distribution or a conditional mean function. Given the wide use of hot deck imputation procedures in statistical matching, they are described at a certain level of detail.

Most of the articles on statistical matching that make use of hot deck methods (e.g. Singh *et al.*, 1993, and references therein) come under the framework illustrated in Remark 1.2. In this case, the two samples A and B are assigned different roles. One sample assumes the role of the *recipient file* (also called the *host* file): the missing items of each record of the recipient file are imputed using records (suitably chosen) from the other sample, the *donor file*.

Remark 2.8 The distinction between recipient and donor files is typical of many statistical matching applications. One of the most important is the Social Policy Simulation Database (SPSD), a microsimulation database created at Statistics Canada (Wolfson *et al.*, 1987). The SPSD uses the Survey of Consumer Finance (SCF) as recipient file. The SCF is successively enriched with information obtained from unemployment insurance claim histories, personal income tax returns, and the Family Expenditure Survey.

The choice of which file should be the recipient and which the donor depends on many factors. The most important are the phenomena under study and the accuracy of information contained in the two files, as stated in Remark 1.2.

Remark 2.9 It is standard practice to assign the role of recipient and donor files also to samples that are equally reliable, in the sense that they can be considered as generated from the same distribution $f(\mathbf{x}, \mathbf{y}, \mathbf{z})$. In this case, another important aspect to consider in assigning the appropriate role to the files is their sample size. For instance, if the number of records in the two files is markedly different, it is common practice to choose the smaller file as the recipient. In fact, if the smaller file was the donor, some records in the donor file would be imputed more than once in the recipient, artificially modifying the variability of the distribution of the imputed variable in the final synthetic file.

Remark 2.10 Let us further explore the meaning of Remarks 1.2 and 2.9 when a synthetic data set is generated with the use of imputation techniques. The matter arises whether it is enough to have a synthetic data set consisting only of the completed A file or if the overall sample $A \cup B$ should be imputed.

Under the i.i.d. assumption of the $n_A + n_B$ units in $A \cup B$, two opposite situations may occur.

(i) A and B are assigned the role of recipient and donor files respectively.

- There is an inefficient use of the available information. In fact, when the synthetic data set is just the completed file A, important sample information on $\mathbf{X}$ in B is discarded.
- Unless limit cases happen, and despite all efforts to produce good imputations, imputed values are generations of an r.v. whose distribution differs from the real (unknown) data generating distribution for finite sample sizes. The larger the donor file, the more accurate the estimated distribution of $\mathbf{Z}$ given $\mathbf{X}$ when consistent estimators are used. This reason justifies the strategy of choosing as recipient file the one with the smaller sample size.

(ii) The overall sample $A \cup B$ can be imputed through hot deck (i.e. both A and B play the role of donor and recipient file in turn).

- In this case the available sample information is fully exploited.
- The effect of the matching noise is magnified (see Section 1.4.3). In particular, when n_B is much larger than n_A, the marginal and conditional $\mathbf{Y}$ distributions can be heavily affected by the matching noise.

As far as the previous cases are concerned, it is not clear which of the two procedures is better. Further research should be devoted to this topic.

When the i.i.d. assumption for the $n_A + n_B$ units in $A \cup B$ can be relaxed, as in Remark 1.2, the recipient file should be held fixed (say, A) and the roles of recipient and donor file cannot be exchanged. In fact, it is enough that A is a set of n_A records generated by $f(\mathbf{x}, \mathbf{y}, \mathbf{z})$ (the distribution to estimate), with $\mathbf{Z}$ missing, and B is a sample of n_B i.i.d. records generated by a distribution $g_{\mathbf{XZ}}(\mathbf{x}, \mathbf{z})$ such that

$$f_{\mathbf{Z}|\mathbf{X}}(\mathbf{z}|\mathbf{x}) = g_{\mathbf{Z}|\mathbf{X}}(\mathbf{z}|\mathbf{x}).$$

When completing A, imputation of A records with live $\mathbf{Z}$ values from B corresponds to mimicking in the completed recipient file the conditional distribution of $\mathbf{Z}$ given $\mathbf{X}$ observed in B. In this case, the alternative of imputing the overall sample $A \cup B$ does not make sense.

From now on, we assume, without loss of generality, that A is the recipient file and B the donor. Hence, the objective is the imputation of $\mathbf{Z}$ in A through the use of the observed units in B. Three hot deck methods have been used in statistical matching (Singh *et al.*, 1993):

(i) *random* hot deck;

(ii) *rank* hot deck;

(iii) *distance* hot deck.

These methods can be considered as the nonparametric counterpart of parametric micro procedures (Section 2.2). Comments on the effectiveness of the hot deck methods in producing a synthetic sample representing a proper sample generated from $f(\mathbf{x}, \mathbf{y}, \mathbf{z})$ are discussed in Section 2.4.4.

2.4.1 Random hot deck

Random hot deck consists in randomly choosing a donor record (in the donor file) for each record in the recipient file. Sometimes the random choice is made within a suitable subset of units in the donor files. In particular, units of both the files are usually grouped into homogeneous subsets according to given common characteristics (units in the same geographical area, individuals with the same demographic characteristics, etc.); we will refer to these subsets as donation classes. Thus, for an individual in a given geographical area, only records in the same area will be considered as possible donors. In general, the donation classes are defined using one or few categorical variables $\mathbf{X}$ chosen within the set of common variables in A and B.

To show how random hot deck works, let us consider a very simple example.

Example 2.6 Let A contain only $n_A = 6$ units, on which three variables have been observed: 'Sex', 'Age' and 'Annual Personal Net Income' (INCOME) (see Table 2.13). Let B contain $n_B = 10$ records, and the observed variables are: 'Sex', 'Age' and 'Annual Personal Expenditure' (EXPENSE) (Table 2.14). Thus we have a set of two common variables $\mathbf{X} = (X_1 =$ 'Sex', $X_2 =$ 'Age'), and two variables not jointly observed: Y = INCOME (thousands of euro) and Z = EXPENSE (thousands of euro).

Let A be the recipient and B the donor file. To each record in A is assigned a donor chosen at random from the 10 units in B. If unit b is assigned to unit a, then the missing Z value in a is imputed with the observed Z in b. The ath record in the final synthetic data set is $(\mathbf{x}_a, y_a, z_b)$. Note that a record in B can be

Table 2.13 List of the units in A

a	X_1	X_2	Y
1	F	27	22
2	M	35	19
3	M	41	47
4	F	61	41
5	F	52	17
6	F	39	26

Table 2.14 List of the units in B

b	X_1	X_2	Z
1	F	54	22
2	M	21	17
3	F	48	15
4	F	33	14
5	M	63	13
6	F	29	15
7	M	36	19
8	M	55	24
9	F	50	26
10	F	27	18

Table 2.15 One realization of the random hot deck method for matching files A and B

a	b donor	X_1^A	X_1^B	X_2^A	X_2^B	Y	Z
1	2	F	M	27	21	22	17
2	8	M	M	35	55	19	24
3	5	M	M	41	63	47	13
4	6	F	F	61	29	41	15
5	4	F	F	52	33	17	14
6	2	F	M	39	21	26	17

assigned more than once to different records in A, i.e. it plays the role of the donor more than once. Theoretically, there are $n_B^{n_A} = 10^6$ possible subsets of records that can be chosen as donors from B, and consequently 10^6 possible distributions for personal expenses. For instance, Table 2.15 reports one of the possible matching outputs.

If the common variable 'Sex' is used to define donation classes (i.e. 'Sex' is used as *cohort* variable), donors in B must be chosen among those within the same recipient gender class. The number of possible donors configurations decreases markedly:

$$(n_{\mathrm{M}}^B)^{n_{\mathrm{M}}^A} + (n_{\mathrm{F}}^B)^{n_{\mathrm{F}}^A} = 6^4 + 4^2 = 1312.$$

Table 2.16 illustrates one of the possible synthetic files that can be obtained by randomly choosing donors within the same gender class.

Remark 2.11 Prediction via random hot deck within donation classes defined through $\mathbf{X}$ (assumed to be categorical) is equivalent to estimating the conditional distribution of $\mathbf{Z}$ given $\mathbf{X}$ in B and drawing an observation from it. When $\mathbf{Z}$ is

Table 2.16 Random hot deck matching within classes of 'Sex'

a	b donor	X_1^A	X_1^B	X_2^A	X_2^B	Y	Z
2	5	M	M	35	63	19	13
3	7	M	M	41	36	47	19
1	3	F	F	27	48	22	15
4	6	F	F	61	29	41	15
5	9	F	F	52	50	17	26
6	3	F	F	39	48	26	15

continuous, the distribution of $\mathbf{Z}$ given $\mathbf{X}$, $F_{\mathbf{Z}|\mathbf{X}}$, is estimated through the empirical cumulative distribution function (2.44), $\hat{F}_{\mathbf{Z}|\mathbf{X}}$. Random drawing of an observation from the donor file within a class $\mathbf{x}$ is equivalent to drawing a value from $\hat{F}_{\mathbf{Z}|\mathbf{X}}$. The same holds when Z is categorical. Instead of the cumulative distribution function, the estimates $\hat{\theta}_{k|i}$, $k = 1, \ldots, K$, should be considered: the random hot deck method coincides with a random draw from that estimated distribution.

When random hot deck is performed without any donation class, it is assumed that $\mathbf{Z}$ and $\mathbf{X}$ are independent (an evaluation of such an assumption is possible in B). Now, instead of $\hat{F}_{\mathbf{Z}|\mathbf{X}}$ and $\hat{\theta}_{k|i}$, the marginal empirical distribution of $\mathbf{Z}$ in B is used for generating the values to impute.

2.4.2 Rank hot deck

When there is one ordinal matching variable X, it can still be used for selecting donors from B to assign to the records in A. In this situation, it is possible to exploit the order relationship between the values of X: this is the case of the rank hot deck method (Singh *et al.* 1990).

The units in both the files are ranked separately according to values of X. If, for simplicity, A is the recipient file and $n_B = kn_A$, with k integer, files are matched by associating records with the same rank. When the files contain a different number of records, matching is carried out by considering the empirical cumulative distribution function of the distribution of X in the recipient file:

$$\hat{F}_X^A(x) = \frac{1}{n_A}\sum_{a=1}^{n_A} I(x_a \leq x), \qquad x \in \mathcal{X},$$

and in the donor file:

$$\hat{F}_X^B(x) = \frac{1}{n_B}\sum_{b=1}^{n_B} I(x_b \leq x), \qquad x \in \mathcal{X}.$$

Then, each $a = 1, \ldots, n_A$ is associated with that record b^* in B such that

$$|\hat{F}_X^A(x_a^A) - \hat{F}_X^B(x_{b^*}^B)| = \min_{1 \leq b \leq n_B} |\hat{F}_X^A(x_a^A) - \hat{F}_X^B(x_b^B)|.$$

Example 2.7 Using the data sets of Example 2.6, if X_2 = Age is used as matching variable, units in sample A are ranked as in Table 2.17. Table 2.18 reports the ranked units in B according to the variable 'Age'. Table 2.19 shows, for each $a \in A$, the nearest record of B with respect to its rank. The final matched file is reported in Table 2.20.

Table 2.17 File A with records ranked by 'Age'

a	X_1	X_2	Y
1	F	27	22
2	M	35	19
6	F	39	26
3	M	41	47
5	F	52	17
4	F	61	41

Table 2.18 File B with records ranked by 'Age'

b	X_1	X_2	Z
2	M	21	17
10	F	27	18
6	F	29	15
4	F	33	14
7	M	36	19
3	F	48	15
9	F	50	26
1	F	54	22
8	M	55	24
5	M	63	13

Table 2.19 The nearest record of B for each record in A according to their ranks

a	$\hat{F}_X^A(x_a^A)$	$\hat{F}_X^B(x_{b*}^B)$
1	1/6	$\hat{F}_X^B(x_{10}^B) = 2/10$
2	2/6	$\hat{F}_X^B(x_6^B) = 3/10$
6	3/6	$\hat{F}_X^B(x_7^B) = 5/10$
3	4/6	$\hat{F}_X^B(x_9^B) = 7/10$
5	5/6	$\hat{F}_X^B(x_1^B) = 8/10$
4	6/6	$\hat{F}_X^B(x_5^B) = 10/10$

Table 2.20 Matched file using rank hot deck matching

a	X_1^A	X_2^A	Y	b donor	Z
1	F	27	22	10	18
2	M	35	19	6	15
6	F	39	26	7	19
3	M	41	47	9	26
5	F	52	17	1	22
4	F	61	41	5	13

2.4.3 Distance hot deck

This type of matching technique was widely used in early applications of statistical matching, e.g. Okner (1972) and Ruggles and Ruggles (1974); see also Rodgers (1984) and references therein. Each record in the recipient file is matched with the closest record in the donor file, according to a distance measure (Appendix C) computed using the matching variables $\mathbf{X}$. For instance, in the simplest case of a single continuous variable X, this means that the donor for the ath record in the recipient file A should be chosen so that:

$$d_{ab^*} = \left|x_a^A - x_{b^*}^B\right| = \min_{1 \leq b \leq n_B} \left|x_a^A - x_b^B\right|. \tag{2.47}$$

In general, when two or more donor records are equally distant from a recipient record, one of them is chosen at random.

Example 2.8 Application of distance hot deck when matching A (Table 2.13) and B (Table 2.14) using 'Age' as matching variable, produces Table 2.21.

Note that there is only one alternative matching output: that in which the donor for unit $a = 5$ is $b = 9$ instead of $b = 1$. In fact, both $b = 1$ and $b = 9$ are at the same distance, in terms of age, from $a = 5$:

$$d_{5,1} = |52 - 54| = |52 - 50| = d_{5,9}.$$

Table 2.21 Matched file obtained by means of nearest neighbour matching

a	X_1^A	X_2^A	Y	b donor	Z
1	F	27	22	10	18
2	M	35	19	7	19
3	M	41	47	7	19
4	F	61	41	5	13
5	F	52	17	1	22
6	F	39	26	7	19

Definition (2.47) is usually called *unconstrained* distance hot deck. It is unconstrained because each record in the donor file B can be used as donor more than once.

Another distance hot deck method is the *constrained* one. This approach allows each record in B to be chosen as donor only once. Constrained hot deck requires that the number of donors is greater than or equal to the number of recipients ($n_A \leq n_B$). In the simplest case of equal number of units in both the files (i.e. $n_A = n_B$), the donor pattern should be such that

$$\sum_{a=1}^{n_A} \sum_{b=1}^{n_B} (d_{ab} w_{ab}) \tag{2.48}$$

is minimized under the following constraints:

$$\sum_{b=1}^{n_B} w_{ab} = 1, \qquad a = 1, \ldots, n_A, \tag{2.49}$$

$$\sum_{a=1}^{n_A} w_{ab} = 1, \qquad b = 1, \ldots, n_B, \tag{2.50}$$

$$w_{ab} \in \{0; 1\},$$

where $w_{ab} = 1$ if the pair (a, b) is matched, and $w_{ab} = 0$ if it is not (Kadane, 1978).

The linear programming problem becomes slightly different when there are more donors than recipients ($n_B > n_A$). In this case the set of constraints becomes:

$$\sum_{b=1}^{n_B} w_{ab} = 1, \qquad a = 1, \ldots, n_A,$$

$$\sum_{a=1}^{n_A} w_{ab} \leq 1, \qquad b = 1, \ldots, n_B,$$

$$w_{ab} \in \{0; 1\}.$$

This system of constraints implies that $\sum_{a=1}^{n_A} \sum_{b=1}^{n_B} w_{ab} = n_A$.

Minimizing the aggregate after-matching distance among the two files corresponds to searching for the best constrained match. From a mathematical viewpoint this optimization problem is a linear programming one: the classical linear assignment problem (Burkard and Derigs 1980). Its solution requires considerable computational effort given the large dimension of the problem.

The main advantage of constrained matching when compared to unconstrained is that the marginal distribution of the imputed variable, $\mathbf{Z}$ in our case, is maintained (perfectly when $n_A = n_B$) in the final synthetic file. A disadvantage is that the average distance of the donor and recipient values of the matching variables **X** is expected to be greater than that in the unconstrained case (and this is the cause of the matching noise in distance hot deck; see Section 2.4.4). Furthermore, the computational cost needed to solve the linear programming problem is heavy.

Example 2.9 Table 2.22 shows the result of constrained matching applied when matching files A and B of Example 2.6 (distances are computed on the variable 'Age' as $d_{ab} = |x_{2,a}^A - x_{2,b}^B|$). Now the aggregate after-matching distance is

$$\sum_{a=1}^{n_A}\sum_{b=1}^{n_B} d_{ab}w_{ab} = 16,$$

which is slightly superior to the value of 13 in the case of unconstrained nearest neighbour matching (Table 2.21).

Example 2.10 Sometimes donor classes are useful, especially when there is a large number of matching variables and computations can be cumbersome. In Example 2.6 (Table 2.16), 'Sex' was used to define the donor classes for the random hot deck method. This is also possible for distance hot deck methods. In this case, distances computed on 'Age' should be restricted only to those units in the two files with the same gender. The results of the unconstrained and constrained distance matching are shown in Tables 2.23 and 2.24 respectively.

Remark 2.12 The distance (unconstrained) hot deck method is the nonparametric regression estimate through the kNN method (Section 2.3), for $k = 1$. More precisely, the distance (unconstrained) hot deck method imputes Z in each record a,

Table 2.22 Matched file obtained by means of constrained nearest neighbour matching

a	X_1^A	X_2^A	Y	b donor	Z
1	F	21	22	10	18
2	M	35	19	4	14
3	M	41	47	3	15
4	F	61	41	5	13
5	F	52	17	1	22
6	F	39	26	7	19

Table 2.23 Matched file obtained by means of unconstrained nearest neighbour matching within the same 'Sex' class

a	X_1^A	X_2^A	b donor	X_1^B	X_2^B	Y	Z
2	M	35	7	M	36	19	19
3	M	41	7	M	36	47	19
1	F	27	10	F	27	22	27
4	F	61	1	F	54	41	22
5	F	52	1	F	54	17	22
6	F	39	4	F	33	26	14

Table 2.24 Matched file obtained by means of constrained nearest neighbour matching within the same 'Sex' class

a	X_1^A	X_2^A	b donor	X_1^B	X_2^B	Y	Z
2	M	35	7	M	36	19	19
3	M	41	8	M	55	47	24
1	F	27	10	F	27	22	27
4	F	61	1	F	54	41	22
5	F	52	9	F	50	17	26
6	F	39	4	F	33	26	14

$a = 1, \ldots, n_A$, with the corresponding value $\hat{r}(x_a)$ of the estimated nonparametric regression function (2.46), with $k = 1$.

Nielsen (2001) suggests the use of imputation by the nonparametric regression function estimated with the kernel method.

Note that this approach is the nonparametric counterpart of the conditional mean matching method of Section 2.2.1. As in the conditional mean matching case, the synthetic file is representative of a model which can be substantially different from the original one. Hence, distance hot deck methods should be used with caution. The discussion of the quality of the synthetic data set and of its ability to preserve the distribution of the imputed variable, i.e. of Z given X in this case, is deferred to Section 2.4.4.

Remark 2.13 One of the most popular distances applied in distance hot deck for statistical matching is the Mahalanobis distance (see Appendix C). Actually, it has been used under a parametric set-up, such as the multinormal, once the variances have been estimated (see Section 2.5 and references therein). The distance hot deck method defined through equation (C.1) is applied, for instance, in Kadane (1978). When X is univariate, Goel and Ramalingam (1989) prove that distance hot deck with distance given by (C.1) coincides with the rank hot deck procedure of Section 2.4.2, also called the *isotonic matching strategy*.

2.4.4 The matching noise

As in Section 2.2.3, one may question whether the data set $A \cup B$ completed with the hot deck methods of Sections 2.4.1, 2.4.2 and 2.4.3 has been actually generated by the true, but unknown, distribution $f(\mathbf{x}, \mathbf{y}, \mathbf{z})$. This is not always the case for finite sample sizes. For the sake of simplicity, let X, Y and Z be univariate, A be the recipient file and B the donor file. The resulting data set is $(x_a, y_a, \tilde{z}_a)$, $a = 1, \ldots, n_A$, where $\tilde{z}_a = z_b$ for a b in B chosen according to a hot deck method. Given that (x_a, y_a) is a sample generated from $f_{XY}(x, y)$, and that the CIA is assumed, it should be proved that $(x_a, \tilde{z}_a)$ is generated from $f_{XZ}(x, z)$ or, in other words, that the distribution of $(X, \tilde{Z})$ equals the distribution of (X, Z). Let $(\tilde{X}, \tilde{Z})$ be the r.v. representing the donor value, conditional on an observed X in A. Then

$$f_{X\tilde{Z}\tilde{X}}(x, z, u) = f_X(x) f_{\tilde{X}|X}(u|x) f_{\tilde{Z}|X\tilde{X}}(z|x, u) = f_X(x) f_{\tilde{X}|X}(u|x) f_{Z|X}(z|u),$$

where the last equality is justified by the fact that, once $\tilde{X}$ is known and equal to u, the corresponding $\tilde{Z}$ is independent of the observed X value in A (records in A and B are observations of independent r.v.s), and is generated according to the distribution of Z given $X = u$ (records in B are generated by identically distributed r.v.s, with distribution $f(x, y, z)$). Consequently, the joint distribution of $(X, \tilde{Z})$ is obtained by marginalization of $\tilde{X}$:

$$f_{X\tilde{Z}}(x, z) = f_X(x) \int_{\mathcal{X}} f_{\tilde{X}|X}(u|x) f_{Z|X}(z|u) \mathrm{d}u. \tag{2.51}$$

In general, $f_{\tilde{X}|X}(u|x)$ is the distribution of the matching noise, and represents the additional variance and distortion due to the noncoincidence between the donor $\tilde{x}_a$ and the recipient x_a value. Note that if the donor and recipient X values coincide with probability one, i.e.

$$f_{\tilde{X}|X}(u|x) = \begin{cases} 1 & \text{when } u = x, \\ 0 & \text{otherwise}, \end{cases} \tag{2.52}$$

for all $x \in \mathcal{X}$, then the matching noise does not exist (as in Rässler 2002, p. 21). This holds, for instance, when X is categorical and conditional random hot deck is applied. More precisely, this procedure is free of matching noise for any sample size. This is a remarkable exception in the panorama of the micro approaches. In comparison, the parametric procedure that most resembles the conditional random hot deck method (i.e. the random draw from an estimated parametric distribution) needs an asymptotic assumption such as consistency of parameter estimators (Section 2.2.3).

However, when X is an absolutely continuous r.v., the event $x_a = x_b$ has probability zero, for any $b = 1, \ldots, n_B$. Hence, for any sample size, the synthetic data set is inevitably affected by matching noise. A preliminary study on the effects of the matching noise is given in Paass (1985). The following results are described in Conti and Scanu (2005).

If a random hot deck method (Section 2.4.1) is used, $f_{\tilde{X}|X}(u|x) = f_X(u)$. Hence,

$$f_{X\tilde{Z}}(x, z) = f_X(x) \int_{\mathcal{X}} f_X(u) f_{Z|X}(z|u) \mathrm{d}u = f_X(x) f_Z(z).$$

Random hot deck, although capable of preserving the marginal Z distribution, forces X and Z to be marginally independent.

If distance hot deck is used (Section 2.4.3), $f_{\tilde{X}|X}(u|x)$ has the following cumulative distribution function:

$$F_{\tilde{X}|X}(u|x_a) = \int_{-\infty}^{u-x_a} P\left(|X - x_a| \geq |t|\right)^{n_B - 1} f_{X-x_a}(t) \mathrm{d}t, \qquad u \in \mathbb{R}. \qquad (2.53)$$

Sometimes numerical approaches should be considered for the evaluation of (2.53) (e.g. when X follows a normal distribution), and consequently for computing the distance between the original distribution function $f_{XZ}(x, z)$ and that affected by the matching noise (2.51).

If the rank hot deck method (Section 2.4.2) is used, assume for simplicity that $n_A = n_B = n$ and let parentheses denote the rank. It follows that $f_{\tilde{X}|X}(u|x_{(a)}) = f_{X_{(a)}}(u)$, i.e. if the recipient record has ath rank in X among the units in A, then the matching noise follows the distribution of the ath ranked X in a sample of n units:

$$f_{\tilde{X}|X}(u|x_{(a)}) = \frac{n!}{(a-1)!(n-a)!} P(X < u)^{a-1} P(X > u)^{n-a} f_X(u).$$

This case may also require numerical approaches in order to compute the distribution of $(X, \tilde{Z})$.

As a matter of fact, all the previous hot deck methods produce an imputed data set which can be considered as generated by a model that differs from the original one for finite sample sizes. When sample sizes are allowed to diverge, both the distance and rank hot deck methods have a pleasant property: the matching noise converges to the distribution (2.52). Again, compare the distance hot deck method to its parametric counterpart, i.e. the conditional mean matching approach. It was shown in Section 2.2.3 that in the normal case the synthetic file could never be representative of the true generating distribution function. On the other hand, the distance hot deck method generates a synthetic file that is asymptotically representative of the true distribution function. Nevertheless, it is still questionable whether imputation by a (nonparametric) conditional mean makes sense. Further research is needed on how to recover a method that, for finite sample sizes, is able to be minimally distant (in average) from the true distribution function.

2.5 Mixed Methods

The previous sections have described the different statistical matching procedures which can be defined respectively in a parametric and a nonparametric setting when considering respectively a micro and a macro approach. Actually, most of the papers on statistical matching apply mixtures of the previous procedures. More precisely, the mixture consists in initially adopting a parametric model, which can be estimated as in Section 2.1, and then obtaining a completed synthetic data set via a nonparametric micro approach such as those in Section 2.4. The idea is to exploit both the following properties: a parametric model is more parsimonious than a nonparametric one, while a nonparametric method is more robust to model misspecification. Additionally, among the nonparametric techniques, hot deck imputes live values, overcoming some of the problems encountered in the parametric micro approach (Section 2.2). Generally speaking, mixed methods may be considered as a two-step procedure: the first step estimates parameters of the parametric model; the second step makes use of one of the hot deck techniques defined in Section 2.4 conditional on the parameters estimated in the first step. Coherently with Section 2.4, file A will be considered as the recipient file and B as the donor file.

2.5.1 Continuous variables

When dealing with continuous variables, a common class of techniques is based on the *predictive mean matching* imputation method; see Rubin (1987) and Little (1988). In the statistical matching context these techniques were first explicitly introduced in Kadane (1978) and Rubin (1986). In general they consist of the following steps.

(i) The regression parameters of $\mathbf{Z}$ on $\mathbf{X}$ (Equation (2.17)) are estimated on file B.

(ii) For each $a = 1, \ldots, n_A$, an intermediate value $\tilde{\mathbf{z}}_a$ is generated, based on the estimated regression function.

(iii) For each $a = 1, \ldots, n_A$, a live value $\mathbf{z}_{b^*}$, with b^* in B, is imputed to the ath record in A through a suitable distance hot deck procedure, taking into account the intermediate value $\tilde{\mathbf{z}}_a$.

Many different approaches have been defined in this context. Following Section 2.1, the first step consists in finding the maximum likelihood estimates of the regression parameters as in Section 2.1.2. In the other two steps various choices are available. The following list is not exhaustive, and is based on Rubin (1986), Singh *et al.* (1993), and Moriarity and Scheuren (2001, 2003). Note that the previous references relate to a more general setting than the CIA and with some minor differences in the techniques; these methods will be further discussed in Chapter 3.

MM1

(a) *Regression step.* Compute intermediate values for the units in A as in Section 2.2.1 via

$$\tilde{\mathbf{z}}_a = \hat{\boldsymbol{\alpha}}_{\mathbf{Z}} + \hat{\boldsymbol{\beta}}_{\mathbf{ZX}}\mathbf{x}_a,$$

for each $a = 1, \ldots, n_A$, and similarly intermediate values for the units in B,

$$\tilde{\mathbf{z}}_b = \hat{\boldsymbol{\alpha}}_{\mathbf{Z}} + \hat{\boldsymbol{\beta}}_{\mathbf{ZX}}\mathbf{x}_b,$$

for each $b = 1, \ldots, n_B$.

(b) *Matching step.* For each $a = 1, \ldots, n_A$, impute the live value $\mathbf{z}_{b^*}$ corresponding to the nearest neighbour b^* in B with respect to a distance between the previously determined intermediate values, $d(\tilde{\mathbf{z}}_a, \tilde{\mathbf{z}}_b)$.

MM2

(a) *Regression step.* Compute intermediate values for the units in A as in Section 2.2.1 via

$$\tilde{\mathbf{z}}_a = \hat{\boldsymbol{\alpha}}_{\mathbf{Z}} + \hat{\boldsymbol{\beta}}_{\mathbf{ZX}}\mathbf{x}_a,$$

for each $a = 1, \ldots, n_A$.

(b) *Matching step.* For each $a = 1, \ldots, n_A$, impute the live value $\mathbf{z}_{b^*}$ corresponding to the nearest neighbour b^* in B with respect to a distance between the previously determined intermediate value and the observed values in B, $d(\tilde{\mathbf{z}}_a, \mathbf{z}_b)$.

MM3

(a) *Regression step.* Compute intermediate values for the units in A as in Section 2.2.2 via

$$\tilde{\mathbf{z}}_a = \hat{\boldsymbol{\alpha}}_{\mathbf{Z}} + \hat{\boldsymbol{\beta}}_{\mathbf{ZX}}\mathbf{x}_a + \mathbf{e}_a,$$

for each $a = 1, \ldots, n_A$, where $\mathbf{e}_a$ is a value generated from a multinormal distribution with zero mean vector and estimated residual variance matrix $\hat{\boldsymbol{\Sigma}}_{\mathbf{ZZ|X}}$.

(b) *Matching step.* For each $a = 1, \ldots, n_A$, impute the live value $\mathbf{z}_{b^*}$ corresponding to the nearest neighbour b^* in B with respect to a distance between the previously determined intermediate value and the observed values in B, $d(\tilde{\mathbf{z}}_a, \mathbf{z}_b)$.

MM4

(a) *Regression step.* Same regression step as MM3.

(b) *Matching step.* For each $a = 1, \ldots, n_A$, impute the live value $\mathbf{z}_{b^*}$ corresponding to the nearest neighbour b^* in B with respect to a distance between the previously determined intermediate value and the observed values in B, $d(\tilde{\mathbf{z}}_a, \mathbf{z}_b)$, and the matching is constrained (see Section 2.4.3).

MM5

(a) *Regression step.* Compute intermediate values for the units in A as in Section 2.2.2 via

$$\tilde{\mathbf{z}}_a = \hat{\boldsymbol{\alpha}}_{\mathbf{Z}} + \hat{\boldsymbol{\beta}}_{\mathbf{ZX}} \mathbf{x}_a + \mathbf{e}_a,$$

for each $a = 1, \ldots, n_A$, and similarly intermediate values for the units in B,

$$\tilde{\mathbf{y}}_b = \hat{\boldsymbol{\alpha}}_{\mathbf{Y}} + \hat{\boldsymbol{\beta}}_{\mathbf{YX}} \mathbf{x}_b + \mathbf{e}_b,$$

for each $b = 1, \ldots, n_B$, where the parameters are estimated as in Section 2.1.2 and $\mathbf{e}_b$ is a random draw from the multinormal distribution with zero mean vector and estimated residual variance matrix $\hat{\boldsymbol{\Sigma}}_{\mathbf{YY|X}}$.

(b) *Matching step.* For each $a = 1, \ldots, n_A$, impute the live value $\mathbf{z}_{b^*}$ corresponding to the nearest neighbour b^* in B with respect to a (constrained) distance between the couples, $d\left((\mathbf{y}_a, \tilde{\mathbf{z}}_a), (\tilde{\mathbf{y}}_b, \mathbf{z}_b)\right)$.

Remark 2.14 Although theoretical results on the properties of these techniques have not yet been defined, they are expected to be affected by two opposite effects.

(i) They can be superior to the ones that use only regression when the model is slightly misspecified, e.g. when the noise is heteroscedastic, as also discussed in a general imputation setting in Little (1988).

(ii) They are affected by the matching noise, in particular when distance hot deck is used (see Section 2.4.4).

Generally speaking, when the samples are small and good matches are hard to find (large matching noise), it could be better to use methods based only on random draws from the estimated distributions (Section 2.2.2); Lazzeroni *et al.* (1990) is relevant to this point. An application in the imputation context can be found in Ezzati-Rice *et al.* (1993). Additional studies are necessary in order to show the importance of issues (i) and (ii).

2.5.2 Categorical variables

As far as categorical variables are concerned, the mixed approach consists of the following two steps.

(a) *Loglinear step*. This mainly involves estimating the expected cell frequencies through the loglinear model corresponding to the CIA (see Appendix B). The most appropriate estimators are those defined in Section 2.1.3. However, when mixed methods are considered, the focus is only on the recipient file A. Hence, inefficient estimators are preferred to maximum likelihood ones. The inefficiency is due to the fact that the marginal distribution of X is forced to be that observed in A. For this purpose, Singh *et al.* (1993) propose the following first step of the mixed procedure.

Let X, Y and Z be three univariate categorical variables as in Section 2.1.3. Singh *et al.* (1988) propose the use of the following raking procedure. Given that the marginal X distribution observed in the two samples A and B is usually different, the observed frequencies in B are modified so that the marginal X distribution coincides with that observed in A. Hence, the difference between the two marginal distributions $n^A_{i..} - n^B_{i..}$, $i = 1, \ldots, I$, is shared between the different Z categories according to their relative conditional frequency:

$$n^{B(0)}_{i.k} = n^B_{i.k} + \left(n^A_{i..} - n^B_{i..}\right) \frac{n^B_{i.k}}{n^B_{i..}} = n^A_{i..} \frac{n^B_{i.k}}{n^B_{i..}}, \quad i = 1, \ldots, I;\ k = 1, \ldots, K.$$

Let $n^{A(0)}_{ij.} = n^A_{ij.}$, $i = 1, \ldots, I$, $j = 1, \ldots, J$. The two contingency tables $\mathbf{n}^{A(0)}$ and $\mathbf{n}^{B(0)}$ are the starting contingency tables of the raking procedure. The cell counts n_{ijk} are first raked on the table $\mathbf{n}^{B(0)}$,

$$n^{(1)}_{ijk} = \frac{n^{B(0)}_{i.k}}{J}, \qquad (i, j, k) \in \mathbf{\Delta},$$

and then the new table $\mathbf{n}^{(1)}$ is raked on the table $\mathbf{n}^{A(0)}$,

$$n^{(2)}_{ijk} = n^{(1)}_{ijk} + \left(n^{A(0)}_{ij.} - n^{(1)}_{ij.}\right) \frac{n^{(1)}_{ijk}}{n^{(1)}_{ij.}} = n^{A(0)}_{ij.} \frac{n^{(1)}_{ijk}}{n^{(1)}_{ij.}}, \qquad (i, j, k) \in \mathbf{\Delta}.$$

The new table $\mathbf{n}^{(2)}$ is such that the marginal (X, Y) and (X, Z) distributions are those of the initial tables $\mathbf{n}^{A(0)}$ and $\mathbf{n}^{B(0)}$. Note that the previous steps are exactly those of the iterative proportional fitting algorithm, see Section B.1. Consequently, if $n^A_{i..} = n^B_{i..}$, $i = 1, \ldots, I$, the raking procedure would produce a table coherent with the ML estimates of the parameters in Section 2.1.3. Singh *et al.* (1993) also propose another modification. Given that the raked estimates may be not integers, they propose rounding the estimates randomly. Let $\tilde{n}_{ijk}$ be the final counts obtained with this procedure.

(b) *Matching step.* A donor hot deck technique from among those described in Section 2.4 is used for the selection of candidate Z values observed in B for each record in A. A candidate value which makes the ath synthetic value equal to (i, j, k) is accepted only if the estimated cell frequency $\tilde{n}_{ijk}$ is not exceeded, otherwise another donor must be searched for the ath record. This constrained imputation is performed in order to make the matched sample meet the relationships estimated by the loglinear model.

An interesting use of this technique is when the original variables (X, Y, Z) to be matched are continuous. In this case, Singh *et al.* (1988) suggest transforming the continuous variables (X, Y, Z) into categorical variables $(X^\bullet, Y^\bullet, Z^\bullet)$. They also assume that $(X^\bullet, Y^\bullet, Z^\bullet)$ follow a loglinear model. Then, according to the estimated probabilities $P(Y^\bullet|X^\bullet)$ and $P(Z^\bullet|X^\bullet)$, the units are assigned to the imputation classes determined by $X^\bullet$. Finally, within the imputation classes, the units are imputed using a hot deck technique.

2.6 Comparison of Some Statistical Matching Procedures under the CIA

How effective is a mixed procedure compared to a parametric one? Is it reasonable to use a synthetic data set obtained via a micro approach for inferences of any kind? These are some of the questions that frequently arise when dealing with statistical matching. This section aims to give some answers through a set of simulations carried out under the CIA.

Let $A \cup B$ be generated by a normal distribution, as in Section 2.1.1. Let us compare the following procedures:

(a) statistical matching via the mixed method as proposed in Moriarity and Scheuren (2003), similar to MM5 (micro approach, Section 2.5.1), with the regression step based on estimates computed with their sample observed counterpart (e.g. estimation of the means with the average of the observed values, of the variances with the observed variances, and so on, as in part (ii) of Remark 2.6);

(b) statistical matching via random draws from an estimated conditional distribution (micro approach, Section 2.2.2);

(c) statistical matching via ML estimation of the parameters (macro approach, steps (i), (ii), and (iii) of Section 2.1.1).

We have considered four different normal models. All the models have a null mean vector

$$\boldsymbol{\mu} = \begin{pmatrix} \mu_X \\ \mu_Y \\ \mu_Z \end{pmatrix} = \begin{pmatrix} 0 \\ 0 \\ 0 \end{pmatrix},$$

and different covariance matrices

$$\mathbf{\Sigma}^{(1)} = \begin{pmatrix} 1 & 0.5 & 0.5 \\ 0.5 & 1 & \rho_{YZ} \\ 0.5 & \rho_{YZ} & 1 \end{pmatrix}, \qquad \mathbf{\Sigma}^{(2)} = \begin{pmatrix} 1 & 0.7 & 0.5 \\ 0.7 & 1 & \rho_{YZ} \\ 0.5 & \rho_{YZ} & 1 \end{pmatrix},$$

$$\mathbf{\Sigma}^{(3)} = \begin{pmatrix} 1 & 0.7 & 0.7 \\ 0.7 & 1 & \rho_{YZ} \\ 0.7 & \rho_{YZ} & 1 \end{pmatrix}, \qquad \mathbf{\Sigma}^{(4)} = \begin{pmatrix} 1 & 0.5 & 0.95 \\ 0.5 & 1 & \rho_{YZ} \\ 0.95 & \rho_{YZ} & 1 \end{pmatrix}.$$

The four covariance matrices, which are also correlation matrices, show increasing correlation for both (X, Y) and (X, Z). The correlation coefficient ρ_{YZ} has been set to 0.35, 0.5 and 0.7.

For each of the previous 12 models, 500 samples $A \cup B$ of 1000 records, $n_A = n_B = 500$, were generated. The three methods were applied under the CIA. In this setting, method (a) corresponds to that suggested by, among others, Moriarity and Scheuren (2003). The regression step is performed as in their paper (see part (ii) of Remark 2.6), hot deck is constrained and both the files are imputed. Method (b) is the one implemented in Rässler (2002), which under the CIA and with large n_A and n_B is very close to the method based on ML estimates. The parameters are estimated essentially following the steps in Section 2.1.1 (recall part (i) of Remark 2.6). Furthermore, $A \cup B$ is completed by imputing Y in A by randomly generating values from the distribution of Y given X, and by imputing Z by randomly generating values from the distribution of Z given X. Final estimates are computed on the overall completed file. Method (c) corresponds to the ML parameter estimates of Section 2.1.1. The experiments are performed in the R environment with *ad hoc* functions. Functions referring to methods (a) and (c) are described in Section E.3, while for method (b) an R version of the code reported in Rässler (2002) is used.

Averages of the parameter estimates with respect to the 500 samples generated for the three methods are shown in Table 2.25. Furthermore, averages of absolute and squared differences between the estimated and the true model parameters are reported (i.e. simulation bias and simulation mean square error).

As a matter of fact, there are no great differences among the different procedures as far as the inestimable parameter σ_{YZ} is concerned. This parameter is obviously well estimated only when the CIA holds: the covariances between Y and Z under the CIA for the four covariance matrices are 0.25, 0.35, 0.49 and 0.475 for $\mathbf{\Sigma}^{(1)}$, $\mathbf{\Sigma}^{(2)}$, $\mathbf{\Sigma}^{(3)}$ and $\mathbf{\Sigma}^{(4)}$ respectively. For these combinations, the mean square error is dramatically lower than the other cases, due to the irrelevance of the bias. Although in all cases the three methods perform similarly, it must be noted that in almost all cases direct computation with the ML approach leads to better results.

The efficiency of the ML approach is particularly noticeable for the estimable parameters. For instance, the mean of Y is much better estimated with method (c) (as suggested in Remark 2.4) than with methods (a) and (b). The mean square errors in the 12 different cases show a gain between 10% and 25% as compared to the corresponding mean square errors of the other two methods. The same also holds

Table 2.25 Averages of the estimates of the covariance of Y and Z, $\bar{\bar{\sigma}}_{YZ}$, simulation bias and simulation mean square error by statistical matching methods (a), (b) and (c)

			Method (a)			Method (b)			Method (c)		
$\mathbf{\Sigma}^{(i)}$	σ_{YZ}	σ_{YZ}^{CIA}	$\bar{\bar{\sigma}}_{YZ}$	|Bias|	MSE	$\bar{\bar{\sigma}}_{YZ}$	|Bias|	MSE	$\bar{\bar{\sigma}}_{YZ}$	|Bias|	MSE
$\mathbf{\Sigma}^{(1)}$	0.25	0.25	0.243 43	0.106 57	0.012 32	0.247 85	0.102 15	0.012 09	0.247 74	0.102 26	0.011 34
$\mathbf{\Sigma}^{(1)}$	0.50	0.25	0.243 58	0.256 42	0.066 71	0.247 81	0.252 19	0.065 24	0.247 76	0.252 24	0.064 51
$\mathbf{\Sigma}^{(1)}$	0.70	0.25	0.243 60	0.456 40	0.209 27	0.247 79	0.452 21	0.206 12	0.247 84	0.452 16	0.205 35
$\mathbf{\Sigma}^{(2)}$	0.35	0.35	0.341 88	0.008 12	0.001 09	0.347 64	0.002 36	0.001 94	0.347 40	0.002 60	0.001 24
$\mathbf{\Sigma}^{(2)}$	0.50	0.35	0.341 61	0.158 39	0.026 10	0.347 56	0.152 44	0.025 16	0.347 38	0.152 62	0.024 54
$\mathbf{\Sigma}^{(2)}$	0.70	0.35	0.341 66	0.358 34	0.129 39	0.347 49	0.352 51	0.126 18	0.347 40	0.352 60	0.125 59
$\mathbf{\Sigma}^{(3)}$	0.35	0.49	0.478 36	0.128 36	0.017 63	0.486 89	0.136 89	0.020 83	0.486 82	0.136 82	0.020 24
$\mathbf{\Sigma}^{(3)}$	0.50	0.49	0.478 14	0.021 86	0.001 62	0.486 84	0.013 16	0.002 26	0.486 82	0.013 18	0.001 71
$\mathbf{\Sigma}^{(3)}$	0.70	0.49	0.476 89	0.223 11	0.050 90	0.485 89	0.214 11	0.047 77	0.487 46	0.212 54	0.046 55
$\mathbf{\Sigma}^{(4)}$	0.35	0.475	0.464 90	0.114 90	0.014 54	0.471 14	0.121 14	0.017 55	0.473 16	0.123 16	0.017 34
$\mathbf{\Sigma}^{(4)}$	0.50	0.475	0.467 04	0.032 96	0.002 21	0.475 80	0.024 20	0.003 16	0.474 78	0.025 22	0.002 60
$\mathbf{\Sigma}^{(4)}$	0.70	0.475	0.465 52	0.234 48	0.056 15	0.471 78	0.228 22	0.054 76	0.474 51	0.225 49	0.052 96

for the other estimable parameters (mean of Z, variances of Z and Y, covariances of (X, Y) and (X, Z)).

2.7 The Bayesian Approach

The Bayesian approach is the usual framework for multiple imputation, which is the method usually considered in order to tackle the identification problem in statistical matching, as discussed in Section 4.8.2. Nevertheless, it is useful to introduce aspects of the Bayesian approach in this section.

Let $\mathbf{Y} = (Y_1, \ldots, Y_Q)$ be an r.v. whose distribution is $f(\mathbf{y}|\boldsymbol{\theta})$. Uncertainty on $\boldsymbol{\theta}$ is formalized by a prior distribution $\pi(\boldsymbol{\theta})$. Given an $n \times Q$ matrix $\mathbf{y}$ whose rows are the observations $\mathbf{y}_i = (y_{i1}, \ldots, y_{iQ})$, $i = 1, \ldots, n$, the Bayes formula allows the posterior distribution of $\boldsymbol{\theta}$ to be computed:

$$\pi(\boldsymbol{\theta}|\mathbf{y}) = \frac{f(\mathbf{y}|\boldsymbol{\theta})\pi(\boldsymbol{\theta})}{\int f(\mathbf{y}|\boldsymbol{\theta})\pi(\boldsymbol{\theta})\mathrm{d}\boldsymbol{\theta}} \propto L(\boldsymbol{\theta}; \mathbf{y})\pi(\boldsymbol{\theta}).$$

The posterior distribution addresses the change of knowledge on $\boldsymbol{\theta}$ before and after the observation of data. Hence, it contains all the information needed to make inference on $\boldsymbol{\theta}$ (macro objective). The information update is essentially the product of the likelihood function and the prior distribution.

When the focus is on predictive inference (see Bernardo and Smith 2000), an important element to consider in order to predict m values for the r.v. $\mathbf{Y}$, i.e. $\hat{\mathbf{y}} = (\mathbf{y}_{n+1}, \ldots, \mathbf{y}_{n+m})$ given that the n units in $\mathbf{y}$ have been observed, is the posterior predictive distribution:

$$f_{\hat{\mathbf{Y}}|\mathbf{Y}}(\hat{\mathbf{y}}|\mathbf{y}) = \int f(\hat{\mathbf{y}}|\mathbf{y}, \boldsymbol{\theta})\pi(\boldsymbol{\theta}|\mathbf{y})\mathrm{d}\boldsymbol{\theta}. \tag{2.54}$$

The r.v.s $\hat{\mathbf{Y}}$ and $\mathbf{Y}$ are generally assumed to be conditionally independent given $\boldsymbol{\theta}$, thus $f(\hat{\mathbf{y}}|\mathbf{y}, \boldsymbol{\theta}) = f(\hat{\mathbf{y}}|\boldsymbol{\theta})$. The posterior predictive distribution is particularly useful when the objective of the statistical matching procedure is micro.

The previous elements can be easily adapted to the statistical matching framework. In order to do this, it is necessary also to consider the r.v. $\mathbf{R}$, as defined in Section 1.3, with probability distribution $h(\mathbf{r}|\boldsymbol{\xi})$ and prior probability distribution $\pi(\boldsymbol{\xi})$. Assuming that the missing data mechanism is at least MAR and that $\boldsymbol{\theta}$ and $\boldsymbol{\xi}$ are independent (distinctness; see Appendix A), the posterior distribution of $\boldsymbol{\theta}$, which is generally what is required, is:

$$\begin{aligned}
\pi(\boldsymbol{\theta}|(\mathbf{x}, \mathbf{y}, \mathbf{z})_{\mathrm{obs}}, \mathbf{r}) &= \int \pi(\boldsymbol{\theta}, \boldsymbol{\xi}|(\mathbf{x}, \mathbf{y}, \mathbf{z})_{\mathrm{obs}}, \mathbf{r})\mathrm{d}\boldsymbol{\xi} \\
&= c^{-1} f((\mathbf{x}, \mathbf{y}, \mathbf{z})_{\mathrm{obs}}|\boldsymbol{\theta})\pi(\boldsymbol{\theta}) \int h(\mathbf{r}|(\mathbf{x}, \mathbf{y}, \mathbf{z})_{\mathrm{obs}}, \boldsymbol{\xi})\pi(\boldsymbol{\xi})\mathrm{d}\boldsymbol{\xi} \\
&\propto L(\boldsymbol{\theta}|(\mathbf{x}, \mathbf{y}, \mathbf{z})_{\mathrm{obs}})\pi(\boldsymbol{\theta}),
\end{aligned} \tag{2.55}$$

where c is the normalizing constant

$$c = \iint f((\mathbf{x}, \mathbf{y}, \mathbf{z})_{\text{obs}}, \mathbf{r}|\boldsymbol{\theta}, \boldsymbol{\xi})\pi(\boldsymbol{\theta})\pi(\boldsymbol{\xi})\mathrm{d}\boldsymbol{\theta}\mathrm{d}\boldsymbol{\xi}.$$

Therefore, all the information on $\boldsymbol{\theta}$ is contained in (2.55), which does not involve the missing data mechanism. By (1.3), we have

$$\pi(\boldsymbol{\theta}|(\mathbf{x}, \mathbf{y}, \mathbf{z})_{\text{obs}}) \propto \pi(\boldsymbol{\theta}) \prod_{a=1}^{n_A} f_{\mathbf{XY}}(\mathbf{x}_a^A, \mathbf{y}_a^A|\boldsymbol{\theta}) \prod_{b=1}^{n_B} f_{\mathbf{XZ}}(\mathbf{x}_b^B, \mathbf{z}_b^B|\boldsymbol{\theta}). \tag{2.56}$$

As a consequence, when the approach is macro, it is necessary to take into account only the observed posterior distribution, because it contains all the information related to the parameters. When the approach is micro, the missing variables should be predicted. For this purpose, the predictive distribution can be adapted to the missing data problem. Splitting each observed record into the observed part $\mathbf{v}$ and the unobserved part $\mathbf{u}$, the posterior predictive distribution is:

$$\begin{aligned} &f_{(\mathbf{X},\mathbf{Y},\mathbf{Z})_{\text{mis}}|(\mathbf{X},\mathbf{Y},\mathbf{Z})_{\text{obs}}}(\mathbf{u}|\mathbf{v}) \\ &= \int f_{(\mathbf{X},\mathbf{Y},\mathbf{Z})_{\text{mis}}|(\mathbf{X},\mathbf{Y},\mathbf{Z})_{\text{obs}}}(\mathbf{u}|\mathbf{v}, \boldsymbol{\theta})\pi(\boldsymbol{\theta}|\mathbf{v})\mathrm{d}\boldsymbol{\theta}. \end{aligned} \tag{2.57}$$

Further details can be found in Schafer (1997) and Rässler (2002).

Under the CIA, the Bayesian approach should fulfil the requirement that the prior distribution of $\boldsymbol{\theta}_{\mathbf{YZ}|\mathbf{X}}$ is concentrated on the situation of conditional independence. When $(\mathbf{X}, \mathbf{Y}, \mathbf{Z})$ is multinormal, this requirement is fulfilled when the prior distribution on the matrix of partial correlations $\boldsymbol{\rho}_{\mathbf{YZ}|\mathbf{X}}$ is equal to the null matrix with probability one. Rubin (1974) shows clearly that the parameters concerning relationships between $\mathbf{Y}$ and $\mathbf{Z}$ given $\mathbf{X}$ are such that their posterior distribution is equal to their prior distribution, due to the lack of joint information on $\mathbf{Y}$ and $\mathbf{Z}$.

In the following example, the multinormal case is analysed.

Example 2.11 Let us consider the case discussed in Section 2.1.2, where the variables $(\mathbf{X}, \mathbf{Y}, \mathbf{Z})$ are multinormal. In the following, the approach described in Rässler (2003) is illustrated.

Let $f_{\mathbf{YZ}|\mathbf{X}}$ be the distribution of interest, i.e. the parameters of interest are those in equations (2.12) and (2.17). For the sake of simplicity, $\boldsymbol{\alpha}$ is included in $\boldsymbol{\beta}$. The parameters are the regression coefficients $\boldsymbol{\beta}_{\mathbf{YX}}$ and $\boldsymbol{\beta}_{\mathbf{ZX}}$ and the covariance matrices $\boldsymbol{\Sigma}_{\mathbf{YY}|\mathbf{X}}$ and $\boldsymbol{\Sigma}_{\mathbf{ZZ}|\mathbf{X}}$. Assume the conventional noninformative prior distributions for these parameters. As far as the covariance matrix $\boldsymbol{\Sigma}_{\mathbf{YZ}|\mathbf{X}}$ is concerned, this is determined by the prior distribution imposed on $\boldsymbol{\rho}_{\mathbf{YZ}|\mathbf{X}}$ due to the CIA, bearing in mind that $\sigma_{Y_q Z_r|\mathbf{X}} = \rho_{Y_q Z_r|\mathbf{X}}\sqrt{\sigma^2_{Y_q|\mathbf{X}}\sigma^2_{Z_r|\mathbf{X}}}$ elementwise. Hence, the prior distribution is assumed to be:

$$\begin{aligned} &\pi(\boldsymbol{\beta}_{\mathbf{YX}}, \boldsymbol{\beta}_{\mathbf{ZX}}, \boldsymbol{\Sigma}_{\mathbf{YY}|\mathbf{X}}, \boldsymbol{\Sigma}_{\mathbf{ZZ}|\mathbf{X}}, \boldsymbol{\rho}_{\mathbf{YZ}|\mathbf{X}}) \\ &= \pi(\boldsymbol{\beta}_{\mathbf{YX}}, \beta_{\mathbf{ZX}})\pi(\boldsymbol{\Sigma}_{\mathbf{YY}|\mathbf{X}}|\boldsymbol{\rho}_{\mathbf{YZ}|\mathbf{X}})\pi(\boldsymbol{\Sigma}_{\mathbf{ZZ}|\mathbf{X}}|\boldsymbol{\rho}_{\mathbf{YZ}|\mathbf{X}})\pi(\boldsymbol{\rho}_{\mathbf{YZ}|\mathbf{X}}) \\ &\propto |\boldsymbol{\Sigma}_{\mathbf{YY}|\mathbf{X}}|^{-(Q+1)/2}|\boldsymbol{\Sigma}_{\mathbf{ZZ}|\mathbf{X}}|^{-(R+1)/2}\pi(\boldsymbol{\rho}_{\mathbf{YZ}|\mathbf{X}}). \end{aligned}$$

Since, for $\boldsymbol{\rho}_{\mathbf{YZ|X}}$ the posterior distribution is equal to the prior distribution, by (2.56) the joint posterior distribution can be written as:

$$\begin{aligned}
&\pi(\boldsymbol{\beta}_{\mathbf{YX}}, \boldsymbol{\beta}_{\mathbf{ZX}}, \boldsymbol{\Sigma}_{\mathbf{YY|X}}, \boldsymbol{\Sigma}_{\mathbf{ZZ|X}}, \boldsymbol{\rho}_{\mathbf{YZ|X}}|A \cup B) \\
&\quad \propto L(\boldsymbol{\beta}_{\mathbf{YX}}, \boldsymbol{\Sigma}_{\mathbf{YY|X}}; A)\pi(\boldsymbol{\Sigma}_{\mathbf{YY|X}}|\boldsymbol{\rho}_{\mathbf{YZ|X}}) \\
&\qquad \times L(\boldsymbol{\beta}_{\mathbf{ZX}}, \boldsymbol{\Sigma}_{\mathbf{ZZ|X}}; B)\pi(\boldsymbol{\Sigma}_{\mathbf{ZZ|X}}|\boldsymbol{\rho}_{\mathbf{YZ|X}})\pi(\boldsymbol{\rho}_{\mathbf{YZ|X}}).
\end{aligned} \tag{2.58}$$

Thus, as Rässler (2002) notes, the problem of specifying the posterior distributions is reduced to a standard problem, where the posterior distributions

$$\pi(\boldsymbol{\Sigma}_{\mathbf{YY|X}}|A), \qquad \pi(\boldsymbol{\Sigma}_{\mathbf{ZZ|X}}|B)$$

follow an inverted Wishart, while the conditional posterior distributions

$$\pi(\boldsymbol{\beta}_{\mathbf{YX}}|\boldsymbol{\Sigma}_{\mathbf{YY|X}}, A), \qquad \pi(\boldsymbol{\beta}_{\mathbf{ZX}}|\boldsymbol{\Sigma}_{\mathbf{ZZ|X}}, B)$$

are multivariate normal; see, for instance, Box and Tiao (1992, p. 439). Note that the posterior distribution $\pi(\boldsymbol{\rho}_{\mathbf{YZ|X}}|A \cup B)$ is equal to the prior distribution $\pi(\boldsymbol{\rho}_{\mathbf{YZ|X}})$ and is assumed to be concentrated on $\boldsymbol{\rho}_{\mathbf{YZ|X}}$ equal to the null matrix.

It is still hard to solve the integral to compute explicitly the posterior predictive distribution. In this case it is possible to draw observations from $f_{(\mathbf{X},\mathbf{Y},\mathbf{Z})_{mis}|(\mathbf{X},\mathbf{Y},\mathbf{Z})_{obs}}(\mathbf{u}|\mathbf{v})$, by the following two steps (Section A.2):

(i) Draw $\tilde{\boldsymbol{\theta}}$ from the observed posterior distribution $\pi(\boldsymbol{\theta}|(\mathbf{x}, \mathbf{y}, \mathbf{z})_{obs})$.

(ii) Draw observations for the missing items $(\mathbf{X}, \mathbf{Y}, \mathbf{Z})_{mis}$ from the distribution $f_{(\mathbf{X},\mathbf{Y},\mathbf{Z})_{mis}|(\mathbf{X},\mathbf{Y},\mathbf{Z})_{obs}}(\mathbf{u}|\mathbf{v}, \tilde{\boldsymbol{\theta}})$.

2.8 Other Identifiable Models

At the beginning of this chapter, it was claimed that statistical matching has primarily focused on the study of an identifiable model for the available data, i.e. the two samples A and B with missing joint information on $\mathbf{Y}$ and $\mathbf{Z}$. The natural model to consider is that described by the CIA. Naturally, all the models included in the CIA are easily estimated from $A \cup B$, e.g. when $\mathbf{X}$, $\mathbf{Y}$ and $\mathbf{Z}$ are pairwise independent. These are not the only identifiable models. In the next paragraphs, two particular models are considered:

(i) a model which assumes (marginal) independence between $\mathbf{Y}$ and $\mathbf{Z}$;

(ii) a model which assumes independence of all the variables $(\mathbf{X}, \mathbf{Y}, \mathbf{Z})$ given an unobserved (latent) variable.

2.8.1 The pairwise independence assumption

In this subsection we will restrict ourselves to the case of univariate, categorical X, Y and Z, although the results can be extended to the multivariate case. Let us consider a model characterized by:

(i) marginal independence between Y and Z;

(ii) a loglinear model for X, Y, and Z with the three-way interaction term equal to zero.

These two assumptions define a model which is estimable (identifiable) for $A \cup B$. In fact, it needs knowledge of the marginal distributions (minimal sufficient statistics) of (X, Y), (X, Z), and (Y, Z), the last of which is estimable thanks to the hypothesis of independence of Y and Z. Furthermore, it is a model which is different from the CIA. In fact, under the two previous assumptions it is possible to estimate a model such that Y and Z are *dependent* given X.

Remark 2.15 This is the other side of the CIA under the so-called Simpson paradox; see Dawid (1979). When analysing a pairwise distribution, the relationship between the two variables may be completely transformed by the use of a third variable, in this case X.

In the case of the CIA, Y and Z are marginally dependent, but actually this dependence is completely explained by a third variable X: Y and Z are assumed independent given X.

On the other hand, it is possible that the marginal (Y, Z) distribution shows independence between the two variables, but actually knowledge of a third variable (i.e. conditioning on it) makes Y and Z dependent.

The peculiar aspect of these models is that they are strictly different, and at the same time estimable for $A \cup B$.

Note that, as for the CIA, assumptions (i) and (ii) define a model which is untestable with the data set $A \cup B$. Consequently, it must be considered as an assumption, henceforth called the *pairwise independence assumption* (PIA).

Differently from the CIA, the sample $A \cup B$ does not allow a maximum likelihood estimate of the parameters θ_{ijk} in closed form under the PIA. In this case, the IPF algorithm (Section B.1) can be used. This procedure first estimates the minimal sufficient tables of the imposed loglinear model, and then applies the IPF algorithm.

(a) Compute $\hat{\theta}_{i..}$ as in (2.33), $i = 1, 2$, $\hat{\theta}_{j|i}$ as in (2.34), $i = 1, 2$, $j = 1, 2$, and $\hat{\theta}_{k|i}$ as in (2.35), $i = 1, 2$, $k = 1, 2, 3$. These estimates are ML estimates of the corresponding parameters under the PIA.

(b) Estimate the parameters of the tables for (X, Y), (X, Z) and (Y, Z) according to the following formulae:

- $\hat{\theta}_{ij.} = \hat{\theta}_{j|i}\hat{\theta}_{i..}$, $i = 1, \ldots, I$, $j = 1, \ldots, J$, for (X, Y);

- $\hat{\theta}_{i.k} = \hat{\theta}_{k|i}\hat{\theta}_{i..}$, $i = 1, \ldots, I$, $k = 1, \ldots, k$, for (X, Z);
- $\hat{\theta}_{.jk} = \hat{\theta}_{.j.}\hat{\theta}_{..k}$, $j = 1, \ldots, J$, $k = 1, \ldots, K$, for (Y, Z), where

$$\hat{\theta}_{.j.} = \sum_{i=1}^{I} \hat{\theta}_{ij.}, \qquad j = 1, \ldots, J,$$

$$\hat{\theta}_{..k} = \sum_{i=1}^{I} \hat{\theta}_{i.k}, \qquad k = 1, \ldots, K.$$

Hence, these are the ML estimates of the minimal sufficient tables (X, Y), (X, Z), and (Y, Z) under assumptions (i) and (ii) of the PIA.

(c) Starting from a contingency table that satisfies the PIA, e.g. a table consisting of 1s, apply the IPF algorithm through iterative adaptation of the frequency estimates $\hat{\theta}_{ijk}$ to the marginal tables of (X, Y), (X, Z) and (Y, Z) obtained in step (b).

Both the ML estimates under the CIA and PIA respectively coincide in some aspects. To be precise, steps (a) and (b) ensure that the ML estimators of $\theta_{i..}$, $\theta_{j|i}$, $\theta_{k|i}$, $\theta_{ij.}$ and $\theta_{i.k}$ are the same under the two distinct models, $i = 1, \ldots, I$, $j = 1, \ldots, J$, $k = 1, \ldots, K$. The difference is in the joint distribution. If, under the CIA, the final ML estimate $\hat{\theta}_{ijk}$ reports marginal dependence between Y and Z (as usually happens unless stricter models hold, such as pairwise independence between X, Y and Z), then under the PIA the final ML estimate should be strictly different, because it must show independence between Y and Z (for the properties of the IPF algorithm, the minimal sufficient tables, and in particular (Y, Z), are preserved in the final estimate).

Remark 2.16 The fact that under the PIA the parameters are estimable with $A \cup B$ and the ML estimate is distinct from the ML estimate under the CIA, may raise suspicion that the PIA can be a competitor of the CIA. Actually this is not true. They are equivalent in the sense that their likelihoods coincide. This fact can be easily seen from the observed likelihood function for sample $A \cup B$, i.e. (2.3), and from the fact that the two ML estimates coincide for the parameters $\theta_{i..}$, $\theta_{j|i}$ and $\theta_{k|i}$. It will be apparent in Chapter 4 that both the estimates are just two solutions of a much wider family of estimates: the likelihood ridge. They have in common that they are both identifiable and estimable for $A \cup B$. In this section we can privilege the CIA or the PIA if we have additional information available.

Remark 2.17 The previous model was considered by Singh *et al.* (1990), although with the following differences.

- Assuming A as the recipient and B as the donor files, they rake B in order to meet the observed X distribution in A. The marginal Y and Z distributions are then used for an additional table, (Y, Z), which is obtained under the hypothesis of independence of Y and Z.

- The joint table for X, Y and Z is obtained by raking with respect to the observed (X, Y) table in A, the raked (X, Z) table in B, and the additional table for (Y, Z). The steps of the raking procedure are exactly those of the IPF algorithm. If the marginal distributions of X in A and B coincide, this procedure coincides with the ML procedure.

- They use mixed methods, as in Section 2.5.

Singh *et al.* (1990) named this approach *RAKEYZ* (the corresponding raking method under the CIA is named *RAKEXYZ*). The previous approach will also be applied in Section 3.7, when external auxiliary information on the marginal table (Y, Z) is available. In this last case, other models than independence of Y and Z can be considered.

Example 2.12 Let X, Y and Z be categorical variables with respectively $I = 2$, $J = 2$, and $K = 3$ categories. Let Tables 2.26 and 2.27 be the observed tables in respectively A and B.

Under the CIA, maximum likelihood estimates of θ_{ijk} are computed as in Section 2.1.3. In other words, compute $\hat{\theta}_{i..}$ as in (2.33), $i = 1, 2$, $\hat{\theta}_{j|i}$ as in (2.34), $i = 1, 2$, $j = 1, 2$, and $\hat{\theta}_{k|i}$ as in (2.35), $i = 1, 2$, $k = 1, 2, 3$, and estimate θ_{ijk} using the formula

$$\hat{\theta}_{ijk} = \hat{\theta}_{j|i}\hat{\theta}_{k|i}\hat{\theta}_{i..}, \qquad i = 1, 2, \ j = 1, 2, \ k = 1, 2, 3.$$

The result is shown in Table 2.28.

In order to estimate parameters under the PIA, it is necessary to consider the ML estimates for tables (X, Y), (X, Z) and (Y, Z), the latter under the assumption

Table 2.26 Contingency table for (X, Y) observed on sample A consisting of $n_A = 48\,000$ units

	$Y = 0$	$Y = 1$
$X = 0$	20 805	16 195
$X = 1$	6660	4340

Table 2.27 Contingency table for (X, Z) observed on sample B consisting of $n_B = 52\,000$ units

	$Z = 0$	$Z = 1$	$Z = 2$
$X = 0$	4766	9336	7898
$X = 1$	12 600	10 216	7184

Table 2.28 Maximum likelihood estimates $\hat{\theta}_{ijk}$ of the (X, Y, Z) distribution from sample $A \cup B$, where A is in Table 2.26 and B is in Table 2.27, under the CIA

	$Y=0$			$Y=1$		
	$Z=0$	$Z=1$	$Z=2$	$Z=0$	$Z=1$	$Z=2$
$X=0$	0.071	0.141	0.119	0.056	0.110	0.093
$X=1$	0.105	0.084	0.059	0.068	0.055	0.039

Table 2.29 Maximum likelihood estimates $\hat{\theta}_{ijk}$ of the (X, Y, Z) distribution from sample $A \cup B$, where A is in Table 2.26 and B is in Table 2.27, under the PIA

	$Y=0$			$Y=1$		
	$Z=0$	$Z=1$	$Z=2$	$Z=0$	$Z=1$	$Z=2$
$X=0$	0.052	0.136	0.144	0.076	0.115	0.068
$X=1$	0.122	0.090	0.026	0.050	0.049	0.062

of independence between Y and Z. The application of the previously described steps (i), (ii) and (iii) give the results in Table 2.29. Note that Tables 2.28 and 2.29 coincide in the marginal parameters for (X, Y) and (X, Z) (differences are due to stopping the IPF after a few iterations).

2.8.2 Finite mixture models

The CIA is a model that assumes independence of $\mathbf{Y}$ and $\mathbf{Z}$ given $\mathbf{X}$. Sometimes it is possible to have pairwise independence between all the variables given a latent variable. This can be modelled through *finite mixture models*.

Finite mixture models are used in different contexts, among them clustering and density estimation; see McLachlan and Peel (2000). Finite mixture models correspond to a particular parameterization of the distributions in $\mathcal{F}$:

$$f(\mathbf{x}, \mathbf{y}, \mathbf{z}; \boldsymbol{\theta}) = \sum_{l=1}^{G} \pi_l f_l(\mathbf{x}, \mathbf{y}, \mathbf{z}; \boldsymbol{\theta}_l), \tag{2.59}$$

i.e. the joint distribution f is obtained combining G different distributions, f_l, $l = 1, \ldots, G$, where $\boldsymbol{\theta}_l$ represents the parameter vector associated with the distribution f_l, and π_l are the mixing proportions such that $\sum_{l=1}^{G} \pi_l = 1$ and $\pi_l \geq 0$ for $l = 1, \ldots, G$. Note that, in this context, there is an additional r.v. L (the component of the mixture $l = 1, \ldots, G$) and this variable is missing in A and B, i.e. it is a latent variable.

Although model (2.59) is more complex than those in Sections 2.1.2 and 2.1.3, it may lead to a substantial simplification of the dependence model among the variables of interest $(\mathbf{X}, \mathbf{Y}, \mathbf{Z})$. This idea was introduced into the statistical matching process by Kamakura and Wedel (1997). They refer to Everitt's (1984) assumption that the interdependence amongst $(\mathbf{X}, \mathbf{Y}, \mathbf{Z})$ is due to their common dependence on a latent variable and that, once determined, the behaviour of the $(\mathbf{X}, \mathbf{Y}, \mathbf{Z})$ is essentially random. This is equivalent to stating that the variables are independent of one another within any class $L = l$, and thus the distribution (2.59) can be formulated as

$$f(\mathbf{x}, \mathbf{y}, \mathbf{z}; \boldsymbol{\theta}) = \sum_{l=1}^{G} \pi_l \prod_{p=1}^{P} f_{X_p;l}(x_p; \boldsymbol{\theta}_l) \prod_{q=1}^{Q} f_{Y_q;l}(y_q; \boldsymbol{\theta}_l) \prod_{r=1}^{R} f_{Z_r;l}(z_r; \boldsymbol{\theta}_l). \quad (2.60)$$

This is a particular form of the CIA: $\mathbf{X}$, $\mathbf{Y}$ and $\mathbf{Z}$ are assumed independent given L.

By (1.3), the likelihood function is

$$L(\boldsymbol{\theta}, \boldsymbol{\pi}; A \cup B) = \prod_{a=1}^{n_A} \left[\sum_{l=1}^{G} \pi_l \prod_{p=1}^{P} f_{X_p;l}(x_{ap}; \boldsymbol{\theta}_l) \prod_{q=1}^{Q} f_{Y_q;l}(y_{aq}; \boldsymbol{\theta}_l) \right]$$
$$\times \prod_{b=1}^{n_B} \left[\sum_{l=1}^{G} \pi_l \prod_{p=1}^{P} f_{X_p;l}(x_{bp}; \boldsymbol{\theta}_l) \prod_{r=1}^{R} f_{Z_q;l}(z_{bq}; \boldsymbol{\theta}_l) \right] \quad (2.61)$$

The ML estimates $(\hat{\boldsymbol{\theta}}, \hat{\boldsymbol{\pi}})$ are generally computed by means of the EM algorithm (Section A.1.2; see also McLachlan and Basford, 1988). Differently from the typical missing data pattern of the statistical matching problem (Table 1.1), this approach is characterized by additional missing data: the unobserved latent variable L. Since the maxima of the observed loglikelihood (2.3) can be obtained in closed form (conditionally on L), the algorithm essentially consists in estimating the probability of belonging to a latent class l, and conditionally on these probabilities computing the estimates of parameters $\boldsymbol{\theta}$.

Kamakura and Wedel (1997) analyse the case where $(\mathbf{X}, \mathbf{Y}, \mathbf{Z})$ are categorical variables, while Rässler and Fleischer (1999) propose an algorithm to deal with categorical and continuous variables simultaneously. In particular, the r.v.s $\mathbf{X}$ are divided into two groups C_x and N_x, where $C_x \cup N_x = \{1, \ldots, P\}$. The r.v.s X_p, $p \in C_x$, are categorical, with probability ${}_p\theta_{i;l}$ for each category $i = 1, \ldots, I_P$. The r.v.s X_p, $p \in N_x$, are normal with mean $\mu_{p;l}$ and variance $\sigma^2_{p;l}$. Similar definitions also hold for $\mathbf{Y}$ and $\mathbf{Z}$.

In this context, the EM algorithm for the computation of the ML estimate of $(\boldsymbol{\theta}, \boldsymbol{\pi})$ adapts the estimates on a set of equations that $(\boldsymbol{\theta}, \boldsymbol{\pi})$ must satisfy. For the sake of simplicity, we report only equations related to $\mathbf{X}$. The others can easily be derived by observing that, when the equations refer to $\mathbf{Y}$, the sums are over the n_A units in A, while, when the equations are relative to $\mathbf{Z}$, the sums are over the n_B

units in B:

$$\hat{\pi}_l = \frac{1}{n_A + n_B}\left[\sum_{a=1}^{n_A} \hat{\pi}_{a;l} + \sum_{b=1}^{n_B} \hat{\pi}_{b;l}\right], l = 1, \ldots, G, \quad (2.62)$$

$$_p\hat{\theta}_{i;l} = \frac{\sum_{a=1}^{n_A} I_i(x_{ap})\hat{\pi}_{a;l} + \sum_{b=1}^{n_B} I_i(x_{bp})\hat{\pi}_{b;l}}{\sum_{a=1}^{n_A} \hat{\pi}_{a;l} + \sum_{b=1}^{n_B} \hat{\pi}_{b;l}}, i = 1, \ldots, I_p, p \in C_x, \quad (2.63)$$

$$\hat{\mu}_{p;l} = \frac{\sum_{a=1}^{n_A} x_{ap}\hat{\pi}_{a;l} + \sum_{b=1}^{n_B} x_{bp}\hat{\pi}_{b;l}}{\sum_{a=1}^{n_A} \hat{\pi}_{a;l} + \sum_{b=1}^{n_B} \hat{\pi}_{b;l}}, p \in N_x, \quad (2.64)$$

$$\hat{\sigma}^2_{p;l} = \frac{\sum_{a=1}^{n_A} (x_{ap} - \hat{\mu}_{p;l})^2\hat{\pi}_{a;l} + \sum_{b=1}^{n_B} (x_{bp} - \hat{\mu}_{p;l})^2\hat{\pi}_{b;l}}{\sum_{a=1}^{n_A} \hat{\pi}_{a;l} + \sum_{b=1}^{n_B} \hat{\pi}_{b;l}}, p \in N_x, \quad (2.65)$$

where $I_i(x)$ is the indicator function taking the value 1 when $x = i$ and 0 otherwise.

The estimated probability that each unit belongs to a group $L = l$, $\pi_{a;l}$, $a = 1, \ldots, n_A$, and $\pi_{b;l}$, $b = 1, \ldots, n_B$, respectively can be obtained by the following equations:

$$\hat{\pi}_{a;l} = \frac{\hat{\pi}_l \prod_{p=1}^{P} f_{X_p;l}(x_{ap}; \hat{\boldsymbol{\theta}}_l) \prod_{q=1}^{Q} f_{Y_q;l}(y_{aq}; \hat{\boldsymbol{\theta}}_l)}{\sum_{l=1}^{L} \hat{\pi}_l \prod_{p=1}^{P} f_{X_p;l}(x_{ap}; \hat{\boldsymbol{\theta}}_l) \prod_{q=1}^{Q} f_{Y_q;l}(y_{aq}; \hat{\boldsymbol{\theta}}_l)}, \quad (2.66)$$

$a = 1, \ldots, n_A$, for the units in A, and

$$\hat{\pi}_{b;l} = \frac{\hat{\pi}_l \prod_{p=1}^{P} f_{X_p;l}(x_{bp}; \hat{\boldsymbol{\theta}}_l) \prod_{r=1}^{R} f_{Z_r;l}(z_{br}; \hat{\boldsymbol{\theta}}_l)}{\sum_{l=1}^{L} \prod_{p=1}^{P} f_{X_p;l}(x_{bp}; \hat{\boldsymbol{\theta}}_l) \prod_{r=1}^{R} f_{Z_r;l}(z_{br}; \hat{\boldsymbol{\theta}}_l)}, \quad (2.67)$$

$b = 1, \ldots, n_B$, for the units in B.

The algorithm to compute the estimates can be summarized in the following steps.

(i) At the first iteration ($h = 0$), initialize (randomly) the posterior probabilities $\hat{\pi}^{(0)}_{i;l}$.

(ii) Compute $\hat{\pi}_l$ using (2.62).

(iii) Compute estimates of the parameters θ using (2.63).

(iv) Compute estimates of the parameters μ using (2.64).

(v) Compute estimates of the parameters σ using (2.65).

(vi) Convergence test: if the change in the loglikelihood from iteration $h - 1$ to h is smaller than a predefined tolerance level, then stop; otherwise go to step (vii).

(vii) Compute estimates of the parameters $\pi_{i;l}$ using (2.66) and (2.67); go to step (ii).

This procedure has been defined when the number of mixture components G is fixed and known. If G is to be estimated, measures such as the Akaike information criterion (AIC), consistent Akaike information criterion (CAIC) or Bayesian information criterion (BIC) can be used (Keribin, 2000).

The cases of only categorical or only continuous (Gaussian) variables, may be handled using just the relevant steps of the EM algorithm previously described.

Remark 2.18 The approach based on finite mixture models assumes that conditioning on a latent variable can simplify the association among $(\mathbf{X}, \mathbf{Y}, \mathbf{Z})$ to independence. As in Remark 2.1, this assumption cannot be tested in the matching context because of a lack of joint information.

Remark 2.19 Once the parameters of the finite mixture models have been estimated (macro objective), a micro objective can be pursued. For instance, by the results of this subsection, it is possible to impute missing values in A and B according to random draws from a finite mixture model. The missing $\mathbf{Z}$ in A and $\mathbf{Y}$ in B are imputed through the following scheme.

(i) Assign the ith observation to the latent class l by sampling from the estimated distribution $\hat{\pi}_{i;l}$, for $l = 1, \ldots, G$.

(ii) Impute the missing variable Z_r for $r = 1, \ldots, R$ according to the distribution $f_{Z_r;l}(z_r; \hat{\boldsymbol{\theta}}_l)$ (or Y_q with the distribution $f_{Y_q;l}(y_q; \hat{\boldsymbol{\theta}}_l)$, for $q = 1, \ldots, Q$).

3

Auxiliary Information

3.1 Different Kinds of Auxiliary Information

The assumption of conditional independence between **Y** and **Z** given **X** cannot be tested from the data sets at hand, A and B, as anticipated in Remark 2.1. The CIA is an assumption, and more often than not an incorrect assumption. A large part of the literature on statistical matching, e.g. Sims (1972), Kadane (1978), Cassel (1983), Rodgers (1984), Paass (1986), Barry (1988), Cohen (1991) and Singh *et al.* (1993), describes the effects of the CIA on the statistical matching output (either micro or macro) when the true model is different. It is easy to understand that estimates of the joint $(\mathbf{X}, \mathbf{Y}, \mathbf{Z})$ distribution will be quite different from the real generating distribution. In discussing this problem some writers talk about the 'bias' of the statistical matching procedures due to the CIA. In fact, the statistical matching procedures are unbiased (or at least suitable) for the conditional independence model. The problem is the misspecification of the model.

When the CIA does not hold, the parameters for the statistical relationship between **Y** and **Z** are inestimable from the data sets A and B. In Chapter 2, two particular distributions were studied in depth: the normal and the multinomial. Let us see in the following what are the inestimable parameters.

Example 3.1 In Section 2.1.1, (X, Y, Z) are univariate normal distributions. In this case, the parameter $\sigma_{YZ|X}$ is inestimable for $A \cup B$, unless strict assumptions are made on that parameter. Under the CIA, it was assumed that $\sigma_{YZ|X} = 0$. As a consequence, the partial regression coefficients of Y on Z and of Z on Y are assumed null.

In a general setting, the regression function (2.4) should assume the form

$$Y = \mu_{Y|XZ} + \epsilon_{Y|XZ} = \mu_Y + \beta_{YX.Z}(X - \mu_X) + \beta_{YZ.X}(Z - \mu_Z) + \epsilon_{Y|XZ},$$

Statistical Matching: Theory and Practice M. D'Orazio, M. Di Zio and M. Scanu

where

$$\beta_{YX.Z} = \frac{\sigma_{XY|Z}}{\sigma^2_{X|Z}},$$

$$\beta_{YZ.X} = \frac{\sigma_{YZ|X}}{\sigma^2_{Z|X}}$$

are the partial regression coefficients of Y on X and Z, and $\epsilon_{Y|XZ}$ is normally distributed with zero mean and residual variance

$$\sigma^2_{Y|XZ} = \sigma^2_Y - (\sigma_{XY}\ \sigma_{YZ}) \begin{pmatrix} \sigma^2_X & \sigma_{XZ} \\ \sigma_{XZ} & \sigma^2_Z \end{pmatrix}^{-1} \begin{pmatrix} \sigma_{XY} \\ \sigma_{YZ} \end{pmatrix}.$$

The same holds for the regression equation (2.6), which in a general setting assumes the form

$$Z = \mu_{Z|XY} + \epsilon_{Z|XY} = \mu_Z + \beta_{ZX.Y}(X - \mu_X) + \beta_{ZY.X}(Y - \mu_Y) + \epsilon_{Z|XY},$$

where the partial regression coefficients are

$$\beta_{ZX.Y} = \frac{\sigma_{XZ|Y}}{\sigma^2_{X|Y}},$$

$$\beta_{ZY.X} = \frac{\sigma_{YZ|X}}{\sigma^2_{Y|X}},$$

and $\epsilon_{Z|XY}$ follows a normal distribution with zero mean and variance

$$\sigma^2_{Z|XY} = \sigma^2_Z - (\sigma_{XZ}\ \sigma_{YZ}) \begin{pmatrix} \sigma^2_X & \sigma_{XY} \\ \sigma_{XY} & \sigma^2_Y \end{pmatrix}^{-1} \begin{pmatrix} \sigma_{XZ} \\ \sigma_{YZ} \end{pmatrix}.$$

The previous arguments carry over to the multivariate case of Section 2.1.2. Under the CIA, the conditional covariance matrix between $\mathbf{Y}$ and $\mathbf{Z}$ given $\mathbf{X}$ is assumed null, so that the covariance matrix $\Sigma_{\mathbf{YZ}}$ can be estimated from $A \cup B$: $\Sigma_{\mathbf{YZ}} = \Sigma_{\mathbf{YX}} \Sigma^{-1}_{\mathbf{XX}} \Sigma_{\mathbf{XZ}}$. Actually, in a general setting the conditional covariance matrix between $\mathbf{Y}$ and $\mathbf{Z}$ given $\mathbf{X}$ should be

$$\Sigma_{\mathbf{YZ|X}} = \Sigma_{\mathbf{YZ}} - \Sigma_{\mathbf{YX}} \Sigma^{-1}_{\mathbf{XX}} \Sigma_{\mathbf{XZ}}$$

(see Anderson, 1984). The regression equations (2.12) and (2.17) now assume the following form:

$$\mathbf{Y} = \boldsymbol{\mu}_{\mathbf{Y}} + \boldsymbol{\beta}_{\mathbf{YX.Z}}(\mathbf{X} - \boldsymbol{\mu}_{\mathbf{X}}) + \boldsymbol{\beta}_{\mathbf{YZ.X}}(\mathbf{Z} - \boldsymbol{\mu}_{\mathbf{Z}}) + \boldsymbol{\epsilon}_{\mathbf{Y|XZ}}, \quad (3.1)$$

$$\mathbf{Z} = \boldsymbol{\mu}_{\mathbf{Z}} + \boldsymbol{\beta}_{\mathbf{ZX.Y}}(\mathbf{X} - \boldsymbol{\mu}_{\mathbf{X}}) + \boldsymbol{\beta}_{\mathbf{ZY.X}}(\mathbf{Y} - \boldsymbol{\mu}_{\mathbf{Y}}) + \boldsymbol{\epsilon}_{\mathbf{Z|XY}}, \quad (3.2)$$

where the matrices of the partial regression coefficients are defined by the equations:

$$\boldsymbol{\beta}_{\mathbf{YX.Z}} = \Sigma_{\mathbf{YX|Z}} \Sigma^{-1}_{\mathbf{XX|Z}}, \qquad \boldsymbol{\beta}_{\mathbf{YZ.X}} = \Sigma_{\mathbf{YZ|X}} \Sigma^{-1}_{\mathbf{ZZ|X}},$$

$$\boldsymbol{\beta}_{\mathbf{ZX.Y}} = \Sigma_{\mathbf{ZX|Y}} \Sigma^{-1}_{\mathbf{XX|Y}}, \qquad \boldsymbol{\beta}_{\mathbf{ZY.X}} = \Sigma_{\mathbf{ZY|X}} \Sigma^{-1}_{\mathbf{YY|X}},$$

and $\boldsymbol{\epsilon}_{\mathbf{Y}|\mathbf{XZ}}$ and $\boldsymbol{\epsilon}_{\mathbf{Z}|\mathbf{XY}}$ are multinormal r.v.s with null mean vector and residual covariance matrices respectively equal to

$$\Sigma_{\mathbf{YY}|\mathbf{XZ}} = \Sigma_{\mathbf{YY}} - (\Sigma_{\mathbf{YX}}\ \Sigma_{\mathbf{YZ}}) \begin{pmatrix} \Sigma_{\mathbf{XX}} & \Sigma_{\mathbf{XZ}} \\ \Sigma_{\mathbf{ZX}} & \Sigma_{\mathbf{ZZ}} \end{pmatrix}^{-1} \begin{pmatrix} \Sigma_{\mathbf{XY}} \\ \Sigma_{\mathbf{ZY}} \end{pmatrix},$$

$$\Sigma_{\mathbf{ZZ}|\mathbf{XY}} = \Sigma_{\mathbf{ZZ}} - (\Sigma_{\mathbf{ZX}}\ \Sigma_{\mathbf{ZY}}) \begin{pmatrix} \Sigma_{\mathbf{XX}} & \Sigma_{\mathbf{XY}} \\ \Sigma_{\mathbf{YX}} & \Sigma_{\mathbf{YY}} \end{pmatrix}^{-1} \begin{pmatrix} \Sigma_{\mathbf{XZ}} \\ \Sigma_{\mathbf{YZ}} \end{pmatrix}.$$

When the variables are categorical, as in Section 2.1.3, the inestimable parameters are $\theta_{jk|i}$. Under the CIA, these parameters were derived from the usual independence formula $\theta_{jk|i} = \theta_{j|i}\theta_{k|i}$.

In order to obtain a *point* estimate of the overall density $f(\mathbf{x}, \mathbf{y}, \mathbf{z})$, given that A and B are not enough, it is necessary to resort to *external auxiliary information.* Singh *et al.* (1993) have identified two different sources of external information:

(i) a third file C where either $(\mathbf{X}, \mathbf{Y}, \mathbf{Z})$ or $(\mathbf{Y}, \mathbf{Z})$ are jointly observed;

(ii) a plausible value of the inestimable parameters of either $(\mathbf{Y}, \mathbf{Z}|\mathbf{X})$ or $(\mathbf{Y}, \mathbf{Z})$.

These sources of information may not be perfect. For instance, C may come from an outdated statistical investigation (e.g. a census or a sample survey) or from a nonstatistical source (an administrative register), or (perhaps the best situation) a supplemental (even small) *ad hoc* survey performed in order to gain information on the inestimable parameters. Possible plausible values for the inestimable parameters may come from the relationship between proxy variables, i.e. jointly observed variables $\mathbf{Y}^{\bullet}$ and $\mathbf{Z}^{\bullet}$ that are expected to be distributed similarly to $\mathbf{Y}$ and $\mathbf{Z}$.

At this stage, it may seem that the use of auxiliary information can definitively solve the statistical matching problem, i.e. the use of an untestable and most of the time unreliable assumption such as the CIA. This is not at all true. Actually, we are still dealing with the following untestable assumption: auxiliary information and the data sets A and B are compatible, i.e. refer to the same (unknown) model. For instance, if C is an outdated data set, this means that the statistical model that has generated C has not changed over time and coincides with that of A and B. More formally, there are three alternatives for the possible sources of information and the inevitable associated hypotheses:

(i) The data set C is a sample of n_C units generated independently from the overall model $f(\mathbf{x}, \mathbf{y}, \mathbf{z})$:

$$(\mathbf{x}_c, \mathbf{y}_c, \mathbf{z}_c), \qquad c = 1, \ldots, n_C.$$

Consequently, inferences can be gained from the overall sample $A \cup B \cup C$, where $\mathbf{Z}$ is missing in A and $\mathbf{Y}$ is missing in B, while C is complete.

(ii) The data set C is a sample of n_C units generated independently from the overall model $f(\mathbf{x}, \mathbf{y}, \mathbf{z})$, with $\mathbf{X}$ missing:

$$(\mathbf{y}_c, \mathbf{z}_c), \qquad c = 1, \ldots, n_C.$$

Consequently, inferences can be obtained from the overall sample $A \cup B \cup C$, where $\mathbf{Z}$ is missing in A, $\mathbf{Y}$ is missing in B, and $\mathbf{X}$ is missing in C.

(iii) The set of distributions $\mathcal{F} = \{f_{\mathbf{XYZ}}\}$ is parametric, i.e. each distribution in $\mathcal{F}$ can be indexed by $\boldsymbol{\theta} \in \Theta$. Auxiliary parametric information (obtained from proxy variables or other sources) allows the restriction of the overall parameter set Θ. Usually, restrictions are on the conditional parameters of $(\mathbf{Y}, \mathbf{Z})$ given $\mathbf{X}$ or on the parameters of the distribution of $(\mathbf{Y}, \mathbf{Z})$.

Note that the first two types of auxiliary information affect the data set used in the estimation process, leaving unchanged the set $\mathcal{F}$ of possible estimates, while the third type restricts $\mathcal{F}$ to a new set of distributions $\mathcal{F}^*$, or in other words restricts the parameter set Θ to Θ^*. In this chapter, the changes to the statistical matching procedures outlined in Chapter 2 due to the use of auxiliary information will be described (Sections 3.2–3.4). It may happen that the assumptions on the supplemental file C are too strict, i.e. it is not possible to hypothesize that the units in C are generated according to the same model as A and B. Nevertheless, it is still possible to use the supplemental file, and mixed procedures can be useful (Section 3.6).

Finally, note that there is another source of auxiliary information that can lead to a unique estimate: a suitable set of coherent logical constraints on the parameter values. However this issue is related to the assessment of the uncertainty of the statistical matching output, and discussion of it is deferred to Chapter 4.

3.2 Parametric Macro Methods

When the model is parametric, estimates may be obtained by maximum likelihood. The density $f(\mathbf{x}, \mathbf{y}, \mathbf{z}; \boldsymbol{\theta})$ can be written in the form

$$f(\mathbf{x}, \mathbf{y}, \mathbf{z}; \boldsymbol{\theta}) = f_{\mathbf{X}}(\mathbf{x}; \boldsymbol{\theta}_{\mathbf{X}}) \, f_{\mathbf{YZ}|\mathbf{X}}\left(\mathbf{y}, \mathbf{z}|\mathbf{x}; \boldsymbol{\theta}_{\mathbf{YZ}|\mathbf{X}}\right), \tag{3.3}$$

where $\mathbf{x} \in \mathcal{X}$, $\mathbf{y} \in \mathcal{Y}$ and $\mathbf{z} \in \mathcal{Z}$, and the parameter set $\Theta = \{\boldsymbol{\theta}\}$ is reparameterized into the two sets $\Theta_{\mathbf{X}} = \{\boldsymbol{\theta}_{\mathbf{X}}\}$, $\Theta_{\mathbf{YZ}|\mathbf{X}} = \{\boldsymbol{\theta}_{\mathbf{YZ}|\mathbf{X}}\}$. The likelihood function depends on the kind of auxiliary information. When recourse can be had to a third file C, the likelihood function will be computed on a sample of $n_A + n_B + n_C$ units. When auxiliary parametric information is used, the likelihood function will be computed on a sample of $n_A + n_B$ units and the ML estimate will be sought in a restricted space, i.e. $\Theta_{\mathbf{YZ}|\mathbf{X}}$ is restricted to the parameter vector suggested by the auxiliary information itself. Consequently, the procedures described in Section 2.1 must be adapted according to the different kinds of auxiliary information available.

3.2.1 The use of a complete third file

Let C be a third file where $(\mathbf{X}, \mathbf{Y}, \mathbf{Z})$ is completely observed on n_C units. As a consequence, let $A \cup B \cup C$ be the overall sample of i.i.d. observations generated from $f(\mathbf{x}, \mathbf{y}, \mathbf{z}; \boldsymbol{\theta})$, where $\mathbf{Z}$ is missing in A and $\mathbf{Y}$ is missing in B. The observed likelihood function can be written in the form

$$
\begin{aligned}
L(\boldsymbol{\theta}; A \cup B \cup C) = & \prod_{a=1}^{n_A} f_{\mathbf{X}}(\mathbf{x}_a; \boldsymbol{\theta}_{\mathbf{X}}) \prod_{b=1}^{n_B} f_{\mathbf{X}}(\mathbf{x}_b; \boldsymbol{\theta}_{\mathbf{X}}) \prod_{c=1}^{n_C} f_{\mathbf{X}}(\mathbf{x}_c; \boldsymbol{\theta}_{\mathbf{X}}) \\
& \times \int_{\mathcal{Z}} \prod_{a=1}^{n_A} f_{\mathbf{YZ}|\mathbf{X}}\left(\mathbf{y}_a, \mathbf{t}|\mathbf{x}_a; \boldsymbol{\theta}_{\mathbf{YZ}|\mathbf{X}}\right) \mathrm{d}\mathbf{t} \\
& \times \int_{\mathcal{Y}} \prod_{b=1}^{n_B} f_{\mathbf{YZ}|\mathbf{X}}\left(\mathbf{t}, \mathbf{z}_b, |\mathbf{x}_b; \boldsymbol{\theta}_{\mathbf{YZ}|\mathbf{X}}\right) \mathrm{d}\mathbf{t} \\
& \times \prod_{c=1}^{n_C} f_{\mathbf{YZ}|\mathbf{X}}\left(\mathbf{y}_c, \mathbf{z}_c|\mathbf{x}_c; \boldsymbol{\theta}_{\mathbf{YZ}|\mathbf{X}}\right). \qquad (3.4)
\end{aligned}
$$

According to Section A.1.2, the likelihood function (3.4) can be factorized into just two factors. One factor is relative to the marginal distribution of $\mathbf{X}$, which is completely observed in $A \cup B \cup C$. The other factor is composed of the joint distribution of $(\mathbf{Y}, \mathbf{Z})$ given $\mathbf{X}$ that cannot be factorized further. Following Rubin (1974), this last factor is irreducible, in the sense that the distribution of $(\mathbf{Y}, \mathbf{Z}|\mathbf{X})$ should be estimated on a partially missing data set, by means of appropriate iterative procedures.

Example 3.2 Let $(\mathbf{X}, \mathbf{Y}, \mathbf{Z})$ be as in Section 2.1.2. The ML estimate of $\boldsymbol{\theta}_{\mathbf{X}}$ is computed as in Section 2.1.2, with the additional help of sample C:

$$
\hat{\mu}_{\mathbf{X}} = \frac{1}{n_A + n_B + n_C} \left(\sum_{a=1}^{n_A} \mathbf{x}_a + \sum_{b=1}^{n_B} \mathbf{x}_b + \sum_{c=1}^{n_C} \mathbf{x}_c \right),
$$

$$
\begin{aligned}
\hat{\Sigma}_{\mathbf{XX}} = & \frac{1}{n_A + n_B + n_C} \sum_{a=1}^{n_A} (\mathbf{x}_a - \hat{\mu}_{\mathbf{X}})(\mathbf{x}_a - \hat{\mu}_{\mathbf{X}})' \\
& + \frac{1}{n_A + n_B + n_C} \sum_{b=1}^{n_B} (\mathbf{x}_b - \hat{\mu}_{\mathbf{X}})(\mathbf{x}_b - \hat{\mu}_{\mathbf{X}})' \\
& + \frac{1}{n_A + n_B + n_C} \sum_{c=1}^{n_C} (\mathbf{x}_c - \hat{\mu}_{\mathbf{X}})(\mathbf{x}_c - \hat{\mu}_{\mathbf{X}})'.
\end{aligned}
$$

The conditional distribution of $(\mathbf{Y}, \mathbf{Z}|\mathbf{X})$ cannot be estimated in closed form, due to the presence of missing items in the overall sample $A \cup B \cup C$. Note that $(\mathbf{Y}, \mathbf{Z}|\mathbf{X})$ follows a multinormal distribution with parameters

$$
\boldsymbol{\theta}_{\mathbf{YZ}|\mathbf{X}} = \left[\begin{pmatrix} \mu_{\mathbf{Y}|\mathbf{X}} \\ \mu_{\mathbf{Z}|\mathbf{X}} \end{pmatrix}, \begin{pmatrix} \Sigma_{\mathbf{YY}|\mathbf{X}} & \Sigma_{\mathbf{YZ}|\mathbf{X}} \\ \Sigma_{\mathbf{ZY}|\mathbf{X}} & \Sigma_{\mathbf{ZZ}|\mathbf{X}} \end{pmatrix} \right].
$$

In this case, it is necessary to resort to iterative procedures, such as the EM algorithm. The presence of file C enables a unique ML estimate of $\boldsymbol{\theta}_{\mathbf{YZ}|\mathbf{X}}$.

Example 3.3 The same situation as outlined in Example 3.2 also holds when (X, Y, Z) are categorical, as in Section 2.1.3. The ML estimate of $\theta_{i..}$, $i = 1, \ldots, I$, is computed from the overall sample $A \cup B \cup C$, i.e.

$$\hat{\theta}_{i..} = \frac{n_{i..}^A + n_{i..}^B + n_{i..}^C}{n_A + n_B + n_C}, \qquad i = 1, \ldots, I.$$

On the other hand, ML estimates $\hat{\theta}_{jk|i}$ of the parameters $\theta_{jk|i}$ should be computed by means of the EM algorithm, and cannot be written in closed form. However, a unique ML estimate is expected.

3.2.2 The use of an incomplete third file

Let C be a third file where $(\mathbf{Y}, \mathbf{Z})$ are completely observed on n_C units, while $\mathbf{X}$ is missing. As a consequence, let $A \cup B \cup C$ be the overall sample of i.i.d. observations generated from $f(\mathbf{x}, \mathbf{y}, \mathbf{z}; \boldsymbol{\theta})$, where $\mathbf{Z}$ is missing in A, $\mathbf{Y}$ is missing in B, and $\mathbf{X}$ is missing in C. The observed likelihood function can be written in the form

$$\begin{aligned} L(\boldsymbol{\theta}; A \cup B \cup C) = & \int_{\mathcal{Z}} \prod_{a=1}^{n_A} f(\mathbf{x}, \mathbf{y}, \mathbf{t}; \boldsymbol{\theta}) d\mathbf{t} \\ & \times \int_{\mathcal{Y}} \prod_{b=1}^{n_B} f(\mathbf{x}, \mathbf{t}, \mathbf{z}; \boldsymbol{\theta}) d\mathbf{t} \\ & \times \int_{\mathcal{X}} \prod_{c=1}^{n_C} f(\mathbf{t}, \mathbf{y}, \mathbf{z}; \boldsymbol{\theta}) d\mathbf{t}. \end{aligned} \tag{3.5}$$

Following Rubin (1974), the likelihood function (3.5) cannot be factorized (see Section A.1.2). As a matter of fact, this is the case when $(\mathbf{X}, \mathbf{Y}, \mathbf{Z})$ is irreducible. An ML estimate of $\boldsymbol{\theta}$ can be obtained by iterative procedures such as the EM algorithm. As a consequence, even the marginal $\mathbf{X}$ parameters cannot be computed in closed form.

If it is possible to obtain an ML estimate of the parameter $\boldsymbol{\theta}$, it might not be unique. In other words, file C might not contain enough information for the model parameters, and it is useless if a point estimate is required. The next two examples illustrate two different situations where file C plays different roles.

Example 3.4 Let $(\mathbf{X}, \mathbf{Y}, \mathbf{Z})$ be multinormal r.v.s as in Section 2.1.2, and A, B and C be three incomplete samples generated by $(\mathbf{X}, \mathbf{Y}, \mathbf{Z})$, respectively of size n_A, n_B and n_C. A property of the multinormal distribution is that it depends only on the marginal and bivariate parameters of the variables (i.e. means, variances and correlations). Hence, $\boldsymbol{\theta}$ is estimable, i.e. a unique ML estimate of $\boldsymbol{\theta}$ is expected, when C is also available. However, it cannot be expressed in closed form, but iterative methods such as the EM algorithm should be applied.

Example 3.5 Let (X, Y, Z) be categorical r.v.s as in Section 2.1.3. The case of an additional incomplete sample C was considered by Paass (1986). In this case, the additional information available through file C is not enough for a unique estimate of $\boldsymbol{\theta}$, and in particular of the conditional parameters $\theta_{jk|i}$, $i = 1, \ldots, I$, $j = 1, \ldots, J$, $k = 1, \ldots, K$, unless particular assumptions hold. As Klevmarken (1986) and Singh *et al.* (1993) noted, it is necessary to assume a loglinear model with the three-way interactions set to zero.

It will be seen in Chapter 4 that the presence of multiple ML solutions defines the uncertainty on the inestimable parameters. Comments on this point are deferred to Chapter 4.

3.2.3 The use of information on inestimable parameters

Since Kadane (1978), it has been apparent that auxiliary parametric information may be useful in avoiding the CIA in the statistical matching problem. Usually, auxiliary parametric information is gained in two ways:

- through previous samples, archives or collection of data;
- from proxy variables.

In both cases, although not perfect, this parametric information is also assumed to hold for the model that has generated the data in samples A and B. Hence, parametric external information plays a very important role in the statistical matching context: it may be used in order to constrain the problem, i.e. reduce the set of possible parameters Θ. Hopefully, these restrictions will lead to a unique ML solution in the statistical matching problem. Unfortunately, this is not always true. In this subsection, it will be shown how parametric information on the inestimable parameters may be used by means of the ML approach. Differences between the ML approach and some of the procedures available in the statistical matching literature will be outlined.

First of all, it is necessary to distinguish between different situations:

(i) where parametric auxiliary information refers to a few parameters of $\boldsymbol{\theta}_{\mathbf{YZ|X}}$, in particular the inestimable ones;

(ii) where parametric auxiliary information refers to a few parameters of $\boldsymbol{\theta}_{\mathbf{YZ}}$, in particular the inestimable ones.

The interaction between unrestricted ML estimates and parametric auxiliary information is the focus of this subsection. All the unrestricted ML estimates when the observed sample is $A \cup B$ can be characterized by the following proposition.

Proposition 3.1 *Let $A \cup B$ be a sample of size $n_A + n_B$ from $f(\mathbf{x}, \mathbf{y}, \mathbf{z}; \theta)$, with* **Z** *missing in A and* **Y** *missing in B. The unrestricted ML estimates $\hat{\boldsymbol{\theta}}$ of $\boldsymbol{\theta}$ are those, and only those, compatible with the following ML estimates:*

$$\hat{\theta}_{\mathbf{X}}, \text{ computed in } A \cup B, \tag{3.6}$$

$$\hat{\boldsymbol{\theta}}_{\mathbf{Y}|\mathbf{X}}, \text{ computed in } A, \tag{3.7}$$

$$\hat{\boldsymbol{\theta}}_{\mathbf{Z}|\mathbf{X}}, \text{ computed in } B. \tag{3.8}$$

Proof. The likelihood function has the following form:

$$\begin{aligned} L(\boldsymbol{\theta}; A \cup B) = & \prod_{a=1}^{n_A} f_{\mathbf{X}}(\mathbf{x}_a; \boldsymbol{\theta}_{\mathbf{X}}) \prod_{b=1}^{n_B} f_{\mathbf{X}}(\mathbf{x}_b; \boldsymbol{\theta}_{\mathbf{X}}) \\ & \times \prod_{a=1}^{n_A} f_{\mathbf{Y}|\mathbf{X}}\left(\mathbf{y}_a|\mathbf{x}_a; \boldsymbol{\theta}_{\mathbf{Y}|\mathbf{X}}\right) \prod_{b=1}^{n_B} f_{\mathbf{Z}|\mathbf{X}}\left(\mathbf{y}_b|\mathbf{x}_b; \boldsymbol{\theta}_{\mathbf{Z}|\mathbf{X}}\right) \end{aligned} \tag{3.9}$$

Hence, conditions (3.6)–(3.8) are sufficient because $\hat{\boldsymbol{\theta}}$ maximizes each component in (3.9). They are also necessary because if a vector $\tilde{\boldsymbol{\theta}}$ is not compatible with at least one of conditions (3.6)–(3.8), then its likelihood cannot be larger than those which fulfil all the conditions.

All the unrestricted ML estimates form the *likelihood ridge*. The presence of a nonunique ML estimate is a clear effect of the unidentifiability of the model. The above proposition implies that all the ML estimates, i.e. all the parameter vectors $\boldsymbol{\theta}$ in the likelihood ridge, are compatible with the parameters estimated under the CIA in Chapter 2. Hence, the model of conditional independence between **Y** and **Z** given **X** with parameter estimates (3.6)–(3.8) is also an ML solution in the likelihood ridge.

Parametric auxiliary information may be either compatible with (3.6)–(3.8) or not. If it is compatible, then the restricted ML solution(s) will still lie in the unrestricted likelihood ridge. If it is not, then the restricted ML solutions will be outside. In the rest of this section these situations will be described when the variables are multinormal, as in Section 2.1.2, and multinomial, as in Section 2.1.3.

Remark 3.1 Note that parametric auxiliary information on $\boldsymbol{\theta}_{\mathbf{YZ}|\mathbf{X}}$ is always compatible with the ML estimates (3.6). Sometimes, it is possible to have knowledge of just one of the parameters of the vector $\boldsymbol{\theta}_{\mathbf{YZ}|\mathbf{X}}$. In the normal case, for instance, knowledge may be confined to the partial correlation coefficient of **Y** and **Z** given **X**. In this case, compatibility of this kind of knowledge is also ensured with (3.7) and (3.8).

On the other hand, information on $\boldsymbol{\theta}_{\mathbf{YZ}}$ may be incompatible with the unrestricted ML estimates (3.6)–(3.8).

3.2.4 The multinormal case

In the multinormal case, the joint distribution of $(\mathbf{X}, \mathbf{Y}, \mathbf{Z})$ is characterized by the parameter

$$\boldsymbol{\theta} = (\boldsymbol{\mu}, \Sigma) = \left[\begin{pmatrix} \mu_{\mathbf{X}} \\ \mu_{\mathbf{Y}} \\ \mu_{\mathbf{Z}} \end{pmatrix}, \begin{pmatrix} \Sigma_{\mathbf{XX}} & \Sigma_{\mathbf{XY}} & \Sigma_{\mathbf{XZ}} \\ \Sigma_{\mathbf{YX}} & \Sigma_{\mathbf{YY}} & \Sigma_{\mathbf{YZ}} \\ \Sigma_{\mathbf{ZX}} & \Sigma_{\mathbf{ZY}} & \Sigma_{\mathbf{ZZ}} \end{pmatrix} \right]. \tag{3.10}$$

In this case, it is enough to gain information on the inestimable parameter, i.e. the one describing the relationship between the not jointly observed variables $\mathbf{Y}$ and $\mathbf{Z}$. In the following, two possible restrictions are considered: the first one is on the matrix of the partial correlation coefficients $\rho_{\mathbf{YZ}|\mathbf{X}}$; the second on the marginal covariance matrix between $\mathbf{Y}$ and $\mathbf{Z}$, $\Sigma_{\mathbf{YZ}}$.

Known partial correlation coefficients

Roughly speaking, this parameter is the most important in the statistical matching problem. In fact, this distribution can be decomposed in the following factorization:

$$f(\mathbf{x}, \mathbf{y}, \mathbf{z}; \boldsymbol{\theta}) = f_{\mathbf{X}}(\mathbf{x}; \boldsymbol{\theta}_{\mathbf{X}}) f_{\mathbf{YZ}|\mathbf{X}}(\mathbf{y}, \mathbf{z}|\mathbf{x}; \boldsymbol{\theta}_{\mathbf{YZ}|\mathbf{X}}). \tag{3.11}$$

In the multinormal case, $\boldsymbol{\theta}_{\mathbf{YZ}|\mathbf{X}}$ contains the partial correlations between $\mathbf{Y}$ and $\mathbf{Z}$ given $\mathbf{X}$, which is the only inestimable parameter for $A \cup B$. The imposed restriction and Proposition 3.1 allow a unique ML estimate of $\boldsymbol{\theta}$, according to the following steps.

(a) The ML estimate (3.6) of the marginal parameters of $\mathbf{X}$ is computed on the overall sample $A \cup B$, as in step (i), Section 2.1.2, i.e. $\hat{\boldsymbol{\mu}}_{\mathbf{X}}$ and $\hat{\Sigma}_{\mathbf{XX}}$.

(b) The ML estimate (3.7) of the partial linear regression parameters of $\mathbf{Y}$ given $\mathbf{X}$ is computed as in step (ii), Section 2.1.2: $\hat{\boldsymbol{\beta}}_{\mathbf{YX}}$, $\hat{\boldsymbol{\alpha}}_{\mathbf{Y}}$, and $\hat{\Sigma}_{\mathbf{YY}|\mathbf{X}}$. The ML estimates of the marginal parameters for (3.10) are obtained through (2.9)–(2.11):

$$\hat{\boldsymbol{\mu}}_{\mathbf{Y}} = \hat{\boldsymbol{\alpha}}_{\mathbf{Y}} + \hat{\boldsymbol{\beta}}_{\mathbf{YX}} \hat{\boldsymbol{\mu}}_{\mathbf{X}},$$
$$\hat{\Sigma}_{\mathbf{YX}} = \hat{\boldsymbol{\beta}}_{\mathbf{YX}} \hat{\Sigma}_{\mathbf{XX}},$$
$$\hat{\Sigma}_{\mathbf{YY}} = \hat{\Sigma}_{\mathbf{YY}|\mathbf{X}} + \hat{\Sigma}_{\mathbf{YX}} \hat{\Sigma}_{\mathbf{XX}}^{-1} \hat{\Sigma}_{\mathbf{XY}}.$$

(c) The ML estimate (3.8) of the partial linear regression parameters of $\mathbf{Z}$ given $\mathbf{X}$ is computed as in step (iii), Section 2.1.2: $\hat{\boldsymbol{\beta}}_{\mathbf{ZX}}$, $\hat{\boldsymbol{\alpha}}_{\mathbf{Z}}$, and $\hat{\Sigma}_{\mathbf{ZZ}|\mathbf{X}}$. The ML estimates of $\boldsymbol{\mu}_{\mathbf{Z}}$, $\Sigma_{\mathbf{ZZ}}$ and $\Sigma_{\mathbf{XZ}}$ are easily obtained through (2.14)–(2.16):

$$\hat{\boldsymbol{\mu}}_{\mathbf{Z}} = \hat{\boldsymbol{\alpha}}_{\mathbf{Z}} + \hat{\boldsymbol{\beta}}_{\mathbf{ZX}} \hat{\boldsymbol{\mu}}_{\mathbf{X}},$$
$$\hat{\Sigma}_{\mathbf{ZX}} = \hat{\boldsymbol{\beta}}_{\mathbf{ZX}} \hat{\Sigma}_{\mathbf{XX}},$$
$$\hat{\Sigma}_{\mathbf{ZZ}} = \hat{\Sigma}_{\mathbf{ZZ}|\mathbf{X}} + \hat{\Sigma}_{\mathbf{ZX}} \hat{\Sigma}_{\mathbf{XX}}^{-1} \hat{\Sigma}_{\mathbf{XZ}}.$$

(d) The constraint imposed on the partial correlation coefficient matrix, i.e. $\rho_{\mathbf{YZ}|\mathbf{X}} = \rho^*_{\mathbf{YZ}|\mathbf{X}}$, and the previously estimated parameters induce the ML estimates of the residual covariances of $\mathbf{Y}$ and $\mathbf{Z}$ given $\mathbf{X}$, i.e. $\hat{\Sigma}_{\mathbf{YZ}|\mathbf{X}}$:

$$\hat{\sigma}_{Y_j Z_k|\mathbf{X}} = \rho^*_{Y_j Z_k|\mathbf{X}} \sqrt{\hat{\sigma}^2_{Y_j|\mathbf{X}} \hat{\sigma}^2_{Z_k|\mathbf{X}}}, \qquad j = 1, \ldots, Q,\ k = 1, \ldots, R.$$

(e) The last parameter to be estimated in (3.10) is $\Sigma_{\mathbf{YZ}}$. Its ML estimate can be obtained through the estimates in (a), (b), (c), and (d):

$$\hat{\Sigma}_{\mathbf{YZ}} = \hat{\Sigma}_{\mathbf{YZ}|\mathbf{X}} + \hat{\Sigma}_{\mathbf{YX}} \hat{\Sigma}^{-1}_{\mathbf{XX}} \hat{\Sigma}_{\mathbf{XZ}}.$$

Note that steps (a)–(c) are determined by Proposition 3.1, while step (d) is the imposed constraint.

A different, but equivalent, approach would consider the following factorization:

$$f(\mathbf{x}, \mathbf{y}, \mathbf{z}; \boldsymbol{\theta}) = f_{\mathbf{X}}(\mathbf{x}; \boldsymbol{\theta}_{\mathbf{X}}) f_{\mathbf{Y}|\mathbf{X}}(\mathbf{y}|\mathbf{x}; \boldsymbol{\theta}_{\mathbf{Y}|\mathbf{X}}) f_{\mathbf{Z}|\mathbf{XY}}(\mathbf{z}|\mathbf{x}, \mathbf{y}; \boldsymbol{\theta}_{\mathbf{Z}|\mathbf{XY}}) \qquad (3.12)$$

instead of (3.11). For the properties of the multinormal distributions, the three distributions on the right-hand side of (3.12) are still multinormal. The ML estimates of $\boldsymbol{\theta}_{\mathbf{X}}$ and $\boldsymbol{\theta}_{\mathbf{Y}|\mathbf{X}}$ are still those outlined in steps (a) and (b) respectively. Step (c) is retained in order to gain information on the ML estimate of the regression parameters of $\mathbf{Z}$ given $\mathbf{X}$. Actually, the parameter $\boldsymbol{\theta}_{\mathbf{Z}|\mathbf{XY}}$ is defined by the mean vector, given by the regression of $\mathbf{Z}$ on $\mathbf{X}$ and $\mathbf{Y}$, and by the covariance matrix, which is the residual covariance matrix of $\mathbf{Z}$ on $\mathbf{X}$ and $\mathbf{Y}$. Under the constraint outlined in step (d), step (e) is substituted by the following.

(f) According to Seber (1977) and Cox and Wermuth (1996), it is possible to obtain the following estimates of the regression parameters of $\mathbf{Z}$ given $\mathbf{X}$ and $\mathbf{Y}$ through the ML estimates computed in (a), (b), (c), and (d):

$$\hat{\boldsymbol{\beta}}_{\mathbf{ZY.X}} = \hat{\Sigma}_{\mathbf{ZY}|\mathbf{X}} \hat{\Sigma}^{-1}_{\mathbf{YY}|\mathbf{X}}, \qquad (3.13)$$

$$\hat{\boldsymbol{\beta}}_{\mathbf{ZX.Y}} = \hat{\boldsymbol{\beta}}_{\mathbf{ZX}} - \hat{\boldsymbol{\beta}}_{\mathbf{ZY.X}} \hat{\boldsymbol{\beta}}_{\mathbf{YX}}, \qquad (3.14)$$

$$\hat{\Sigma}_{\mathbf{ZZ}|\mathbf{XY}} = \hat{\Sigma}_{\mathbf{ZZ}|\mathbf{X}} - \hat{\Sigma}_{\mathbf{ZY}|\mathbf{X}} \hat{\Sigma}^{-1}_{\mathbf{YY}|\mathbf{X}} \hat{\Sigma}_{\mathbf{YZ}|\mathbf{X}}.$$

Hence, by (3.2), the conditional distribution of $\mathbf{Z}$ given $\mathbf{X}$ and $\mathbf{Y}$ is multinormal, with mean vector

$$\hat{\boldsymbol{\mu}}_{\mathbf{Z}} + \hat{\Sigma}_{\mathbf{ZX}|\mathbf{Y}} \hat{\Sigma}^{-1}_{\mathbf{XX}|\mathbf{Y}} (\mathbf{X} - \hat{\boldsymbol{\mu}}_{\mathbf{X}}) + \hat{\Sigma}_{\mathbf{ZY}|\mathbf{X}} \hat{\Sigma}^{-1}_{\mathbf{YY}|\mathbf{X}} (\mathbf{Y} - \hat{\boldsymbol{\mu}}_{\mathbf{Y}})$$

and covariance matrix $\hat{\Sigma}_{\mathbf{ZZ}|\mathbf{XY}}$.

Remark 3.2 Auxiliary parametric information on the partial correlation coefficients of $\mathbf{Y}$ and $\mathbf{Z}$ given $\mathbf{X}$ has been used by Rubin (1986) and Rässler (2002), with minor differences between them; see also Rässler (2003). In particular, it has been used in the so-called RIEPS method (frequentist regression imputation with random residuals). Although RIEPS has a micro objective, the strength of this method is that imputations are performed once suitable parameter estimates have been obtained. Steps (a)–(d) and (f) are very similar to those outlined in Rässler's papers, and differ from Rubin's in minor respects. However, some differences with the ML approach outlined in the previous section can be detected.

First of all, both Rubin and Rässler assume $\mathbf{X}$ as given, and consequently they do not apply step (a). Secondly, the covariance matrices in steps (a), (b) and (c) are computed with the least squares method instead of ML; see part (i) of Remark 2.6. Finally, Rässler estimates the residual covariance matrices by means of the sum of the squared residuals between the observed and the regressed items. This estimate is equivalent to those outlined in steps (b) and (c), i.e. as far as the partial regressions of $\mathbf{Y}$ on $\mathbf{X}$ and $\mathbf{Z}$ on $\mathbf{X}$ are concerned. However, the residual covariance matrix of $\mathbf{Z}$ given $\mathbf{X}$ and $\mathbf{Y}$ in step (f) cannot be equivalently obtained through the sum of squared residuals. For instance, in order to compute the residual with respect to the estimated regression function in step (f), Rässler completes file B with the estimated regression function of $\mathbf{Y}$ on $\mathbf{X}$ and $\mathbf{Z}$. Hence, the predicted $\mathbf{Z}$ values lack all the variability of the unobserved and imputed $\mathbf{Y}$ values.

Known marginal covariances

Assume that $\Sigma_{\mathbf{YZ}}$ is known and fixed at a value $\Sigma^*_{\mathbf{YZ}}$. Given the property that the unconstrained ML estimates of (3.10) are coherent with the CIA estimates, there are two possibilities.

(i) The estimates of the estimable parameters of (3.10) obtained in steps (a), (b) and (c) on page 73 are coherent with the imposed constraint. In other words, the covariance matrix

$$\hat{\Sigma} = \begin{pmatrix} \hat{\Sigma}_{\mathbf{XX}} & \hat{\Sigma}_{\mathbf{XY}} & \hat{\Sigma}_{\mathbf{XZ}} \\ \hat{\Sigma}_{\mathbf{YX}} & \hat{\Sigma}_{\mathbf{YY}} & \Sigma^*_{\mathbf{YZ}} \\ \hat{\Sigma}_{\mathbf{ZX}} & \Sigma^*_{\mathbf{ZY}} & \hat{\Sigma}_{\mathbf{ZZ}} \end{pmatrix} \tag{3.15}$$

is positive semidefinite. Note that this solution is one of the unconstrained ML estimates.

(ii) The matrix (3.15) is negative definite. This means that the imposed $\Sigma^*_{\mathbf{YZ}}$ is not compatible with the unconstrained ML estimates of the other parameters, i.e. it is incompatible with the CIA. In this case, the restricted ML solution is not in the likelihood ridge. Suitable transformations of the EM algorithm in order to take the constraint into account are necessary.

Remark 3.3 This constraint has been applied in a more general framework in Moriarity and Scheuren (2001, 2003, 2004), following ideas of Kadane (1978); see Section 4.8.1. They pursue a micro objective, once suitable models for the imputation of the missing items is obtained. Note that Kadane (1978) suggests the use of consistent estimators for the estimable parameters. However, Moriarity and Scheuren adopt consistent but not ML estimators; see issue (ii) in Remark 2.6. Their procedure follows these steps.

(a) Estimate all the parameters of $\boldsymbol{\theta}$ except for $\Sigma_{\mathbf{YZ}}$ from the available samples through the usual sample estimates. In particular, the following covariance matrix is considered:

$$\mathbf{S} = \begin{pmatrix} \mathbf{S}_{\mathbf{XX};A\cup B} & \mathbf{S}_{\mathbf{XY};A} & \mathbf{S}_{\mathbf{XZ};B} \\ \mathbf{S}_{\mathbf{YX};A} & \mathbf{S}_{\mathbf{YY};A} & \\ \mathbf{S}_{\mathbf{ZX};B} & & \mathbf{S}_{\mathbf{ZZ};B} \end{pmatrix}.$$

These are unbiased and consistent covariance estimates, but not ML ones.

(b) Substitute the inestimable matrix $\Sigma_{\mathbf{YZ}}$ with a compatible matrix $\Sigma^*_{\mathbf{YZ}}$, such that $\mathbf{S}$ is positive semidefinite.

As a matter of fact, their approach is not exactly a constrained approach, given that they verify the compatibility of $\Sigma^*_{\mathbf{YZ}}$ after the other parameters are estimated. However, they suggest a micro statistical matching approach, (Moriarity and Scheuren, 2001) able to preserve the imposed constraint, and consequently they treat the imposed $\Sigma^*_{\mathbf{YZ}}$ as a constraint. Their micro approach estimates the regression parameters through $\mathbf{S}$ and the sample averages of $\mathbf{X}$, $\mathbf{Y}$ and $\mathbf{Z}$, although they claim that some adjustments are necessary (for the variance of $\mathbf{X}$). For instance, when computing the residual covariance matrix $\hat{\Sigma}_{\mathbf{ZZ}|\mathbf{XY}}$ for the regression of $\mathbf{Z}$ on $\mathbf{Y}$ and $\mathbf{X}$ that is used to impute a residual noise to the regressed values, they suggest the following estimate:

$$\hat{\hat{\Sigma}}_{\mathbf{ZZ}|\mathbf{XY}} = \mathbf{S}_{\mathbf{ZZ};B} - \left(\mathbf{S}_{\mathbf{ZX};B}\ \Sigma^*_{\mathbf{ZY}}\right) \begin{pmatrix} \mathbf{S}_{\mathbf{XX};A} & \mathbf{S}_{\mathbf{XY};A} \\ \mathbf{S}_{\mathbf{YX};A} & \mathbf{S}_{\mathbf{YY};A} \end{pmatrix}^{-1} \begin{pmatrix} \mathbf{S}_{\mathbf{ZX};B} \\ \Sigma^*_{\mathbf{ZY}} \end{pmatrix}.$$

The lack of coherence between $\mathbf{S}$ and the regression estimates will be a problem in their approach (residual covariance matrices which are negative definite). Actually, the use of the ML approach is able to avoid this problem.

3.2.5 Comparison of different regression parameter estimators through simulation

Different ways of estimating the regression parameters have been introduced. They mainly reduce to the approaches based on ML, that described in Moriarity and Scheuren (2001) (MS), and that used in Rubin (1986) and detailed in Rässler (2002) for the RIEPS matching method (RegRieps). Note that, as far as the last two approaches are concerned, this section investigates only the preliminary parameter estimates that the previous authors subsequently use in the relevant micro matching procedures.

The differences among these approaches have been underlined in Remarks 2.5, 2.6, 3.2 and 3.3. Regression parameters estimated via ML are expected to perform better than those estimated by the RegRieps method, improving the estimate of the residual variance of the regression. They are also expected to be superior to those estimated by MS, especially when the variables are highly correlated, as already anticipated in Remark 2.4.

D'Orazio *et al.* (2005a) describe an extensive simulation study. Twelve different trivariate normal distributions with common mean and variance vectors

$$\boldsymbol{\mu} = \begin{pmatrix} \mu_X \\ \mu_Y \\ \mu_Z \end{pmatrix} = \begin{pmatrix} 0 \\ 0 \\ 0 \end{pmatrix}, \qquad \boldsymbol{\sigma} = \begin{pmatrix} \sigma_X^2 \\ \sigma_Y^2 \\ \sigma_Z^2 \end{pmatrix} = \begin{pmatrix} 1 \\ 1 \\ 1 \end{pmatrix},$$

and different correlation matrices

$$\boldsymbol{\rho}^{(1)} = \begin{pmatrix} 1 & 0.5 & 0.5 \\ 0.5 & 1 & \rho_{YZ} \\ 0.5 & \rho_{YZ} & 1 \end{pmatrix}, \qquad \boldsymbol{\rho}^{(2)} = \begin{pmatrix} 1 & 0.7 & 0.5 \\ 0.7 & 1 & \rho_{YZ} \\ 0.5 & \rho_{YZ} & 1 \end{pmatrix},$$

$$\boldsymbol{\rho}^{(3)} = \begin{pmatrix} 1 & 0.7 & 0.7 \\ 0.7 & 1 & \rho_{YZ} \\ 0.7 & \rho_{YZ} & 1 \end{pmatrix}, \qquad \boldsymbol{\rho}^{(4)} = \begin{pmatrix} 1 & 0.5 & 0.95 \\ 0.5 & 1 & \rho_{YZ} \\ 0.95 & \rho_{YZ} & 1 \end{pmatrix},$$

are considered. The four correlation matrices show increasing correlation for both (X, Y) and (X, Z). The correlation coefficient ρ_{YZ} is set to 0.35, 0.5 and 0.7, imposing increasing correlation also for the pair (Y, Z).

A random sample of 1000 observations is drawn from each population. Then the sample is split randomly into two samples of 500 observations, and in the one the variable Z is deleted in order to create file A, while in the other the variable discarded is Y (file B). The three estimation procedures (MS, RegRieps, ML) are applied to the files A and B, provided that a value $\rho^*_{YZ|X}$ for the parameter $\rho_{YZ|X}$ (or equivalently ρ^*_{YZ} for the parameter ρ_{YZ}) is postulated (three levels including the CIA are supplied). This task is iterated 1000 times for each population, and the simulation MSE, bias, and variance of the estimates with respect to the true parameters are computed.

For the sake of simplicity, in this section only results related to the two cases with normal distributions with correlation matrices $\boldsymbol{\rho}^{(1)}$ with $\rho_{YZ} = 0.35$, and $\boldsymbol{\rho}^{(4)}$ with $\rho_{YZ} = 0.5$ are reported. The corresponding residual variances are $\sigma^2_{Y|XZ} = 0.736\,667$ and $\sigma^2_{Z|XY} = 0.736\,667$ for the first case, and $\sigma^2_{Y|XZ} = 0.743\,590$ and $\sigma^2_{Z|XY} = 0.096\,667$ for the second.

Tables 3.1 and 3.2 show the average of the estimates over 1000 iterations, and the average of squared differences between estimated and true population parameters (MSE). As expected, ML estimates are better, and the gain is more noticeable when the correlation is high. For instance, Table 3.2 shows that the improvement obtained by using ML instead of MS is more noticeable for ρ_{XZ} (X is highly correlated with Z) instead of ρ_{XY}.

Table 3.1 Averages of the estimates of μ_Y, μ_Z, ρ_{XY}, ρ_{XZ}, $\sigma^2_{Y|XZ}$, $\sigma^2_{Z|XY}$ and the simulation MSE by the MS, RegRieps and ML methods with correlation matrix $\rho^{(1)}$ and $\rho_{YZ} = 0.35$

	$\rho^*_{YZ} = 0.25$ (CIA)			$\rho^*_{YZ} = 0.5125$			$\rho^*_{YZ} = 0.775$		
	MS	RegRieps	ML	MS	RegRieps	ML	MS	RegRieps	ML
$\hat{\mu}_Y$	−0.000 02	0.000 49	0.000 49	−0.002 94	−0.002 71	−0.002 71	−0.001 52	−0.000 63	−0.000 63
MSE($\hat{\mu}_Y$)	0.001 83	0.001 53	0.001 53	0.001 99	0.001 76	0.001 76	0.002 17	0.001 88	0.001 88
$\hat{\mu}_Z$	0.000 34	−0.000 22	−0.000 22	−0.000 71	−0.000 87	−0.000 87	0.000 21	−0.000 66	−0.000 66
MSE($\hat{\mu}_Z$)	0.001 95	0.001 71	0.001 71	0.001 98	0.001 72	0.001 72	0.002 10	0.001 80	0.001 80
$\hat{\rho}_{XY}$	0.500 15	0.499 75	0.500 50	0.500 29	0.499 56	0.500 31	0.499 70	0.498 99	0.499 74
MSE($\hat{\rho}_{XY}$)	0.001 45	0.001 04	0.001 04	0.001 44	0.001 01	0.001 01	0.001 46	0.001 07	0.001 07
$\hat{\rho}_{XZ}$	0.500 82	0.500 29	0.501 04	0.499 64	0.499 40	0.500 15	0.499 71	0.499 62	0.500 36
MSE($\hat{\rho}_{XZ}$)	0.001 45	0.001 09	0.001 09	0.001 51	0.001 03	0.001 03	0.001 47	0.001 11	0.001 11
$\hat{\sigma}^2_{Y\|XZ}$	0.747 18	0.759 03	0.748 53	0.653 50	0.589 43	0.655 14	0.379 08	0.216 06	0.381 68
MSE($\hat{\sigma}^2_{Y\|XZ}$)	0.003 15	0.002 93	0.002 50	0.008 99	0.024 22	0.008 48	0.128 52	0.273 16	0.126 59
$\hat{\sigma}^2_{Z\|XY}$	0.747 57	0.759 56	0.749 06	0.653 35	0.588 99	0.654 78	0.377 50	0.211 73	0.379 92
MSE($\hat{\sigma}^2_{Z\|XY}$)	0.002 88	0.002 71	0.002 29	0.008 80	0.024 17	0.008 44	0.129 63	0.277 75	0.127 87

Table 3.2 Averages of the estimates of μ_Y, μ_Z, ρ_{XY}, ρ_{XZ}, $\sigma^2_{Y|XZ}$, $\sigma^2_{Z|XY}$ and the simulation MSE by the MS, RegRieps and ML methods with correlation matrix $\rho^{(4)}$ and $\rho_{XY} = 0.5$

	$\rho^*_{YZ} = 0.475$ (CIA)			$\rho^*_{YZ} = 0.5696$			$\rho^*_{YZ} = 0.6637$		
	MS	RegRieps	ML	MS	RegRieps	ML	MS	RegRieps	ML
$\hat{\mu}_Y$	0.001 40	0.000 80	0.000 80	0.001 07	0.000 63	0.000 63	0.000 54	0.000 47	0.000 47
MSE($\hat{\mu}_Y$)	0.001 94	0.001 66	0.001 66	0.001 89	0.001 70	0.001 70	0.001 84	0.001 62	0.001 62
$\hat{\mu}_Z$	0.000 43	0.001 53	0.001 53	−0.001 48	−0.000 49	−0.000 49	0.000 54	0.000 69	0.000 69
MSE($\hat{\mu}_Z$)	0.002 03	0.001 09	0.001 09	0.002 10	0.001 05	0.001 05	0.001 10	0.001 08	0.001 08
$\hat{\rho}_{XY}$	0.499 75	0.499 61	0.500 36	0.498 14	0.498 15	0.498 90	0.498 82	0.499 04	0.499 79
MSE($\hat{\rho}_{XY}$)	0.001 37	0.001 07	0.001 07	0.001 55	0.001 15	0.001 15	0.001 43	0.001 03	0.001 03
$\hat{\rho}_{XZ}$	0.950 02	0.949 81	0.949 10	0.950 25	0.949 81	0.950 00	0.950 59	0.949 84	0.950 02
MSE($\hat{\rho}_{XZ}$)	0.000 54	0.000 01	0.000 01	0.000 57	0.000 01	0.000 01	0.000 54	0.000 01	0.000 01
$\hat{\sigma}^2_{Y\|XZ}$	0.728 82	0.957 49	0.745 70	0.635 40	0.993 41	0.657 68	0.360 24	0.983 22	0.380 83
MSE($\hat{\sigma}^2_{Y\|XZ}$)	0.003 63	0.049 55	0.002 20	0.017 22	0.067 32	0.009 01	0.165 34	0.062 81	0.132 12
$\hat{\sigma}^2_{Z\|XY}$	0.095 49	0.305 35	0.097 28	0.083 87	0.811 54	0.085 33	0.054 76	2.556 29	0.049 53
MSE($\hat{\sigma}^2_{Z\|XY}$)	0.004 50	0.045 25	0.000 04	0.004 25	0.538 66	0.000 16	0.004 59	6.218 98	0.002 23

Table 3.3 Average of the estimates of ρ_{YZ}, and its simulation bias and variance by the MS, RegRieps and ML methods

	$\rho^*_{YZ} = 0.25$ (CIA)			$\rho^*_{YZ} = 0.5125$			$\rho^*_{YZ} = 0.775$		
	MS	RegRieps	ML	MS	RegRieps	ML	MS	RegRieps	ML
$\hat{\rho}_{YZ}$	0.250 00	0.250 08	0.250 83	0.512 50	0.511 78	0.512 26	0.775 00	0.773 93	0.774 15
Bias($\hat{\rho}_{YZ}$)	−0.100 00	−0.099 92	−0.099 17	0.162 50	0.161 78	0.162 26	0.425 00	0.423 93	0.424 15
Var($\hat{\rho}_{YZ}$)	0.000 00	0.000 56	0.000 56	0.000 00	0.000 23	0.000 24	0.000 00	0.000 06	0.000 06
	$\rho^*_{YZ} = 0.475$ (CIA)			$\rho^*_{YZ} = 0.5696$			$\rho^*_{YZ} = 0.6637$		
	MS	RegRieps	ML	MS	RegRieps	ML	MS	RegRieps	ML
$\hat{\rho}_{YZ}$	0.475 00	0.474 56	0.475 36	0.569 65	0.567 95	0.568 53	0.663 97	0.663 45	0.663 82
Bias($\hat{\rho}_{YZ}$)	−0.025 00	−0.025 44	−0.024 64	0.069 65	0.067 95	0.068 53	0.163 97	0.163 45	0.163 82
Var($\hat{\rho}_{YZ}$)	0.000 00	0.000 99	0.000 99	0.000 00	0.000 89	0.000 89	0.000 10	0.000 67	0.000 67

The results show a generally better behaviour on the part of the ML approach. To be precise, it is generally superior to the MS estimation for all those parameters that can be estimated without the use of auxiliary information. Furthermore, the residual variances $\sigma^2_{Y|XZ}$ and $\sigma^2_{Y|XZ}$ are better estimated with ML than with either MS or RegRieps. Concerning RegRieps, in some cases this improvement is fairly large. These considerations are true for all the experiments performed on the 12 populations.

A further remark concerns the inestimable parameter ρ_{YZ} (see Table 3.3). All the methods reproduce exactly the postulated value, i.e. $\rho^*_{YZ|X}$ in ML and RegRieps and the corresponding ρ^*_{YZ} in MS (note that ML can work with both the constraints, while MS and RegRieps have been designed for just one of the two constraints; it was just our choice to consider $\rho_{YZ|X}$ as the postulated value in ML). As a result, the simulation MSE for MS consists only of the simulation bias, while the simulation MSE for ML and RegRieps is also affected by sample variability induced by the estimates $\hat{\rho}_{YX}$ and $\hat{\rho}_{ZX}$ needed to go from the postulated $\rho^*_{YZ|X}$ to $\hat{\rho}_{YZ}$.

The experiments are performed in the R environment with *ad hoc* functions. Functions referring to MS and ML methods are described in Section E.3, while for RegRieps an R version of the code reported in Rässler (2002) is used. A detailed description of the results can be found in D'Orazio *et al.* (2005a).

3.2.6 The multinomial case

Let (X, Y, Z) be the multinomial r.v. of Section 2.1.3. In this case, parametric auxiliary information might not be as useful as in the normal case. Auxiliary information can assume the following form:

(i) $\theta^*_{jk|i}$, $j = 1, \ldots, J$, $k = 1, \ldots, K$, $i = 1, \ldots, I$;

(ii) $\theta^*_{.jk}$, $j = 1, \ldots, J$, $k = 1, \ldots, K$.

The first case is rather rare: the whole conditional distribution of (Y, Z) given X should be known in advance. If such information is trustworthy, it is possible to disregard the available sample information on Y and Z in favour of the auxiliary information itself.

If auxiliary information (ii) is available, not all the models for (X, Y, Z) admit a unique ML estimate. Following Proposition 3.1, the marginal parameters $\theta_{i..}$, $i = 1, \ldots, I$, can be estimated as in (2.33) by $\hat{\theta}_{i..}$, $i = 1, \ldots, I$. Note that the unrestricted likelihood ridge (Proposition 3.1) is characterized by the conditional distributions (2.34) and (2.35). In order to estimate the parameter $\theta_{jk|i}$ of the conditional (Y, Z) distribution given X, the (linear) equality constraint to impose is:

$$\sum_{i=1}^{I} \theta_{jk|i}\hat{\theta}_{i..} = \theta^*_{.jk}, \qquad j = 1, \ldots, J, \; k = 1, \ldots, K. \tag{3.16}$$

If

$$\sum_{k=1}^{K} \theta^*_{.jk} = \sum_{i=1}^{I} \hat{\theta}_{i..}\hat{\theta}_{j|i}, \qquad j = 1, \ldots, J, \tag{3.17}$$

$$\sum_{j=1}^{J} \theta^*_{.jk} = \sum_{i=1}^{I} \hat{\theta}_{i..}\hat{\theta}_{k|i}, \qquad k = 1, \ldots, K, \tag{3.18}$$

then there is compatibility between constraints (3.16) and the likelihood ridge. Hence, the parameter vectors $\boldsymbol{\theta}$ in the likelihood ridge fulfilling (3.16) are the ML estimates restricted through the use of the available parametric auxiliary information. Note that, similarly to Example 3.5, knowledge of the parameters of the marginal (Y, Z) distribution does not lead to a unique solution. In fact, the sample and auxiliary information is only on the bivariate distributions (X, Y), (X, Z) (obtained through Proposition 3.1) and (Y, Z) (through auxiliary information). Hence, a loglinear model for (X, Y, Z) is identified when the three-way interaction parameters are set to zero. In this case, and given compatibility of the ML estimates $\hat{\theta}_{ij.}$ and $\hat{\theta}_{i.k}$ with $\theta^*_{.jk}$, a solution is offered by the IPF algorithm. It basically consists of the steps (a), (b) and (c) of Section 2.8.1, where the estimate $\hat{\theta}_{.jk}$ under the PIA is substituted by the auxiliary information $\theta^*_{.jk}$.

3.3 Parametric Predictive Approaches

As in Section 2.2, when a unique ML solution is available, the estimated joint distribution may be used to generate a complete synthetic data set. The procedures are identical, i.e. conditional mean matching and random draw. Actually, the additional information that is being used in this chapter affects only the parameter estimation phase. Neither the conditional mean matching nor the random draw procedure need consider additional information once the overall distribution $f(\mathbf{x}, \mathbf{y}, \mathbf{z}; \boldsymbol{\theta})$ has been estimated.

The comments on the predictive procedures (Section 2.2.3) still hold here.

Conditional mean matching determines a synthetic data set that is quite different from that under investigation. For instance, when $(\mathbf{X}, \mathbf{Y}, \mathbf{Z})$ are multinormal r.v.s as in Section 2.1.2, and A is completed through the conditional mean matching method, the missing $\mathbf{Z}$ is substituted by the r.v. $\tilde{\mathbf{Z}} = E(\mathbf{Z}|\mathbf{X}, \mathbf{Y})$, which is a linear combination of $\mathbf{X}$ and $\mathbf{Y}$:

$$\tilde{\mathbf{Z}} = \hat{\boldsymbol{\mu}}_{\mathbf{Z}} + \hat{\Sigma}_{\mathbf{ZX}|\mathbf{Y}}\hat{\Sigma}^{-1}_{\mathbf{XX}|\mathbf{Y}}\left(\mathbf{X} - \hat{\boldsymbol{\mu}}_{\mathbf{X}}\right) + \hat{\Sigma}_{\mathbf{ZY}|\mathbf{X}}\hat{\Sigma}^{-1}_{\mathbf{YY}|\mathbf{X}}\left(\mathbf{Y} - \hat{\boldsymbol{\mu}}_{\mathbf{Y}}\right). \tag{3.19}$$

Let us assume that sample sizes are large enough, so that ML estimates of the parameters almost coincide with their true values. As Kadane (1978) notes, filling in missing $\mathbf{Z}$ values in the n_A rows of A with the previous formula leads to a synthetic sample generated from a multinormal r.v. with mean $(\boldsymbol{\mu}_{\mathbf{X}}, \boldsymbol{\mu}_{\mathbf{Y}}, \boldsymbol{\mu}_{\mathbf{Z}})$ and variance matrix

$$\begin{pmatrix} \Sigma_{\mathbf{XX}} & \Sigma_{\mathbf{XY}} & \mathbf{T}_1' \\ \Sigma_{\mathbf{YX}} & \Sigma_{\mathbf{YY}} & \mathbf{T}_2' \\ \mathbf{T}_1 & \mathbf{T}_2 & \mathbf{T}_3 \end{pmatrix},$$

where

$$\mathbf{T}_1 = \Sigma_{\mathbf{ZX}|\mathbf{Y}} \Sigma_{\mathbf{XX}|\mathbf{Y}}^{-1} \Sigma_{\mathbf{XX}} + \Sigma_{\mathbf{ZY}|\mathbf{X}} \Sigma_{\mathbf{YY}|\mathbf{X}}^{-1} \Sigma_{\mathbf{YX}},$$

$$\mathbf{T}_2 = \Sigma_{\mathbf{ZX}|\mathbf{Y}} \Sigma_{\mathbf{XX}|\mathbf{Y}}^{-1} \Sigma_{\mathbf{XY}} + \Sigma_{\mathbf{ZY}|\mathbf{X}} \Sigma_{\mathbf{YY}|\mathbf{X}}^{-1} \Sigma_{\mathbf{YY}}$$

and

$$\mathbf{T}_3 = \Sigma_{\mathbf{ZX}|\mathbf{Y}} \Sigma_{\mathbf{XX}|\mathbf{Y}}^{-1} \Sigma_{\mathbf{XX}} \Sigma_{\mathbf{XX}|\mathbf{Y}}^{-1} \Sigma_{\mathbf{XZ}|\mathbf{Y}} + \Sigma_{\mathbf{ZY}|\mathbf{X}} \Sigma_{\mathbf{YY}|\mathbf{X}}^{-1} \Sigma_{\mathbf{YY}} \Sigma_{\mathbf{YY}|\mathbf{X}}^{-1} \Sigma_{\mathbf{YZ}|\mathbf{X}}$$
$$+ \Sigma_{\mathbf{ZX}|\mathbf{Y}} \Sigma_{\mathbf{XX}|\mathbf{Y}}^{-1} \Sigma_{\mathbf{XY}} \Sigma_{\mathbf{YY}|\mathbf{X}}^{-1} \Sigma_{\mathbf{YZ}|\mathbf{X}} + \Sigma_{\mathbf{ZY}|\mathbf{X}} \Sigma_{\mathbf{YY}|\mathbf{X}}^{-1} \Sigma_{\mathbf{YX}} \Sigma_{\mathbf{XX}|\mathbf{Y}}^{-1} \Sigma_{\mathbf{XZ}|\mathbf{Y}}.$$

Again, as in Section 2.2.3, this distribution is singular, given that $\tilde{\mathbf{Z}}$ is a linear combination of $\mathbf{X}$ and $\mathbf{Y}$.

On the other hand, it is expected that draws from the conditional distribution better preserve the joint multivariate distribution, at least for large samples when parameters are estimated consistently. Again, assume that $(\mathbf{X}, \mathbf{Y}, \mathbf{Z})$ are multivariate normal distributions. For the sake of simplicity, assume that the aim is the completion of A, i.e. $\mathbf{Z}$ should be predicted. Prediction is still based on (3.19), but an additional term $\mathbf{e}_{\mathbf{Z}|\mathbf{XY}}$ is added:

$$\tilde{\mathbf{Z}} = \hat{\mu}_{\mathbf{Z}} + \hat{\Sigma}_{\mathbf{ZX}|\mathbf{Y}} \hat{\Sigma}_{\mathbf{XX}|\mathbf{Y}}^{-1} (\mathbf{X} - \hat{\mu}_{\mathbf{X}}) + \hat{\Sigma}_{\mathbf{ZY}|\mathbf{X}} \hat{\Sigma}_{\mathbf{YY}|\mathbf{X}}^{-1} (\mathbf{Y} - \hat{\mu}_{\mathbf{Y}}) + \mathbf{e}_{\mathbf{Z}|\mathbf{XY}}.$$

In order to be consistent with the $(\mathbf{X}, \mathbf{Y}, \mathbf{Z})$ distribution, $\mathbf{e}_{\mathbf{Z}|\mathbf{XY}}$ is drawn from a multivariate normal distribution with null mean vector and covariance matrix $\hat{\Sigma}_{\mathbf{ZZ}|\mathbf{XY}}$ (step (f) in Section 3.2.4).

3.4 Nonparametric Macro Methods

When it is not possible to hypothesize a parametric model for the multivariate distribution of $(\mathbf{X}, \mathbf{Y}, \mathbf{Z})$, nonparametric methods are usually considered. However, it does not appear to be easy to apply nonparametric procedures in the present context. Actually, the CIA allows the use of nonparametric methods on completely observed data subsets (Section 2.3). When the CIA does not hold, distributions must inevitably be estimated on partially observed data sets. When auxiliary information consists of a third sample C, there are as usual two different situations.

The first consists in a completely observed C. The marginal $\mathbf{X}$ distribution may be estimated by nonparametric methods (based on kernels or kNN methods) on the overall sample $A \cup B \cup C$, which is completely observed as far as $\mathbf{X}$ is concerned. Consequently, the methods described in Section 2.3 can be considered. Those methods cannot be directly applied for the estimation of the distribution of $(\mathbf{Y}, \mathbf{Z}|\mathbf{X})$ given that both $\mathbf{Y}$ and $\mathbf{Z}$ are partially missing in $A \cup B \cup C$.

The second consists in a partially observed sample C, with $\mathbf{X}$ missing. In this case, the overall distribution $(\mathbf{X}, \mathbf{Y}, \mathbf{Z})$ should be estimated by means of nonparametric methods on a sample $A \cup B \cup C$ which is only partially observed.

Actually, nonparametric methods are usually applied when samples are complete. Cheng and Chu (1996) prove consistency of kernel estimators of distribution functions when a data set is only partially observed. Another useful reference is Nielsen (2001), whose objective is conditional mean imputation, and consequently may turn out to be useful for a nonparametric micro approach.

Nonparametric methods have also been proposed by Paass (1986). Paass considers the case of a partially observed sample C, and proposes the use of kNN in the IPF algorithm (when variables are categorical) or in the EM algorithm (when variables are continuous). However, the author states that convergence properties of the proposed algorithm have not yet been established.

3.5 The Nonparametric Micro Approach with Auxiliary Information

The hot deck methods exploiting auxiliary information of a third file C reporting micro data on $(\mathbf{X}, \mathbf{Y}, \mathbf{Z})$ (or $(\mathbf{Y}, \mathbf{Z})$) can be represented as a two-step procedure. Let A be the recipient and B the donor files.

(i) For each $a = 1, \ldots, n_A$, a live value $\mathbf{z}_{c^*}$, with c^* in C, is imputed to the ath record in A through one of the hot deck procedures of Section 2.4. Note that, when distance hot deck methods are used, the following procedures are applied:

   (a) when C contains information on variables $(\mathbf{X}, \mathbf{Y}, \mathbf{Z})$, the distance is computed with respect to $(\mathbf{X}, \mathbf{Y})$;

   (b) when C contains information on variables $(\mathbf{Y}, \mathbf{Z})$, the distance is computed only with respect to $\mathbf{Y}$.

(ii) For each $a = 1, \ldots, n_A$, impute the final live value $\mathbf{z}_{b^*}$ corresponding to the nearest neighbour b^* in B with respect to a distance that also considers the previously determined intermediate values $d\left((\mathbf{x}_a, \mathbf{z}_{c^*}), (\mathbf{x}_b, \mathbf{z}_b)\right)$.

This technique, proposed by Singh *et al.* (1993), is a simplified version of the more general technique introduced by Paass (1986) involving nonparametric techniques like the ones based on kNN methods.

This way of using auxiliary information exploits the relationship between $\mathbf{X}$, $\mathbf{Y}$, and $\mathbf{Z}$ observed in the third data file C.

Remark 3.4 If C is assumed to be a sample from $f(\mathbf{x}, \mathbf{y}, \mathbf{z})$, it is sufficient to perform only the first step. In other words, similarly to Section 2.4.1, random hot deck is equivalent to drawing observations from the probability distribution

$F_{\mathbf{Z}|\mathbf{Y},\mathbf{X}}(\mathbf{z}, |\mathbf{y}, \mathbf{x})$, estimated through the empirical distribution computed on C. Furthermore, similarly to Section 2.4.3, distance hot deck is equivalent to imputing values according to the conditional mean matching approach, when $E(\mathbf{Z}|\mathbf{Y}, \mathbf{X})$ is estimated on C through the kNN method. Nevertheless, Singh *et al.* (1993) introduce this algorithm because they suppose that the auxiliary information C is not completely reliable, and thus they try to robustify the nonparametric matching procedure by using C in order to exploit information regarding $(\mathbf{Z}|\mathbf{X}, \mathbf{Y})$. Since the imputed value from C is not coming from $f(\mathbf{x}, \mathbf{y}, \mathbf{z})$, they look for a similar value in B.

3.6 Mixed Methods

When auxiliary information is available, most of the techniques are essentially based on those of Section 2.5. In fact, they still consist of two main steps:

(a) model parameter estimation;

(b) use of hot deck techniques conditional on the first step.

As usual, differences among the techniques depend on the nature of auxiliary information. The distinction is between auxiliary information:

- at a micro level, i.e. a file C;
- in parametric form for the key parameters of the model under study (e.g. partial correlation coefficients $\rho_{\mathbf{YZ}|\mathbf{X}}$ for the normal case);
- in parametric form, but on parameters that are not directly key (e.g. information on the distribution of a categorization of $(\mathbf{X}, \mathbf{Y}, \mathbf{Z})$, when $(\mathbf{X}, \mathbf{Y}, \mathbf{Z})$ are continuous).

The important case where a complete auxiliary file C is available but not reliable belongs to the last situation, unless $(\mathbf{X}, \mathbf{Y}, \mathbf{Z})$ are coarsely partitioned.

3.6.1 Continuous variables

A first important source of auxiliary information is when values for the conditional parameters of $\boldsymbol{\theta}_{\mathbf{ZY}|\mathbf{X}}$ (or, similarly, the parameters $\boldsymbol{\theta}_{\mathbf{ZY}}$) are known. This information can also be deduced by a reliable third data set C.

Under this hypothesis, the methods listed in Section 2.5.1 change accordingly. All the parameter estimates reported in the following list refer to Section 3.2.4.

MM1*

(a) *Regression step 1.* Compute primary intermediate values for the units in A as in (3.19):

$$\tilde{\mathbf{z}}_a = \hat{\boldsymbol{\mu}}_{\mathbf{Z}} + \hat{\Sigma}_{\mathbf{ZX}|\mathbf{Y}} \hat{\Sigma}_{\mathbf{XX}|\mathbf{Y}}^{-1} (\mathbf{x}_a - \hat{\boldsymbol{\mu}}_{\mathbf{X}}) + \hat{\Sigma}_{\mathbf{ZY}|\mathbf{X}} \hat{\Sigma}_{\mathbf{YY}|\mathbf{X}}^{-1} (\mathbf{y}_a - \hat{\boldsymbol{\mu}}_{\mathbf{Y}})$$

for each $a = 1, \ldots, n_A$. Similarly, compute primary intermediate values for the units in B:

$$\tilde{\mathbf{y}}_b = \hat{\boldsymbol{\mu}}_{\mathbf{Y}} + \hat{\Sigma}_{\mathbf{YX|Z}} \hat{\Sigma}^{-1}_{\mathbf{XX|Z}} (\mathbf{x}_b - \hat{\boldsymbol{\mu}}_{\mathbf{X}}) + \hat{\Sigma}_{\mathbf{ZY|X}} \hat{\Sigma}^{-1}_{\mathbf{ZZ|X}} (\mathbf{z}_b - \hat{\boldsymbol{\mu}}_{\mathbf{Z}})$$

for each $b = 1, \ldots, n_B$.

(b) *Regression step 2*. Compute final intermediate values for the variable Y for the units in A as in (3.19), taking into account the primary intermediate values $\tilde{\mathbf{z}}$:

$$\tilde{\tilde{\mathbf{y}}}_a = \hat{\boldsymbol{\mu}}_{\mathbf{Y}} + \hat{\Sigma}_{\mathbf{YX|Z}} \hat{\Sigma}^{-1}_{\mathbf{XX|Z}} (\mathbf{x}_a - \hat{\boldsymbol{\mu}}_{\mathbf{X}}) + \hat{\Sigma}_{\mathbf{ZY|X}} \hat{\Sigma}^{-1}_{\mathbf{ZZ|X}} (\tilde{\mathbf{z}}_a - \hat{\boldsymbol{\mu}}_{\mathbf{Z}})$$

for each $a = 1, \ldots, n_A$. Similarly, compute intermediate values for Z for the units in B:

$$\tilde{\tilde{\mathbf{z}}}_b = \hat{\boldsymbol{\mu}}_{\mathbf{Z}} + \hat{\Sigma}_{\mathbf{ZX|Y}} \hat{\Sigma}^{-1}_{\mathbf{XX|Y}} (\mathbf{x}_b - \hat{\boldsymbol{\mu}}_{\mathbf{X}}) + \hat{\Sigma}_{\mathbf{ZY|X}} \hat{\Sigma}^{-1}_{\mathbf{YY|X}} (\tilde{\mathbf{y}}_b - \hat{\boldsymbol{\mu}}_{\mathbf{Y}})$$

for each $b = 1, \ldots, n_B$.

(c) *Matching step*. For each $a = 1, \ldots, n_A$, impute the live value $\mathbf{z}_{b^*}$ corresponding to the nearest neighbour b^* in B with respect to a distance between the previously determined intermediate values, $d(\tilde{\mathbf{z}}_a, \tilde{\tilde{\mathbf{z}}}_b)$.

MM2*

(a) *Regression step*. Compute primary intermediate values for the units in A as in (3.19):

$$\tilde{\mathbf{z}}_a = \hat{\boldsymbol{\mu}}_{\mathbf{Z}} + \hat{\Sigma}_{\mathbf{ZX|Y}} \hat{\Sigma}^{-1}_{\mathbf{XX|Y}} (\mathbf{x}_a - \hat{\boldsymbol{\mu}}_{\mathbf{X}}) + \hat{\Sigma}_{\mathbf{ZY|X}} \hat{\Sigma}^{-1}_{\mathbf{YY|X}} (\mathbf{y}_a - \hat{\boldsymbol{\mu}}_{\mathbf{Y}})$$

for each $a = 1, \ldots, n_A$.

(b) *Matching step*. For each $a = 1, \ldots, n_A$, impute the live value $\mathbf{z}_{b^*}$ corresponding to the nearest neighbour b^* in B with respect to a distance between the previously determined intermediate value and the observed values in B, $d(\tilde{\mathbf{z}}_a, \mathbf{z}_b)$.

MM3*

(a) *Regression step*. Compute primary intermediate values for the units in A via

$$\tilde{\mathbf{z}}_a = \hat{\boldsymbol{\mu}}_{\mathbf{Z}} + \hat{\Sigma}_{\mathbf{ZX|Y}} \hat{\Sigma}^{-1}_{\mathbf{XX|Y}} (\mathbf{x}_a - \hat{\boldsymbol{\mu}}_{\mathbf{X}}) + \hat{\Sigma}_{\mathbf{ZY|X}} \hat{\Sigma}^{-1}_{\mathbf{YY|X}} (\mathbf{y}_a - \hat{\boldsymbol{\mu}}_{\mathbf{Y}}) + \mathbf{e}_{\mathbf{Z|XY}}$$

for each $a = 1, \ldots, n_A$, where $\mathbf{e}_{\mathbf{Z|XY}}$ is a random draw from a multinormal distribution with zero mean vector and covariance matrix $\hat{\Sigma}_{\mathbf{ZZ|XY}}$.

(b) *Matching step*. For each $a = 1, \ldots, n_A$, impute the live value $\mathbf{z}_{b^*}$ corresponding to the nearest neighbour b^* in B with respect to a distance between the previously determined intermediate value and the observed values in B, $d(\tilde{\mathbf{z}}_a, \mathbf{z}_b)$.

MM4*

(a) *Regression step.* As MM3*.

(b) *Matching step.* For each $a = 1, \ldots, n_A$, impute the live value $\mathbf{z}_{b^*}$ corresponding to the nearest neighbour b^* in B with respect to a distance between the previously determined intermediate value and the observed values in B, $d(\tilde{\mathbf{z}}_a, \mathbf{z}_b)$, and the matching is constrained (see Section 2.4.3).

MM5*

(a) *Regression step.* Compute intermediate values for the units in A via

$$\tilde{\mathbf{z}}_a = \hat{\boldsymbol{\mu}}_{\mathbf{Z}} + \hat{\Sigma}_{\mathbf{ZX}|\mathbf{Y}} \hat{\Sigma}^{-1}_{\mathbf{XX}|\mathbf{Y}} (\mathbf{x}_a - \hat{\boldsymbol{\mu}}_{\mathbf{X}}) + \hat{\Sigma}_{\mathbf{ZY}|\mathbf{X}} \hat{\Sigma}^{-1}_{\mathbf{YY}|\mathbf{X}} (\mathbf{y}_a - \hat{\boldsymbol{\mu}}_{\mathbf{Y}})$$

for each $a = 1, \ldots, n_A$, and similarly intermediate values for the units in B via

$$\tilde{\mathbf{y}}_b = \hat{\boldsymbol{\mu}}_{\mathbf{Y}} + \hat{\Sigma}_{\mathbf{YX}|\mathbf{Z}} \hat{\Sigma}^{-1}_{\mathbf{XX}|\mathbf{Z}} (\mathbf{x}_b - \hat{\boldsymbol{\mu}}_{\mathbf{X}}) + \hat{\Sigma}_{\mathbf{ZY}|\mathbf{X}} \hat{\Sigma}^{-1}_{\mathbf{ZZ}|\mathbf{X}} (\mathbf{z}_b - \hat{\boldsymbol{\mu}}_{\mathbf{Z}})$$

for each $b = 1, \ldots, n_B$.

(b) *Matching step.* For each $a = 1, \ldots, n_A$, impute the live value $\mathbf{z}_{b^*}$ corresponding to the nearest neighbour b^* in B with respect to a distance $d\left((\mathbf{x}_a, \mathbf{y}_a, \tilde{\mathbf{z}}_a), (\mathbf{x}_b, \tilde{\mathbf{y}}_b, \mathbf{z}_b)\right)$. Matching is constrained to the couple $(\mathbf{Y}, \mathbf{Z})$ after the regression step. Kadane (1978) suggests the use of the Mahalanobis distance.

MM6*

(a) *Regression step.* Compute intermediate values for the units in A via

$$\tilde{\mathbf{z}}_a = \hat{\boldsymbol{\mu}}_{\mathbf{Z}} + \hat{\Sigma}_{\mathbf{ZX}|\mathbf{Y}} \hat{\Sigma}^{-1}_{\mathbf{XX}|\mathbf{Y}} (\mathbf{x}_a - \hat{\boldsymbol{\mu}}_{\mathbf{X}}) + \hat{\Sigma}_{\mathbf{ZY}|\mathbf{X}} \hat{\Sigma}^{-1}_{\mathbf{YY}|\mathbf{X}} (\mathbf{y}_a - \hat{\boldsymbol{\mu}}_{\mathbf{Y}}) + \mathbf{e}_{\mathbf{Z}|\mathbf{XY}}$$

for each $a = 1, \ldots, n_A$, where $\mathbf{e}_{\mathbf{Z}|\mathbf{XY}}$ is a random draw from a multinormal distribution with zero mean vector and estimated residual variance matrix $\hat{\Sigma}_{\mathbf{ZZ}|\mathbf{XY}}$, and similarly intermediate values for the units in B via

$$\tilde{\mathbf{y}}_b = \hat{\boldsymbol{\mu}}_{\mathbf{Y}} + \hat{\Sigma}_{\mathbf{YX}|\mathbf{Z}} \hat{\Sigma}^{-1}_{\mathbf{XX}|\mathbf{Z}} (\mathbf{x}_b - \hat{\boldsymbol{\mu}}_{\mathbf{X}}) + \hat{\Sigma}_{\mathbf{ZY}|\mathbf{X}} \hat{\Sigma}^{-1}_{\mathbf{ZZ}|\mathbf{X}} (\mathbf{z}_b - \hat{\boldsymbol{\mu}}_{\mathbf{Z}}) + \mathbf{e}_{\mathbf{Y}|\mathbf{XZ}}$$

for each $b = 1, \ldots, n_B$, where $\mathbf{e}_{\mathbf{Y}|\mathbf{XZ}}$ is a random draw from the multinormal distribution with zero mean vector and covariance matrix $\hat{\Sigma}_{\mathbf{YY}|\mathbf{XZ}}$.

(b) *Matching step.* For each $a = 1, \ldots, n_A$, impute the live value $\mathbf{z}_{b^*}$ corresponding to the nearest neighbour b^* in B with respect to a distance between the couples, $d\left((\mathbf{y}_a, \tilde{\mathbf{z}}_a), (\tilde{\mathbf{y}}_b, \mathbf{z}_b)\right)$. The matching is constrained to the couple $(\mathbf{Y}, \mathbf{Z})$ after the regression step.

Remark 3.5 Moriarity and Scheuren (2003) find through empirical studies that method MM6* is preferable both for retaining the postulated value $\Sigma^*_{\mathbf{YZ}}$ and for the estimation of $\Sigma_{\mathbf{XZ}}$ and $\Sigma_{\mathbf{XY}}$. They also point out that MM1* appears to be the worst method because of the use of an intermediate regressed value (regression step 2 of the algorithm) that causes biased estimates of the covariance matrix. Note, finally, that all their results are based not on ML estimates but on the estimators described in Remark 3.3.

3.6.2 Comparison between some mixed methods

In Section 3.2.5 a comparison among different estimation procedures in the multinormal case was carried out. The method based on ML estimates was shown to be the best. Hence, its use is expected also to improve the mixed methods proposed in Moriarity and Scheuren (2001, 2003). In this subsection a comparison between MM6* based on the ML and the method using at the first step the estimation suggested by Moriarity and Scheuren (MSMM6*) is carried out.

The simulation setting is similar to that described in Section 3.2.5. Twelve different trivariate normal distributions with mean and variance vectors

$$\boldsymbol{\mu} = \begin{pmatrix} \mu_X \\ \mu_Y \\ \mu_Z \end{pmatrix} = \begin{pmatrix} 0 \\ 0 \\ 0 \end{pmatrix}, \qquad \boldsymbol{\sigma} = \begin{pmatrix} \sigma_X^2 \\ \sigma_Y^2 \\ \sigma_Z^2 \end{pmatrix} = \begin{pmatrix} 1 \\ 1 \\ 1 \end{pmatrix},$$

and different correlation matrices

$$\boldsymbol{\rho}^{(1)} = \begin{pmatrix} 1 & 0.5 & 0.5 \\ 0.5 & 1 & \rho_{YZ} \\ 0.5 & \rho_{YZ} & 1 \end{pmatrix}, \qquad \boldsymbol{\rho}^{(2)} = \begin{pmatrix} 1 & 0.7 & 0.5 \\ 0.7 & 1 & \rho_{YZ} \\ 0.5 & \rho_{YZ} & 1 \end{pmatrix},$$

$$\boldsymbol{\rho}^{(3)} = \begin{pmatrix} 1 & 0.7 & 0.7 \\ 0.7 & 1 & \rho_{YZ} \\ 0.7 & \rho_{YZ} & 1 \end{pmatrix}, \qquad \boldsymbol{\rho}^{(4)} = \begin{pmatrix} 1 & 0.5 & 0.95 \\ 0.5 & 1 & \rho_{YZ} \\ 0.95 & \rho_{YZ} & 1 \end{pmatrix},$$

are considered. The four correlation matrices show increasing correlation for both (X, Y) and (X, Z). The correlation coefficient ρ_{YZ} is set to 0.35, 0.5 and 0.7, imposing increasing correlation also for the pair (Y, Z).

A random sample of 1000 observations is drawn from each population. Then the sample is split randomly into two samples of 500 observations. In the one, Z is deleted in order to create file A, while in the other the discarded variable is Y (file B). The two estimation procedures (MS, ML) are applied on the files A and B, provided that a value $\rho^*_{YZ|X}$ for the parameter $\rho_{YZ|X}$ (or equivalently ρ^*_{YZ} for the parameter ρ_{YZ}) is postulated (three levels including the CIA are supplied). Finally, the files A and B are filled out in turn, considering first A as recipient and B as donor set and then vice versa.

This procedure is iterated 500 times for each population. At the end of each iteration, two different sets of parameter estimates are obtained referring respectively to A and B after the completion. Finally, the average of the estimates over the 500 iterations, and the average of squared differences between estimated and true population parameters (MSE) are computed.

In general, apart the estimates of the correlations involving missing variables, all the other parameter estimates exhibit the same behaviour. As far as $\boldsymbol{\rho}$ is concerned, MM6* is superior to the mixed method MSMM6*. Tables 3.4 and 3.5 report the estimated ρ_{XZ} and its mean square error (computed on the completed file A), when the normal distribution has correlation $\boldsymbol{\rho}^{(1)}$ with $\rho_{YZ} = 0.35$, and $\boldsymbol{\rho}^{(4)}$ with $\rho_{YZ} = 0.5$. Analogously, Tables 3.6 and 3.7 report the average of the estimated ρ_{XY} and the corresponding simulation mean square error (computed on the completed file B).

Table 3.4 Average of the estimates of ρ_{XZ} and corresponding simulation MSE by the MM6* and MSMM6* methods computed on file A, for the normal case with correlation matrix $\boldsymbol{\rho}^{(1)}$ and $\rho_{YZ} = 0.35$

	$\rho^*_{YZ} = 0.5125$		$\rho^*_{YZ} = 0.7$	
	MSMM6*	MM6*	MSMM6*	MM6*
$\hat{\rho}_{XZ}$	0.494 442	0.493 765	0.495 296	0.493 619
MSE($\hat{\rho}_{XZ}$)	0.002 894	0.001 901	0.002 724	0.002 106

Table 3.5 Average of the estimates of ρ_{XZ} and corresponding simulation MSE by the MM6* and MSMM6* methods computed on file A, for the normal case with correlation matrix $\boldsymbol{\rho}^{(4)}$ and $\rho_{YZ} = 0.5$

	$\rho^*_{YZ} = 0.5020$		$\rho^*_{YZ} = 0.6913$	
	MSMM6*	MM6*	MSMM6*	MM6*
$\hat{\rho}_{XZ}$	0.935 823	0.936 732	0.929 878	0.937 357
MSE($\hat{\rho}_{XZ}$)	0.001 775	0.000 213	0.001 444	0.000 204

Table 3.6 Average of the estimates of ρ_{XY} and corresponding simulation MSE by the MM6* and MSMM6* methods computed on file B, for the normal case with correlation matrix $\boldsymbol{\rho}^{(1)}$ and $\rho_{YZ} = 0.35$.

	$\rho^*_{YZ} = 0.5125$ (CIA)		$\rho^*_{YZ} = 0.7$	
	MSMM6*	MM6*	MSMM6*	MM6*
$\hat{\rho}_{XY}$	0.489 020	0.498 084	0.493 255	0.492 788
MSE($\hat{\rho}_{XY}$)	0.003 054	0.001 751	0.002 838	0.001 851

Table 3.7 Average of the estimates of ρ_{XY} and corresponding simulation MSE by the MM6* and MSMM6* methods computed on file B, for the normal case with correlation matrix $\boldsymbol{\rho}^{(4)}$ and $\rho_{YZ} = 0.5$.

	$\rho^*_{YZ} = 0.5020$ (CIA)		$\rho^*_{YZ} = 0.6913$	
	MSMM6*	MM6*	MSMM6*	MM6*
$\hat{\rho}_{XY}$	0.490 243	0.493 423	0.492 882	0.491 190
MSE($\hat{\rho}_{XY}$)	0.002 655	0.001 604	0.002 367	0.001 633

Table 3.8 Average of the estimates of ρ_{YZ} by the MM6* and MSMM6* methods computed on files A and B, for the normal case with correlation matrix $\boldsymbol{\rho}^{(1)}$ and $\rho_{YZ} = 0.35$.

	$\hat{\rho}^A_{YZ}$		$\hat{\rho}^B_{YZ}$	
ρ^*_{YZ}	MSMM6*	MM6*	MSMM6*	MM6*
0.5125	0.502 837	0.505 099	0.502 837	0.505 099
0.7	0.689 044	0.690 247	0.689 044	0.690 247

Table 3.9 Average of the estimates of ρ_{YZ} by the MM6* and MSMM6* methods computed on files A and B, for the normal case with correlation matrix $\boldsymbol{\rho}^{(4)}$ and $\rho_{YZ} = 0.5$.

	$\hat{\rho}^A_{YZ}$		$\hat{\rho}^B_{YZ}$	
ρ^*_{YZ}	MSMM6*	MM6*	MSMM6*	MM6*
0.5020	0.493 970	0.494 918	0.493 970	0.494 918
0.6913	0.672 673	0.679 379	0.672 673	0.679 379

A final consideration relates to the reproduction of the postulated value ρ^*_{YZ}. Actually, since no observations with both Y and Z are available, the methods should reproduce ρ^*_{YZ}. As shown in Tables 3.8 and 3.9, although both the methods tend to reproduce ρ^*_{YZ}, the method based on ML (MM6*) has (averaged) estimates always closer to ρ^*_{YZ}.

The experiments are performed in the R environment with the functions described in Section E.3 and by setting the argument `macro = FALSE`.

A detailed description of the experiments is given in D'Orazio *et al.* (2005a).

3.6.3 Categorical variables

Let us suppose that the variables (X, Y, Z) are categorical. Then the mixed method consists of the following steps.

(a) *Estimation step*. Estimate θ_{ijk} by ML as in Sections 3.2.1 and 3.2.2, according to the nature of C.

(b) *Matching step*. For each $a = 1, \ldots, n_A$, find a value $\mathbf{z}_{b^*}$ in B through the random hot deck technique. This value is used for the imputation if the corresponding frequency of the cell (X, Y, Z) in A is not larger than the frequency of the same cell estimated in (a). If this constraint is not satisfied, a second donor in B is considered and so on, until the constraint is fulfilled.

This method was first defined in Singh *et al.* (1993) without the use of ML estimates. Their approach assumes A as the recipient file, and consequently the method is forced to have the observed table (X, Y) in A as fixed. Under this approach, the estimation phase (a) should be modified according to the nature of C via the following raking procedure.

Raking step of (X, Y, Z)

(i) Transform the frequencies of (X, Z) in B so that the marginal distribution of X in B is equal to the marginal distribution of X in A, i.e.

$$n_{i.k}^{B(0)} = \frac{n_{i.k}^B}{n_{i..}^B} n_{i..}^A \quad i = 1, \ldots, I; \; k = 1, \ldots, K.$$

(ii) Transform the frequencies of (X, Y, Z) in C in a such a way that their marginal distributions (X, Z) and (X, Y) are equal to the corresponding distributions of $B(0)$ and A, i.e.

$$n_{ijk}^{C(0)} = \frac{n_{ijk}^C}{n_{i.k}^C} n_{i.k}^{B(0)} \quad i = 1, \ldots, I; \; j = 1, \ldots, J; \; k = 1, \ldots, K,$$

$$n_{ijk}^{C(1)} = \frac{n_{ijk}^{C(0)}}{n_{ij.}^{C(0)}} n_{ij.}^A \quad i = 1, \ldots, I; \; j = 1, \ldots, J; \; k = 1, \ldots, K.$$

After this raking the final data set with the frequencies $n_{ijk}^{C(1)}$, $i = 1, \ldots, I$, $j = 1, \ldots, J$, $k = 1, \ldots, K$, is such that the (Y, Z) and (X, Y, Z) associations of the (X, Y, Z) table from C are preserved. Moreover, the (X, Y) and (X, Z) associations reproduce those observed in A and B respectively.

When the third data set C has only the variables (Y, Z) the raking procedure is slightly different.

Raking step of (Y, Z)

(i) Transform the frequencies of (X, Z) in B so that the marginal distribution of X in B is equal to the marginal distribution of X in A, i.e.

$$n_{i.k}^{B(0)} = \frac{n_{i.k}^B}{n_{i..}^B} n_{i..}^A \quad i = 1, \ldots, I; \; k = 1, \ldots, K.$$

(ii) Transform the frequencies of (Y, Z) in C so that the marginal distributions of Y and Z are equal to the marginal distributions of Y and Z in A and B respectively, i.e.

$$n_{.jk}^{C(0)} = \frac{n_{.jk}^{C}}{n_{.j.}^{C}} n_{.j.}^{A} \quad j = 1, \ldots, J;\ k = 1, \ldots, K,$$

$$n_{.jk}^{C(1)} = \frac{n_{.jk}^{C(0)}}{n_{..k}^{C(0)}} n_{..k}^{B(0)} \quad j = 1, \ldots, J;\ k = 1, \ldots, K.$$

(iii) Transform a three-dimensional table of ones, say $C(2)$, so that its bivariate marginal distributions will be equal to those obtained in the previous steps, i.e.

$$n_{ijk}^{C(h+2)} = \frac{n_{ijk}^{C(h+1)}}{n_{ij.}^{C(h+1)}} n_{ij.}^{A} \quad i = 1, \ldots, I;\ j = 1, \ldots, J;\ k = 1, \ldots, K,$$

$$n_{ijk}^{C(h+3)} = \frac{n_{ijk}^{C(h+2)}}{n_{i.k}^{C(h+2)}} n_{i.k}^{B(0)} \quad i = 1, \ldots, I;\ j = 1, \ldots, J;\ k = 1, \ldots, K,$$

$$n_{ijk}^{C(h+4)} = \frac{n_{ijk}^{C(h+3)}}{n_{.jk}^{C(h+3)}} n_{.jk}^{C(1)} \quad i = 1, \ldots, I;\ j = 1, \ldots, J;\ k = 1, \ldots, K;$$

where the last step must be iterated for $h = 1, 2, \ldots$, until convergence is reached. Since the values obtained may be noninteger, a rounding procedure must be adopted.

3.7 Categorical Constrained Techniques

The techniques described in this section are a constrained version of the nonparametric and mixed methods in Sections 2.4, 2.5, 3.5 and 3.6. Their purpose is to deal with continuous variables (X, Y, Z) and with a partial use of the auxiliary information in C. In particular, continuous variables (X, Y, Z) are also categorized with a coarse partition of their ranges, say $(X^\bullet, Y^\bullet, Z^\bullet)$. The basic idea is to preserve as much as possible the relationships among the variables at categorical level. The introduction of categorical constraints is expected to make the estimation of the joint distribution of the synthetic completed file more robust with respect to the imperfect nature of auxiliary information. This case was introduced and investigated in Singh *et al.* (1993). A first group of methods requires that the auxiliary file C contains micro data and the frequency table for $(X^\bullet, Y^\bullet, Z^\bullet)$, while a second group needs only the frequency distribution of $(X^\bullet, Y^\bullet, Z^\bullet)$. These methods also require that C is raked according to the appropriate raking step of Section 3.6.3: rake the observed distribution of $(X^\bullet, Y^\bullet, Z^\bullet)$ in C in order to meet the marginal distributions of $(X^\bullet, Y^\bullet)$ and $(X^\bullet, Z^\bullet)$ observed respectively in A and B.

Let us suppose that the aim is the reconstruction of A, and that the frequency distribution for the categorical variables $(X^{\bullet}, Y^{\bullet}, Z^{\bullet})$ is available.

3.7.1 Auxiliary micro information and categorical constraints

If the auxiliary information is available and reliable at micro level, the techniques to be used are the hot deck and mixed methods as described in Sections 3.5 and 3.6 respectively. They are detailed in the following list.

MM2*.CAT

(a) *Regression step.* Compute primary intermediate values for the units in A as in (3.19),

$$\tilde{\mathbf{z}}_a = \hat{\boldsymbol{\mu}}_{\mathbf{Z}} + \hat{\Sigma}_{\mathbf{ZX}|\mathbf{Y}} \hat{\Sigma}_{\mathbf{XX}|\mathbf{Y}}^{-1} (\mathbf{x}_a - \hat{\boldsymbol{\mu}}_{\mathbf{X}}) + \hat{\Sigma}_{\mathbf{ZY}|\mathbf{X}} \hat{\Sigma}_{\mathbf{YY}|\mathbf{X}}^{-1} (\mathbf{y}_a - \hat{\boldsymbol{\mu}}_{\mathbf{Y}})$$

for each $a = 1, \ldots, n_A$.

(b) *Matching step.* For each $a = 1, \ldots, n_A$, find a value $\mathbf{z}_{b^*}$ corresponding to the nearest neighbour b^* in B with respect to a distance between the previously determined intermediate value and the observed values in B, $d(\tilde{\mathbf{z}}_a, \mathbf{z}_b)$. This proposed value is used for the imputation if the frequency of the corresponding cell $(X^{\bullet}, Y^{\bullet}, Z^{\bullet})$ in A is not larger than the frequency of the same cell in C. If this constraint is not satisfied, the second nearest neighbour in B is considered. If this value also does not fulfil the constraint, the third nearest neighbour in B is considered, and so on until the constraint is fulfilled.

HOD*.CAT

(a) *Hot deck step.* Compute primary intermediate values $\tilde{\mathbf{z}}_a$ for the units in A based on the use of hot deck methods with auxiliary information, as discussed in Section 3.5; for each $a = 1, \ldots, n_A$, find a live value $\mathbf{z}_{c^*}$ corresponding to the nearest neighbour c^* in C with respect to a distance between the observed values, $d((\mathbf{x}_a, \mathbf{y}_a), (\mathbf{x}_c, \mathbf{y}_c))$. This value is used for the imputation only if the frequency of the corresponding cell $(X^{\bullet}, Y^{\bullet}, Z^{\bullet})$ in A is not larger than the frequency of the same cell in C. When this constraint is not satisfied, the second nearest neighbour in C is considered, and so on until the constraint is fulfilled.

(b) *Matching step.* For each $a = 1, \ldots, n_A$, impute the live value $\mathbf{z}_{b^*}$ corresponding to the nearest neighbour b^* in B with respect to a distance between the previously determined intermediate value and the observed values in B, $d((\mathbf{x}, \tilde{\mathbf{z}}_a), (\mathbf{x}_b, \mathbf{z}_b))$.

An alternative method to HOD*.CAT might be that of making the first step without constraints but then using them in the second step. However Singh *et al.* (1993) state that this alternative performs poorly.

3.7.2 Auxiliary information in the form of categorical constraints

What distinguishes the auxiliary information of the following techniques is that, differently from the previous section, only the categorical distribution of the variables $(X^\bullet, Y^\bullet, Z^\bullet)$ is needed. This means that the available information is not enough to eliminate the CIA completely, but only to relax it. In fact, in the case of regression we are not able to estimate the parameters of the regression of Z on X and Y, since we have no information on $\rho_{YZ|X}$, and in the case of hot deck techniques we do not have a micro data set C available for a first step of imputation of a preliminary value. The only option is to use the categorical distribution of $(X^\bullet, Y^\bullet, Z^\bullet)$ of the external sources of information in order to apply the statistical matching techniques introduced under the CIA in Chapter 2, but with the categorical constraints induced by the distribution of the external source $(X^\bullet, Y^\bullet, Z^\bullet)$.

After the raking step described in Section 3.6.3, either one of the following two procedures can be applied.

MM2.CAT

(a) *Regression step.* Compute intermediate values $\tilde{z}$ for the units in A by the method of Section 2.1.2,

$$\tilde{\mathbf{z}}_a = \bar{\mathbf{z}}_B + \mathbf{S}_{\mathbf{ZX};B}\mathbf{S}_{\mathbf{XX};B}^{-1}(\mathbf{x}_a - \bar{\mathbf{x}}_B)$$

for each $a = 1, \ldots, n_A$.

(b) *Matching step.* For each $a = 1, \ldots, n_A$, find a value $\mathbf{z}_{b^*}$ corresponding to the nearest neighbour b^* in B with respect to a distance between the observed values of X and Z in A and B, $d((\mathbf{x}_a, \tilde{\mathbf{z}}_a), (\mathbf{x}_b, \mathbf{z}_b))$. This value is used for the imputation if the frequency of the corresponding cell $(X^\bullet, Y^\bullet, Z^\bullet)$ in A is not larger than the frequency of the same cell in C. If this constraint is not satisfied, the second nearest neighbour in B is used and so on, until the constraint is fulfilled.

HOD.CAT

Matching step. For each $a = 1, \ldots, n_A$, find a value $\mathbf{z}_{b^*}$ corresponding to the nearest neighbour b^* in B with respect to a distance between the observed values of X in A and B, $d(\mathbf{x}_a, \mathbf{x}_b)$. This value is used for the imputation if the frequency of the corresponding cell $(X^\bullet, Y^\bullet, Z^\bullet)$ in A is not larger than the frequency of the same cell in C. If this constraint is not satisfied, the second nearest neighbour in B is considered and so on, until the constraint is fulfilled.

Remark 3.6 There is an interesting finding in Singh *et al.* (1993) and Filippello *et al.* (2004). They state that estimators perform better by using only information concerning $(\mathbf{Y}, \mathbf{Z})$ also when auxiliary information is available at micro level on all the variables $(\mathbf{X}, \mathbf{Y}, \mathbf{Z})$. Filippello *et al.* (2004), studying the impact of the sample size of C, find that this is particularly true when n_C is small. The statement that the use of less information gives better results for small sample sizes may appear surprising. On the other hand, it is generally true that the more unnecessarily complex a model is the poorer is the performance of inferences with respect to simpler but adequate models; see Agresti (1990, p. 182). In our case, the use of just $(\mathbf{Y}, \mathbf{Z})$ corresponds to estimating a loglinear model without the triple interaction term, which on the other hand is taken into account by using all the variables $(\mathbf{X}, \mathbf{Y}, \mathbf{Z})$.

3.8 The Bayesian Approach

The Bayesian approach is particularly useful for handling auxiliary information. However, in the context of statistical matching, it loses its main characteristic: auxiliary information on the parameters of interest cannot be updated by the information contained in $A \cup B$, as already discussed in Section 2.7 and clearly explained in Rubin (1974). Applications of Bayesian methods using auxiliary information when dealing with continuous variables are presented in Rässler (2002, 2003), although the goal of the techniques introduced in those papers is mainly the application of multiple imputation methods.

Rässler (2003) studies the case where an auxiliary data set C is available. A first way of using this information is to take the value $\boldsymbol{\rho}^*_{YZ|X}$ estimated on C for the prior conditional correlation, which means that in Example 2.11 the prior distribution for the conditional correlation is

$$\pi(\boldsymbol{\rho}_{YZ|X} = \boldsymbol{\rho}^*_{YZ|X}) = 1 \tag{3.20}$$

Hence, the posterior distribution on all the parameters is given by (2.58) with $\pi(\boldsymbol{\rho}_{\mathbf{YZ|X}})$ given by (3.20).

In the same paper it is also shown how to treat this kind of auxiliary information when using the data augmentation algorithm (see Section A.2) as introduced in Schafer (1997) or when an iterative imputation method–multivariate imputation by chained equations (MICE), proposed by Van Buuren and Oudshoorn (1999, 2000)–is applied.

4

Uncertainty in Statistical Matching

4.1 Introduction

The main characteristic of the statistical matching problem is the lack of joint information on the variables of interest. In order to overcome this problem, two solutions were considered in Chapters 2 and 3. In the first case, a particular assumption (the CIA) was adopted in order to yield an identifiable model for $A \cup B$. In the second case, auxiliary information was introduced and models other than those defined by the CIA became identifiable. It is, however, possible that neither case is appropriate: the CIA may be a misspecified assumption, and auxiliary information may be not available. This situation yields a kind of *uncertainty* on the model of $(\mathbf{X}, \mathbf{Y}, \mathbf{Z})$: while, given a sample, standard statistical problems are characterized by a unique estimate of the model parameters (e.g. the parameter estimate $\hat{\boldsymbol{\theta}}$ which maximizes the likelihood function), in the statistical matching problem sample information is unable to distinguish between a set, sometimes a very large set, of possible parameters. As a result, statistical matching techniques should aim at:

- a set of equally plausible parameter estimates, when the objective is macro;
- a collection of synthetic data sets, under the different equally plausible parameter estimates, when the objective is micro.

In the rest of this section, only the first issue will be considered. The second will be recalled in Section 4.8.2, but will not be analysed further.

Uncertainty defined as a set of equally plausible values is a characteristic of a more general problem: the presence of missing items in a data set. When a data set is only partially observed, the statistical model is identifiable if the MAR assumption

Statistical Matching: Theory and Practice M. D'Orazio, M. Di Zio and M. Scanu

holds (see Section A.1.1). Manski (1995) illustrates the situation where the MAR assumption is not considered and how to find an interval for all the plausible estimates, according to the different plausible models for the generation of missing data. When all the models for missingness are assumed to be plausible, this problem has been referred to as the *identification problem*. Manski explains what the interval of plausible estimates means, comparing this result with a very common interval procedure in statistics: confidence intervals. Confidence intervals at a $100\,(1-\alpha)\%$ of confidence are related to a sampling based concept of uncertainty, the concept of sampling variability: $100(1-\alpha)\%$ of the samples generated according to the sampling mechanisms contain the true, but unknown, parameter. On the other hand, all the plausible estimates gained from a sample (i.e. *given* a sample) are related to what Manski calls a population concept: given a sample, what can be learned about the *population* (i.e. model) parameters when the sample does not contain important information for estimating the model, because it is only partially observed, and when no assumptions are assumed to hold for the missingness mechanism. In a sense, the concept of uncertainty due to the identification problem is *model based*: the set of plausible models $\mathcal{F}$ for $(\mathbf{X}, \mathbf{Y}, \mathbf{Z})$ is allowed to be larger than the one, $\mathcal{F}^* \subset \mathcal{F}$, that the sample is able to estimate, if strict hypotheses (as MAR) are assumed to hold.

Example 4.1 Manski (1995) illustrates the above ideas with a very simple example. This example is further simplified by Rässler (2002, p. 8), as follows.

Let X be a dichotomous r.v., $P(X=1)=\theta$, $P(X=0)=1-\theta$, and R the corresponding indicator of missingness, $R=1$ if X is observed and 0 otherwise. Let (x_i, r_i), $i=1,\ldots,n$, be an i.i.d. sample of size n from (X, R). Assume that the missing data mechanism is MAR (i.e. in this case MCAR, given that X has no covariates). Then R and X are independent, this model is identifiable for the available data set, and θ can be estimated from the observed data. In particular, there is a unique maximum likelihood estimator of θ:

$$\hat{\theta} = \frac{\sum_{i=1}^{n} I_{1,1}(x_i, r_i)}{\sum_{i=1}^{n} I_1(r_i)}.$$

When MAR does not hold, i.e. under a MNAR assumption, R and X are dependent:

$$\theta = P(R=1)P(X=1|R=1) + P(R=0)P(X=1|R=0).$$

All the previous probabilities but $P(X=1|R=0)$ can be estimated from the sample. Note that $P(X=1|R=0)$ defines which MNAR model holds between X and R, i.e. the statistical relationship between the variable of interest and the missing data mechanism. Given that the data set does not help in choosing a particular MNAR model among all the possible MNAR models (this information is clearly not available), all these MNAR models are equally plausible. Hence, the interval

$$0 \leq P(X=1|R=0) \leq 1$$

describes uncertainty. As a result, θ is allowed to take values in the following interval:

$$P(R=1)P(X=1|R=1) \leq \theta \leq P(R=1)P(X=1|R=1) + P(R=0).$$

Considering the ML estimators of $P(R=1)$, $P(R=0)$, and $P(X=1|R=1)$, which are unique in our sample and equal to

$$\hat{P}(R=1) = \frac{\sum_{i=1}^{n} I_1(r_i)}{n},$$

$$\hat{P}(R=0) = 1 - \hat{P}(R=1),$$

$$\hat{P}(X=1|R=1) = \frac{\sum_{i=1}^{n} I_{1,1}(x_i, r_i)}{\sum_{i=1}^{n} I_1(r_i)},$$

the ML estimates of θ are in the following interval:

$$\hat{P}(R=1)\hat{P}(X=1|R=1) \leq \hat{\theta} \leq \hat{P}(R=1)\hat{P}(X=1|R=1) + \hat{P}(R=0).$$

Rässler (2002) correctly links the statistical matching problem with the identification problem described by Manski. It must be noted that, although both problems are related to the fact that the samples are only partially observed, there is an important difference between the two problems. In statistical matching, when the units in A and B are i.i.d., the missingness mechanism is MCAR, as already illustrated in Section 1.3: this is not an assumption, but a consequence of the data generating process. Nevertheless, the identification problem is still present: A and B, even under the MCAR model, are unable to identify the overall model for $(\mathbf{X}, \mathbf{Y}, \mathbf{Z})$, due to the absence of joint observation of $\mathbf{Y}$ and $\mathbf{Z}$, given $\mathbf{X}$. As a consequence, the missing data generation model is held fixed in the statistical matching problem, and the set of all plausible models is studied considering all the relationship models between $\mathbf{Y}$ and $\mathbf{Z}$ given $\mathbf{X}$. The identification problem in terms of *model parameter uncertainty* is the particular form of uncertainty for the statistical matching problem.

Remark 4.1 When, as in Remark 1.2, the two samples A and B are not identically distributed, due to differences in the reference times of the respective surveys, or in methods of data collection, and so on, the identification problem is much more complex: it is a combination of the model parameter uncertainty and of the missingness model uncertainty. Actually, all the papers (cited above) on the assessment of uncertainty for the statistical matching problem refer (explicitly or not) only to the notion of model parameter uncertainty.

Kadane (1978) was the first to describe model parameter uncertainty in the statistical matching problem. His ideas have been analysed and deepened by Moriarity and Scheuren (2001). Uncertainty is assessed by checking for all the parameters compatible with the estimates of the estimable parameters when the variables are normal. Rubin (1986) uses multiple imputation ideas in order to obtain both a collection of synthetic data sets and a set of plausible estimates. Rässler (2002) defines a fully Bayesian technique that leads to a proper multiple imputation approach. Again, both Rubin and Rässler refer to the normal case. A multiple imputation procedure for categorical variables is considered by Kamakura and Wedel (1997). Finally, a likelihood based approach is described in D'Orazio *et al.* (2005b) for categorical variables.

In this chapter, a likelihood based definition of uncertainty, as in D'Orazio *et al.* (2005b), is given. First of all, the case of complete, perfect knowledge of the $(\mathbf{X}, \mathbf{Y})$ and $(\mathbf{X}, \mathbf{Z})$ distributions is illustrated (Section 4.2). The properties of the set of equally plausible values are analysed in Section 4.3. These properties allow the definition of some measures of uncertainty. The normal and multinomial cases are studied in depth (Sections 4.3.1 and 4.3.2). When complete, perfect knowledge of the $(\mathbf{X}, \mathbf{Y})$ and $(\mathbf{X}, \mathbf{Z})$ distributions is substituted by the sample $A \cup B$, uncertainty measures are estimated by maximum likelihood (Section 4.4). The possibility of reducing uncertainty by means of structural constraints is discussed in Section 4.5. Finally, a review of the other approaches for the estimation of uncertainty is presented in Section 4.8.

Remark 4.2 Sometimes the assessment of uncertainty is termed 'sensitivity analysis'. We agree with Horowitz and Manski (2000) that this usage is inappropriate. Sensitivity analysis is focused on the evaluation of results when a limited range of alternative assumptions are assumed. The assessment of uncertainty evaluates all the possible results for all the alternative models that can be assumed.

Remark 4.3 This chapter considers only the parametric case, i.e. the set of models $\mathcal{F}$ for $(\mathbf{X}, \mathbf{Y}, \mathbf{Z})$ is described by the set of parameters $\boldsymbol{\theta} \in \Theta$, $\Theta \subset \mathbb{R}^T$, for T finite. The assessment of uncertainty in the nonparametric case remains an unsolved problem.

4.2 A Formal Definition of Uncertainty

The statistical matching problem is inevitably characterized by uncertainty. Even in the optimal case of complete knowledge of the $(\mathbf{X}, \mathbf{Y})$ and $(\mathbf{X}, \mathbf{Z})$ distributions, it is not possible to draw unique and certain conclusions on the overall distribution $(\mathbf{X}, \mathbf{Y}, \mathbf{Z})$.

Let $f(\mathbf{x}, \mathbf{y}, \mathbf{z}; \boldsymbol{\theta}^*)$ be the distribution of $(\mathbf{X}, \mathbf{Y}, \mathbf{Z})$, where $\boldsymbol{\theta}^*$, the true but unknown parameter, belongs to Θ, the natural parameter space for $\boldsymbol{\theta}$. If nothing else is known, the value $\boldsymbol{\theta}^*$ of $\boldsymbol{\theta}$ is *completely* uncertain, and Θ describes its uncertainty.

Now, let the partial distributions of $(\mathbf{X}, \mathbf{Y})$ and $(\mathbf{X}, \mathbf{Z})$ be perfectly known, i.e. the partial parameters $\boldsymbol{\theta}_{\mathbf{XY}}$ and $\boldsymbol{\theta}_{\mathbf{XZ}}$ are assumed to be equal to respectively $\boldsymbol{\theta}^*_{\mathbf{XY}}$ and $\boldsymbol{\theta}^*_{\mathbf{XZ}}$ (with the obvious compatibility assumption that the marginal $\mathbf{X}$ distribution, i.e. $\boldsymbol{\theta}^*_{\mathbf{X}}$, is the same in both cases). In this framework, $\boldsymbol{\theta}$ becomes a bit less uncertain:

it can be just one of those parameters in Θ compatible with the constraints $\boldsymbol{\theta}_{\mathbf{XY}} = \boldsymbol{\theta}^*_{\mathbf{XY}}$ and $\boldsymbol{\theta}_{\mathbf{XZ}} = \boldsymbol{\theta}^*_{\mathbf{XZ}}$. More formally, uncertainty on $\boldsymbol{\theta}$ is described by the subset $\Theta^{\mathrm{SM}} \subset \Theta$ whose $\boldsymbol{\theta}$ satisfy the constraints

$$\int_{\mathcal{Z}} f(\mathbf{x}, \mathbf{y}, \mathbf{z}; \boldsymbol{\theta}) d\mathbf{z} = f_{\mathbf{XY}}(\mathbf{x}, \mathbf{y}; \boldsymbol{\theta}^*_{\mathbf{XY}}), \tag{4.1}$$

$$\int_{\mathcal{Y}} f(\mathbf{x}, \mathbf{y}, \mathbf{z}; \boldsymbol{\theta}) d\mathbf{y} = f_{\mathbf{XZ}}(\mathbf{x}, \mathbf{z}; \boldsymbol{\theta}^*_{\mathbf{XZ}}). \tag{4.2}$$

Example 4.2 Let (X, Y, Z) be univariate normal r.v.s with parameter

$$\boldsymbol{\theta}^* = \left(\boldsymbol{\mu}^*, \boldsymbol{\Sigma}^*\right) = \left[\begin{pmatrix} \mu^*_X \\ \mu^*_Y \\ \mu^*_Z \end{pmatrix}, \begin{pmatrix} \sigma^{*2}_X & \sigma^*_{XY} & \sigma^*_{XZ} \\ \sigma^*_{XY} & \sigma^{*2}_Y & \sigma^*_{YZ} \\ \sigma^*_{XZ} & \sigma^*_{YZ} & \sigma^{*2}_Z \end{pmatrix}\right],$$

where $\boldsymbol{\theta}^* \in \Theta$, and Θ is composed of all those $\boldsymbol{\theta}$ whose components are in the intervals

$$-\infty < \mu_X < \infty, \quad -\infty < \mu_Y < \infty, \quad -\infty < \mu_Z < \infty, \tag{4.3}$$

$$0 \leq {\sigma_X}^2 < \infty, \quad 0 \leq {\sigma_Y}^2 < \infty, \quad 0 \leq {\sigma_Z}^2 < \infty, \tag{4.4}$$

$$-\infty < \sigma_{XY} < \infty, \quad -\infty < \sigma_{XZ} < \infty, \quad -\infty < \sigma_{YZ} < \infty, \tag{4.5}$$

and where

$$\boldsymbol{\Sigma} \text{ is positive semidefinite.} \tag{4.6}$$

Constraint (4.6) is crucial in the present context. Once the marginal distributions (X, Y) and (X, Z), i.e. the marginal parameters

$$\boldsymbol{\theta}^*_{XY} = \left(\boldsymbol{\mu}^*_{XY}, \Sigma^*_{XY}\right) = \left[\begin{pmatrix} \mu^*_X \\ \mu^*_Y \end{pmatrix}, \begin{pmatrix} \sigma^{*2}_X & \sigma^*_{XY} \\ \sigma^*_{XY} & \sigma^{*2}_Y \end{pmatrix}\right]$$

and

$$\boldsymbol{\theta}^*_{XZ} = \left(\boldsymbol{\mu}^*_{XZ}, \Sigma^*_{XZ}\right) = \left[\begin{pmatrix} \mu^*_X \\ \mu^*_Z \end{pmatrix}, \begin{pmatrix} \sigma^{*2}_X & \sigma^*_{XZ} \\ \sigma^*_{XZ} & \sigma^{*2}_Z \end{pmatrix}\right],$$

are known, the parameter space Θ reduces dramatically. In fact, all the intervals in (4.3)–(4.5) degenerate to a single value defined by $\boldsymbol{\theta}^*_{XY}$ or $\boldsymbol{\theta}^*_{XZ}$, with the exception of σ_{YZ}. This parameter is not specified by $\boldsymbol{\theta}^*_{XY}$ and $\boldsymbol{\theta}^*_{XZ}$, and need only be compatible with them. In other words, all the values of $\sigma_{YZ} \in (-\infty, +\infty)$ satisfying constraint (4.6) are plausible given $\boldsymbol{\theta}^*_{XY}$ and $\boldsymbol{\theta}^*_{XZ}$. Hence, as already discussed in Kadane (1978), Θ^{SM} contains all those $\boldsymbol{\theta} \in \Theta$ with the already fixed parameters and all those σ_{YZ} satisfying the inequality

$$\begin{vmatrix} \sigma^{*2}_X & \sigma^*_{XY} & \sigma^*_{XZ} \\ \sigma^*_{XY} & \sigma^{*2}_Y & \sigma_{YZ} \\ \sigma^*_{XZ} & \sigma_{YZ} & \sigma^{*2}_Z \end{vmatrix} \geq 0. \tag{4.7}$$

The former results can be rewritten and more easily interpreted in terms of the equivalent parameterization of Θ with the matrix of the correlation coefficients $\boldsymbol{\rho}$:

$$\boldsymbol{\theta} = (\boldsymbol{\mu}, \boldsymbol{\sigma}, \boldsymbol{\rho}) = \left[\begin{pmatrix} \mu_X \\ \mu_Y \\ \mu_Z \end{pmatrix}, \begin{pmatrix} {\sigma_X}^2 \\ {\sigma_Y}^2 \\ {\sigma_Z}^2 \end{pmatrix}, \begin{pmatrix} 1 & \rho_{XY} & \rho_{XZ} \\ \rho_{XY} & 1 & \rho_{YZ} \\ \rho_{XZ} & \rho_{YZ} & 1 \end{pmatrix} \right].$$

In this case, Θ allows the parameters to take values according to (4.3) and (4.4). Inequalities (4.5) are substituted by the obvious

$$-1 \leq \rho_{XY} \leq 1, \ -1 \leq \rho_{XZ} \leq 1, \ -1 \leq \rho_{YZ} \leq 1, \tag{4.8}$$

while condition (4.6) becomes

$$\boldsymbol{\rho} \text{ is positive semidefinite.} \tag{4.9}$$

When $\boldsymbol{\theta}_{XY} = \boldsymbol{\theta}^*_{XY}$ and $\boldsymbol{\theta}_{XZ} = \boldsymbol{\theta}^*_{XZ}$, ρ_{YZ} is the only uncertain parameter, i.e. the only parameter whose interval of values is not concentrated at a point. Condition (4.9), reduces to

$$\begin{vmatrix} 1 & \rho^*_{XY} & \rho^*_{XZ} \\ \rho^*_{XY} & 1 & \rho_{YZ} \\ \rho^*_{XZ} & \rho_{YZ} & 1 \end{vmatrix} \geq 0,$$

and induces ρ_{YZ} to take values in the interval bounded by

$$\rho^*_{XY}\rho^*_{XZ} \pm \sqrt{\left(1 - \rho^{*2}_{XY}\right)\left(1 - \rho^{*2}_{XZ}\right)}. \tag{4.10}$$

Note that, under the CIA, the corresponding parameter

$$\rho^{\mathrm{CIA}}_{YZ} = \rho^*_{XY}\rho^*_{XZ}$$

is located at the midpoint of the interval (4.10), as remarked in Moriarity and Scheuren (2001).

Example 4.3 Let $(\mathbf{X}, \mathbf{Y}, \mathbf{Z})$ be multinormal r.v.s, as in Section 2.1.2. For the sake of simplicity, in this case only the equivalent parameterization will be considered:

$$\boldsymbol{\theta}^* = \left(\boldsymbol{\mu}^*, \boldsymbol{\sigma}^*, \boldsymbol{\rho}^*\right),$$

where

$$\boldsymbol{\mu}^* = \begin{pmatrix} \boldsymbol{\mu}^*_{\mathbf{X}} \\ \boldsymbol{\mu}^*_{\mathbf{Y}} \\ \boldsymbol{\mu}^*_{\mathbf{Z}} \end{pmatrix}, \tag{4.11}$$

$$\boldsymbol{\sigma}^* = \begin{pmatrix} \boldsymbol{\sigma}^*_{\mathbf{X}} \\ \boldsymbol{\sigma}^*_{\mathbf{Y}} \\ \boldsymbol{\sigma}^*_{\mathbf{Z}} \end{pmatrix}, \tag{4.12}$$

$$\rho^* = \begin{pmatrix} \rho^*_{\mathbf{XX}} & \rho^*_{\mathbf{XY}} & \rho^*_{\mathbf{XZ}} \\ \rho^*_{\mathbf{YX}} & \rho^*_{\mathbf{YY}} & \rho^*_{\mathbf{YZ}} \\ \rho^*_{\mathbf{ZX}} & \rho^*_{\mathbf{ZY}} & \rho^*_{\mathbf{ZZ}} \end{pmatrix}, \tag{4.13}$$

with $\boldsymbol{\sigma}_{\mathbf{W}}$ the vector of the variances of $\mathbf{W}$, and $\boldsymbol{\rho}_{\mathbf{WV}}$ correlation matrix of $\mathbf{W}$ with $\mathbf{V}$. The natural parameter space $\Theta = \{(\boldsymbol{\mu}, \boldsymbol{\sigma}, \boldsymbol{\rho})\}$, describing complete uncertainty when nothing else is known, consists of all those parameters such that:

(i) each mean in $\boldsymbol{\mu}$ is a real number;

(ii) each variance in $\boldsymbol{\sigma}$ is a nonnegative number;

(iii) $\boldsymbol{\rho}$ is symmetric and positive semidefinite, with 1 in its principal diagonal and the other values in $[-1, 1]$.

Let $\boldsymbol{\theta}_{\mathbf{XY}} = \boldsymbol{\theta}^*_{\mathbf{XY}}$ and $\boldsymbol{\theta}_{\mathbf{XZ}} = \boldsymbol{\theta}^*_{\mathbf{XZ}}$, i.e. let all the parameters in (4.11)–(4.13), with the exception of $\boldsymbol{\rho}^*_{\mathbf{YZ}}$, be perfectly known. This information restricts Θ to the set of parameters:

$$\begin{cases} \boldsymbol{\mu} = \boldsymbol{\mu}^*, & \boldsymbol{\sigma} = \boldsymbol{\sigma}^*, \\ \boldsymbol{\rho}_{\mathbf{XX}} = \boldsymbol{\rho}^*_{\mathbf{XX}}, & \boldsymbol{\rho}_{\mathbf{YY}} = \boldsymbol{\rho}^*_{\mathbf{YY}}, \quad \boldsymbol{\rho}_{\mathbf{ZZ}} = \boldsymbol{\rho}^*_{\mathbf{ZZ}}, \\ \boldsymbol{\rho}_{\mathbf{XY}} = \boldsymbol{\rho}^*_{\mathbf{XY}}, & \boldsymbol{\rho}_{\mathbf{XZ}} = \boldsymbol{\rho}^*_{\mathbf{XZ}}, \end{cases} \tag{4.14}$$

and

$$\begin{pmatrix} \boldsymbol{\rho}^*_{\mathbf{XX}} & \boldsymbol{\rho}^*_{\mathbf{XY}} & \boldsymbol{\rho}^*_{\mathbf{XZ}} \\ \boldsymbol{\rho}^*_{\mathbf{YX}} & \boldsymbol{\rho}^*_{\mathbf{YY}} & \boldsymbol{\rho}_{\mathbf{YZ}} \\ \boldsymbol{\rho}^*_{\mathbf{ZX}} & \boldsymbol{\rho}_{\mathbf{ZY}} & \boldsymbol{\rho}^*_{\mathbf{ZZ}} \end{pmatrix} \text{ is positive semidefinite.} \tag{4.15}$$

Again, uncertainty induced by complete knowledge of the marginal distributions $(\mathbf{X}, \mathbf{Y})$ and $(\mathbf{X}, \mathbf{Z})$ is described by (4.15), i.e. by all $\boldsymbol{\rho}_{\mathbf{YZ}}$ compatible with the imposed parameter knowledge. Uncertainty over the other parameters degenerates to the point values (4.14).

Note that, if only $\mathbf{X}$ is multivariate, while both Z and Y are univariate r.v.s, it is possible to define explicitly the bounds of the plausible values of the inestimable parameter $\rho_{YZ|\mathbf{X}}$. Its bounds are given by (see Moriarity and Scheuren, 2001, and references therein)

$$\rho_{YZ|\mathbf{X}} \in \left[C - \sqrt{D}, C + \sqrt{D}\right], \tag{4.16}$$

where

$$C = \sum_{p_1, p_2=1}^{P} \rho_{X_{p_1}Y} \rho_{X_{p_2}Z} \rho^{(-1)}_{X_{p_1}X_{p_2}},$$

$$D = \left(1 - \sum_{p_1, p_2=1}^{P} \rho_{X_{p_1}Y} \rho_{X_{p_2}Y} \rho^{(-1)}_{X_{p_1}X_{p_2}}\right) \left(1 - \sum_{p_1, p_2=1}^{P} \rho_{X_{p_1}Z} \rho_{X_{p_2}Z} \rho^{(-1)}_{X_{p_1}X_{p_2}}\right)$$

and $\rho^{(-1)}_{X_{p_1}X_{p_2}}$ denotes the (p_1, p_2)th element of the inverse of the correlation matrix of $\mathbf{X}$, $\boldsymbol{\rho}_{\mathbf{XX}}$.

For multivariate $\mathbf{Y}$ and/or $\mathbf{Z}$, computation of the bounds is more cumbersome. Moriarity and Scheuren (2001) suggest the use of a recursion formula for partial correlations. Equivalently, those bounds can be obtained through (4.15). For instance, if X and Y are univariate and $\mathbf{Z} = (Z_1, Z_2)$ is bivariate, (4.15) is fulfilled when

$$\begin{vmatrix} 1 & \rho^*_{XY} & \rho^*_{XZ_1} \\ \rho^*_{XY} & 1 & \rho_{YZ_1} \\ \rho^*_{XZ_1} & \rho_{YZ_1} & 1 \end{vmatrix} \geq 0, \tag{4.17}$$

$$\begin{vmatrix} 1 & \rho^*_{XY} & \rho^*_{XZ_1} & \rho^*_{XZ_2} \\ \rho^*_{XY} & 1 & \rho_{YZ_1} & \rho_{YZ_2} \\ \rho^*_{XZ_1} & \rho_{YZ_1} & 1 & \rho^*_{Z_1Z_2} \\ \rho^*_{XZ_2} & \rho_{YZ_2} & \rho^*_{Z_1Z_2} & 1 \end{vmatrix} \geq 0. \tag{4.18}$$

Constraint (4.17) implies the usual bounds for ρ_{YZ_1}, i.e. it should be in the interval bounded by

$$\rho_{XY}\rho_{XZ_1} \pm \sqrt{1-\rho^2_{XY}}\sqrt{1-\rho^2_{XZ_1}}. \tag{4.19}$$

Having fixed ρ_{YZ_1} in the previous interval, constraint (4.18) implies that ρ_{YZ_2} should satisfy the inequality

$$E\rho^2_{YZ_2} + F(\rho_{YZ_1})\rho_{YZ_2} + G(\rho_{YZ_1}) \geq 0,$$

where

$$\begin{aligned} E &= \rho^2_{XZ_1} - 1, \\ F(\rho_{YZ_1}) &= 2\rho_{YZ_1}\rho_{Z_1Z_2} + 2\rho_{XY}\rho_{XZ_2} - 2\rho_{XY}\rho_{XZ_1}\rho_{Z_1Z_2} - \rho_{XZ_1}\rho_{XZ_2}\rho_{YZ_1} \\ &\quad -\rho_{XZ_1}\rho_{XZ_2}\rho_{YZ_1}, \\ G(\rho_{YZ_1}) &= 1 - \rho^2_{Z_1Z_2} - \rho^2_{YZ_1} - \rho^2_{XY} - \rho^2_{XZ_1} - \rho^2_{XZ_2} + \rho^2_{XY}\rho^2_{Z_1Z_2} \\ &\quad + \rho^2_{XZ_2}\rho^2_{YZ_1} + 2\rho^2_{XZ_1}\rho^2_{XY}\rho^2_{YZ_1} + 2\rho^2_{XZ_1}\rho^2_{XZ_2}\rho^2_{Z_1Z_2} \\ &\quad - 2\rho^2_{XY}\rho^2_{YZ_1}\rho^2_{XZ_2}\rho^2_{Z_1Z_2}. \end{aligned}$$

Hence, for each ρ_{YZ_1} in (4.19), ρ_{YZ_2} can take values in the interval bounded by

$$\frac{-F(\rho_{YZ_1}) \pm \sqrt{F(\rho_{YZ_1})^2 - 4EG(\rho_{YZ_1})}}{2E}. \tag{4.20}$$

The combination of correlations coefficients in the intervals (4.19) and (4.20) defines uncertainty in this case.

Example 4.4 As in Section 2.1.3, let (X, Y, Z) be a multinomial r.v. with true, but unknown, parameters

$$\theta^*_{ijk} = P(X = i, Y = j, Z = k), \qquad i, j, k \in \Delta, \tag{4.21}$$

with $\Delta = \{(i, j, k) : i = 1, \ldots, I, j = 1, \ldots, J, k = 1, \ldots, K\}$. The natural parameter space,

$$\Theta = \left\{\boldsymbol{\theta} : \theta_{ijk} \geq 0; \sum_{i,j,k} \theta_{ijk} = 1\right\}, \tag{4.22}$$

describes complete parameter uncertainty.

Let the marginal distributions for (X, Y) and (X, Z) be perfectly known:

$$\theta^*_{ij.}, \qquad i = 1, \ldots, I, \; j = 1, \ldots, J, \tag{4.23}$$

$$\theta^*_{i.k}, \qquad i = 1, \ldots, I, \; k = 1, \ldots, K. \tag{4.24}$$

Information in (4.23) and (4.24) shrinks the set of possible distributions from the natural parameter space Θ to

$$\Theta^{\mathrm{SM}} = \left\{\boldsymbol{\theta} \in \Theta : \begin{array}{ll} \sum_k \theta_{ijk} = \theta^*_{ij.} & i = 1, \ldots, I, \; j = 1, \ldots, J \\ \sum_j \theta_{ijk} = \theta^*_{i.k} & i = 1, \ldots, I, \; k = 1, \ldots, K \end{array}\right\}. \tag{4.25}$$

The set of parameters (4.25) describes uncertainty connected to the statistical matching problem under complete knowledge on the (X, Y) and (X, Z) distributions.

4.3 Measures of Uncertainty

When $\boldsymbol{\theta}_{\mathbf{XY}} = \boldsymbol{\theta}^*_{\mathbf{XY}}$ and $\boldsymbol{\theta}_{\mathbf{XZ}} = \boldsymbol{\theta}^*_{\mathbf{XZ}}$, the restricted parameter set $\Theta^{\mathrm{SM}} \subset \Theta$ describes uncertainty on the true, but unknown, parameter $\boldsymbol{\theta}^*$. In particular, it is possible to define measures of uncertainty by studying Θ^{SM}. In this section, some of these measures are illustrated.

Being in a parametric set-up, let Θ be a T-dimensional space, i.e. let $\boldsymbol{\theta}$ be a vector of T components θ_t, $t = 1, \ldots, T$. For each $t = 1, \ldots, T$, the new, restricted space Θ^{SM} defines two values, $\theta_t^{\mathrm{L}} \leq \theta_t^{\mathrm{U}}$, such that if $\boldsymbol{\theta} \in \Theta^{\mathrm{SM}}$ then

$$\theta_t^{\mathrm{L}} \leq \theta_t \leq \theta_t^{\mathrm{U}}, \qquad t = 1, \ldots, T.$$

In some cases it is possible to say that

$$\boldsymbol{\theta} \in \Theta^{\mathrm{SM}} \text{ if and only if } \theta_t^{\mathrm{L}} \leq \theta_t \leq \theta_t^{\mathrm{U}}, \qquad t = 1, \ldots, T. \tag{4.26}$$

In other words, the uncertainty space of values for each parameter is in an interval.

Remark 4.4 An uncertainty space Θ^{SM} as in (4.26) is appealing because it is easy to analyse. Θ^{SM} is determined by the system of equations (4.1) and (4.2). An attractive possibility is to reduce the set of solutions of the system (4.1) and (4.2) to its normal form. Θ^{SM} is in its normal form when there exist functions $a_t(.)$ and $b_t(.)$, $t = 1, \ldots, T$, such that $\boldsymbol{\theta} \in \Theta^{\mathrm{SM}}$ if and only if

$$\begin{cases} a_1(\boldsymbol{\theta}^*_{\mathbf{XY}}, \boldsymbol{\theta}^*_{\mathbf{XZ}}) \leq \theta_1 \leq b_1(\boldsymbol{\theta}^*_{\mathbf{XY}}, \boldsymbol{\theta}^*_{\mathbf{XZ}}), \\ a_2(\boldsymbol{\theta}^*_{\mathbf{XY}}, \boldsymbol{\theta}^*_{\mathbf{XZ}}, \theta_1) \leq \theta_2 \leq b_2(\boldsymbol{\theta}^*_{\mathbf{XY}}, \boldsymbol{\theta}^*_{\mathbf{XZ}}, \theta_1), \\ a_3(\boldsymbol{\theta}^*_{\mathbf{XY}}, \boldsymbol{\theta}^*_{\mathbf{XZ}}, \theta_1, \theta_2) \leq \theta_3 \leq b_3(\boldsymbol{\theta}^*_{\mathbf{XY}}, \boldsymbol{\theta}^*_{\mathbf{XZ}}, \theta_1, \theta_2), \\ \ldots, \\ a_T(\boldsymbol{\theta}^*_{\mathbf{XY}}, \boldsymbol{\theta}^*_{\mathbf{XZ}}, \theta_1, \ldots, \theta_{T-1}) \leq \theta_T \leq b_T(\boldsymbol{\theta}^*_{\mathbf{XY}}, \boldsymbol{\theta}^*_{\mathbf{XZ}}, \theta_1, \ldots, \theta_{T-1}). \end{cases} \tag{4.27}$$

From the examples in Section 4.2 it can be derived that both the multinormal and the multinomial cases admit Θ^{SM} in normal form. This will be more apparent in Sections 4.3.1 and 4.3.2.

When $\theta_t^{\mathrm{L}} = \theta_t^{\mathrm{U}}$, uncertainty on θ_t reduces to zero, i.e. a unique value of θ_t is compatible with $\boldsymbol{\theta}^*_{\mathbf{XY}}$ and $\boldsymbol{\theta}^*_{\mathbf{XZ}}$:

$$\theta_t^* = \theta_t^{\mathrm{L}} = \theta_t^{\mathrm{U}}.$$

Otherwise, the parameter θ_t is uncertain. A natural measure of uncertainty for the statistical matching problem, when complete marginal knowledge is available, is

$$\theta_t^{\mathrm{U}} - \theta_t^{\mathrm{L}}, \qquad t = 1, \ldots, T, \tag{4.28}$$

i.e. the range of the intervals of uncertainty when (4.26) holds. The wider the interval, the more uncertain θ_t is. This measure has been used in statistical matching, especially in the univariate normal case (i.e. the interval (4.10)). Rässler (2002) gives a formula for an overall measure of uncertainty for the multivariate normal case. This measure can be generalized by computing

$$\sum_{t=1}^{T} \left(\theta_t^{\mathrm{U}} - \theta_t^{\mathrm{L}}\right),$$

and dividing it by the number of uncertain parameters, i.e. those parameters such that $\theta_t^{\mathrm{L}} \neq \theta_t^{\mathrm{U}}$. It will be shown that uncertainty measured by (4.28) is suitable for the univariate normal case (Section 4.3.1) and for the dichotomous case when the r.v.s are categorical (Section 4.3.2). In general, more refined measures should be used. These measures study the characteristics of the set Θ^{SM} more in depth. For the sake of simplicity, let the set of parameters in Θ^{SM} correspond to the population of allowable parameters. Each unit in this population, i.e. each element $\boldsymbol{\theta} \in \Theta^{\mathrm{SM}}$, is associated with one, and only one, allowable distribution. Hence, loosely speaking, the frequency distribution of the population of parameters in Θ^{SM} is uniform, according to the following distribution:

$$m(\boldsymbol{\theta}) = \begin{cases} 1/\int_{\Theta^{\mathrm{SM}}} d\boldsymbol{\theta} & \boldsymbol{\theta} \in \Theta^{\mathrm{SM}}, \\ 0 & \text{otherwise.} \end{cases}$$

Note that this density is defined only when $\int_{\Theta^{SM}} d\boldsymbol{\theta}$, the volume of Θ^{SM}, is a real number, which is the case in most of our problems once $\boldsymbol{\theta}^*_{\mathbf{XY}}$ and $\boldsymbol{\theta}^*_{\mathbf{XZ}}$ have been imposed. The distribution $m(\boldsymbol{\theta})$, the *uncertainty distribution*, is the key element for assessing uncertainty on each parameter θ_t, $t = 1, \ldots, T$. In particular, uncertainty on the marginal parameter θ_t is defined through marginalization of $m(\boldsymbol{\theta})$ with respect to the other parameters:

$$m(\theta_t) = \int_{\theta_1^L}^{\theta_1^U} \cdots \int_{\theta_{t-1}^L}^{\theta_{t-1}^U} \int_{\theta_{t+1}^L}^{\theta_{t+1}^U} \cdots \int_{\theta_T^L}^{\theta_T^U} m(\boldsymbol{\theta}) d\theta_1 \ldots d\theta_{t-1} d\theta_{t+1} \ldots d\theta_T. \quad (4.29)$$

This marginal distribution, the *marginal uncertainty distribution* on θ_t, might not be uniform. Hence, a reduction in the uncertainty on θ_t can be induced while also leaving its limits θ_t^U and θ_t^L unchanged, and reducing the dispersion of $m(\theta_t)$. The dispersion of $m(\theta_t)$ can be studied by traditional descriptive measures, among them the standard deviation, the coefficient of variation and the interquartile range.

Let $\bar{\theta}_t$ be the expectation of θ_t with respect to the marginal uncertainty density $m(\theta_t)$:

$$\bar{\theta}_t = \int_{\theta_t^L}^{\theta_t^U} \theta_t m(\theta_t) d\theta_t. \quad (4.30)$$

The parameter $\bar{\theta}_t$ can be considered as a representative value of the uncertainty distribution of θ_t. In some cases, when variability of $m(\theta_t)$ is not too large, instead of an uncertainty distribution, the unique value $\bar{\theta}_t$ can be considered. In some cases, such as the normal case, $\bar{\theta}_t$ is exactly the value induced by the CIA. The properties of the mean value with respect to $m(\theta_t)$ in the case of multinormal and multinomial distributions will be further discussed in Sections 4.3.1 and 4.5.1.

Remark 4.5 The idea of studying the uncertainty of each single parameter θ_t through the evaluation of the dispersion of the corresponding marginal uncertainty distribution $m(\theta_t)$ relies on the following concept. If this distribution is very concentrated, then many $\boldsymbol{\theta}$ in Θ^{SM} have a value for θ_t belonging to a much narrower interval than (θ_t^L, θ_t^U).

This aspect of uncertainty of θ_t may suggest different decisions than just a point estimate. If the decision is in terms of an interval $(a, b) \subset (\theta_t^L, \theta_t^U)$, it is important to compare the error frequency

$$\int_{\theta_t^L}^{a} m(\theta_t)\, d\theta_t + \int_{b}^{\theta_t^U} m(\theta_t)\, d\theta_t \quad (4.31)$$

with the gain given by the reduction of the interval width

$$1 - (b - a)/(\theta_t^U - \theta_t^L). \quad (4.32)$$

Once α is fixed, the shortest interval (a, b) is determined by the region $\Gamma = \{\theta : m(\theta_t) \geq c\}$ where c is chosen such that $1 - \alpha = \int_\Gamma m(\theta_t)\, d\theta_t$.

4.3.1 Uncertainty in the normal case

Let (X, Y, Z) be univariate normal r.v.s. Assume complete information on the marginal distributions (X, Y) and (X, Z) in terms of the marginal parameters $\boldsymbol{\theta}^*_{XY}$ and $\boldsymbol{\theta}^*_{XZ}$ is provided. For all the parameters but ρ_{YZ}, the uncertainty density is a degenerate distribution, i.e. their lower and upper bounds θ_t^{L} and θ_t^{U} coincide:

$$m(\mu_X) = \begin{cases} 1 & \mu_X = \mu_X^*, \\ 0 & \text{otherwise}, \end{cases} \tag{4.33}$$

$$m(\mu_Y) = \begin{cases} 1 & \mu_Y = \mu_Y^*, \\ 0 & \text{otherwise}, \end{cases} \tag{4.34}$$

$$m(\mu_Z) = \begin{cases} 1 & \mu_Z = \mu_Z^*, \\ 0 & \text{otherwise}, \end{cases} \tag{4.35}$$

$$m(\sigma_X^2) = \begin{cases} 1 & \sigma_X^2 = \sigma_X^{*2}, \\ 0 & \text{otherwise}, \end{cases} \tag{4.36}$$

$$m(\sigma_Y^2) = \begin{cases} 1 & \sigma_Y^2 = \sigma_Y^{*2}, \\ 0 & \text{otherwise}, \end{cases} \tag{4.37}$$

$$m(\sigma_Z^2) = \begin{cases} 1 & \sigma_Z^2 = \sigma_Z^{*2}, \\ 0 & \text{otherwise}, \end{cases} \tag{4.38}$$

$$m(\rho_{XY}) = \begin{cases} 1 & \rho_{XY} = \rho_{XY}^*, \\ 0 & \text{otherwise}, \end{cases} \tag{4.39}$$

$$m(\rho_{XZ}) = \begin{cases} 1 & \rho_{XZ} = \rho_{XZ}^*, \\ 0 & \text{otherwise}. \end{cases} \tag{4.40}$$

The uncertainty distribution for ρ_{YZ} takes values in the interval (4.10), i.e.

$$\rho_{YZ}^{\mathrm{L}} = \rho_{XY}^* \rho_{XZ}^* - \sqrt{\left(1 - \rho_{XY}^{*2}\right)\left(1 - \rho_{XZ}^{*2}\right)}, \tag{4.41}$$

$$\rho_{YZ}^{\mathrm{U}} = \rho_{XY}^* \rho_{XZ}^* + \sqrt{\left(1 - \rho_{XY}^{*2}\right)\left(1 - \rho_{XZ}^{*2}\right)}. \tag{4.42}$$

As far as the distribution $m(\rho_{YZ})$ is concerned, it is a uniform distribution between ρ_{YZ}^{L} and ρ_{YZ}^{U}. In fact, for each ρ_{YZ} in that interval, constraints (4.33)–(4.40) induce a unique joint normal density for (X, Y, Z). Hence,

$$m\left(\rho_{YZ}\right) = \begin{cases} \left(\rho_{YZ}^{\mathrm{U}} - \rho_{YZ}^{\mathrm{L}}\right)^{-1} & \rho_{YZ} \in \left(\rho_{YZ}^{\mathrm{L}}, \rho_{YZ}^{\mathrm{U}}\right), \\ 0 & \text{otherwise.} \end{cases}$$

Example 4.5 Let $\boldsymbol{\theta}^*_{XY}$ and $\boldsymbol{\theta}^*_{XZ}$ define the following correlation matrix:

$$\boldsymbol{\rho} = \begin{pmatrix} 1 & 0.66 & 0.63 \\ 0.66 & 1 & \rho_{YZ} \\ 0.63 & \rho_{YZ} & 1 \end{pmatrix}. \tag{4.43}$$

Then the bounds (4.41) and (4.42) are

$$\rho^{\mathrm{L}}_{YZ} = -0.1676, \qquad \rho^{\mathrm{U}}_{YZ} = 0.9992.$$

The distribution $m(\rho_{YZ})$ is described in Figure 4.1.

Note that, under the CIA, the corresponding parameter is the midpoint of the interval bounded by (4.41) and (4.42):

$$\rho^{\mathrm{CIA}}_{YZ} = 0.66 \times 0.63 = 0.4158.$$

Given that the uncertainty density $m(\rho_{YZ})$ is symmetric, ρ^{CIA}_{YZ} is also the mean and median of ρ_{YZ} with respect to $m(\rho_{YZ})$.

In a multivariate setting there are some differences. Again, all the parameters but $\boldsymbol{\rho}_{\mathbf{YZ}}$ are deduced by knowledge of the marginal parameters, $\boldsymbol{\theta}^*_{\mathbf{XY}}$ and $\boldsymbol{\theta}^*_{\mathbf{XZ}}$. The uncertainty distribution for $\boldsymbol{\rho}_{\mathbf{YZ}}$, $m(\boldsymbol{\rho}_{\mathbf{YZ}})$, is still uniformly distributed, but the uncertainty distributions for each parameter $\rho_{\mathbf{YZ}}$ are no longer uniformly distributed. This is due to constraint (4.15).

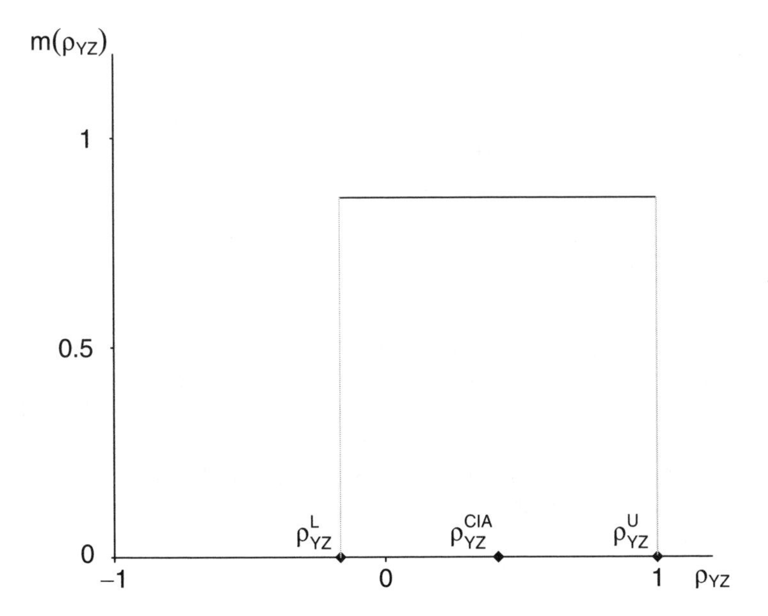

Figure 4.1 Marginal uncertainty distribution of ρ_{YZ} when $\boldsymbol{\rho}$ is (4.43)

Example 4.6 Assume that $(X, Y, \mathbf{Z}) = (X, Y, Z_1, Z_2)$ are the variables of interest, normally distributed with parameter $\boldsymbol{\theta}$. Let all but ρ_{YZ_1} and ρ_{YZ_2} be fixed through complete knowledge of the marginal parameters $\boldsymbol{\theta}_{XY} = \boldsymbol{\theta}^*_{XY}$ and $\boldsymbol{\theta}_{X\mathbf{Z}} = \boldsymbol{\theta}^*_{X\mathbf{Z}}$. Then, Θ^{SM} is composed of those $(\rho_{YZ_1}, \rho_{YZ_2})$ with ρ_{YZ_1} in the interval (4.19) and ρ_{YZ_2} in the interval (4.20). In particular, each pair $(\rho_{YZ_1}, \rho_{YZ_2}) \in \Theta^{\mathrm{SM}}$, together with the parameters $\boldsymbol{\theta}^*_{XY}$ and $\boldsymbol{\theta}^*_{X\mathbf{Z}}$, defines one multinormal density for $(X, Y, \mathbf{Z})$ compatible with the imposed knowledge, $\boldsymbol{\theta}^*_{XY}$ and $\boldsymbol{\theta}^*_{X\mathbf{Z}}$. Hence, the uncertainty density for the pair $(\rho_{YZ_1}, \rho_{YZ_2})$ is uniform:

$$m(\rho_{YZ_1}, \rho_{YZ_2}) = \begin{cases} c & (\rho_{YZ_1}, \rho_{YZ_2}) \in \Theta^{\mathrm{SM}}, \\ 0 & \text{otherwise}, \end{cases}$$

where c is the constant

$$c = \frac{1}{\int_{\Theta^{\mathrm{SM}}} du dv}.$$

The marginal uncertainty density for ρ_{YZ_1} is obtained through marginalization of $m(\rho_{YZ_1}, \rho_{YZ_2})$ with respect to ρ_{YZ_2}. Note that the value of $m(\rho_{YZ_1})$ is given by the width of the interval (4.20). The same procedure applies to the marginal uncertainty density for ρ_{YZ_2}.

For the sake of simplicity, assume that $\boldsymbol{\rho}$, given knowledge of $\boldsymbol{\theta}_{XY}$ and $\boldsymbol{\theta}_{X\mathbf{Z}}$, is given by

$$\boldsymbol{\rho} = \begin{pmatrix} 1 & 0.66 & 0.63 & 0.62 \\ 0.66 & 1 & \rho_{YZ_1} & \rho_{YZ_2} \\ 0.63 & \rho_{YZ_1} & 1 & 0.77 \\ 0.62 & \rho_{YZ_2} & 0.77 & 1 \end{pmatrix}. \tag{4.44}$$

The set of the admissible values for the pairs $(\rho_{YZ_1}, \rho_{YZ_2})$ is shown in Figure 4.2. Note that, although correlation coefficients can assume values between -1 and 1, knowledge of $\boldsymbol{\theta}_{XY}$ and $\boldsymbol{\theta}_{X\mathbf{Z}}$ restricts the admissible values for $(\rho_{YZ_1}, \rho_{YZ_2})$ in the subspace Θ^{SM}. They are bounded by

$$\rho^{\mathrm{L}}_{YZ_1} = -0.167, \qquad \rho^{\mathrm{U}}_{YZ_1} = 0.999,$$
$$\rho^{\mathrm{L}}_{YZ_2} = -0.180, \qquad \rho^{\mathrm{U}}_{YZ_2} = 0.998.$$

The distribution $m(\rho_{YZ_1}, \rho_{YZ_2})$ is constant in Θ^{SM}. Hence, the marginal uncertainty distribution of $m(\rho_{YZ_1})$ is proportional to the width of the interval of plausible values of ρ_{YZ_2} given ρ_{YZ_1}. For instance, Figure 4.2 shows the range of the plausible values of ρ_{YZ_2} when $\rho_{YZ_1} = 0.7$. Roughly speaking, $m(\rho_{YZ_1})$ shows how many normal distributions in Θ are compatible with $\boldsymbol{\theta}^*_{XY}$ and $\boldsymbol{\theta}^*_{X\mathbf{Z}}$ when ρ_{YZ_1} is given, i.e. how many normal distributions with the given ρ_{YZ_1} are in Θ^{SM}. Actually, this range changes according to ρ_{YZ_1}, as shown in (4.20). Hence:

$$m(\rho_{YZ_1}) \approx \begin{cases} \frac{\sqrt{F(\rho_{YZ_1})^2 - 4EG(\rho_{YZ_1})}}{E} & \rho^{\mathrm{L}}_{YZ_1} \leq \rho_{YZ_1} \leq \rho^{\mathrm{U}}_{YZ_1}, \\ 0 & \text{otherwise}. \end{cases}$$

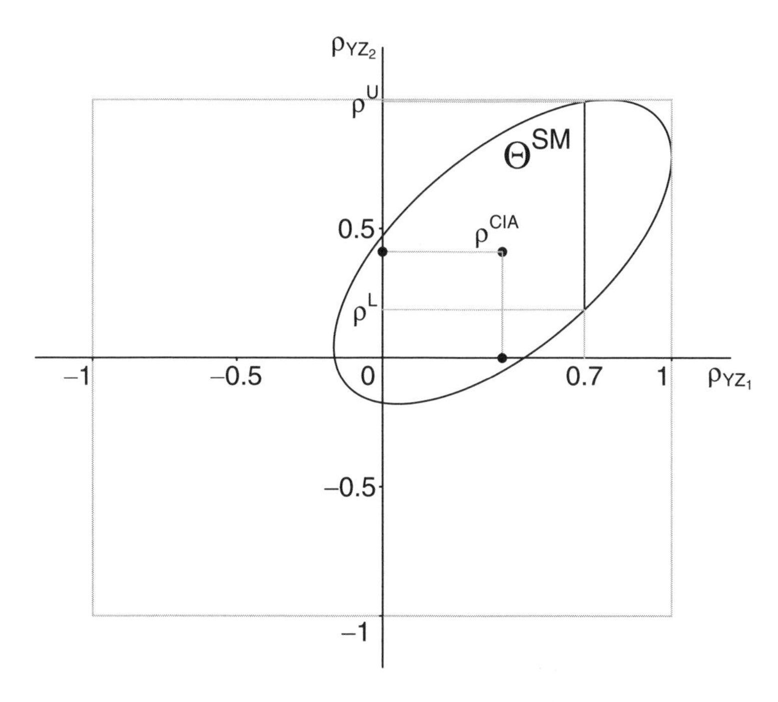

Figure 4.2 Set of admissible values for $(\rho_{YZ_1}, \rho_{YZ_2})$ given $\boldsymbol{\rho}^*$ as in (4.44). $\boldsymbol{\rho}^{\mathrm{CIA}}$ is the point $(\rho_{YZ_1}^{\mathrm{CIA}}, \rho_{YZ_2}^{\mathrm{CIA}})$ under the CIA between Y and $\mathbf{Z}$ given X. ρ^{L} and ρ^{U} are the extremes of the admissible values for ρ_{YZ_2} given $\rho_{YZ_1} = 0.7$

Note that, when ρ_{YZ_1} is equal to one of its extremes, there is only one compatible ρ_{YZ_2}. Figure 4.3 shows $m(\rho_{YZ_1})$ for this example. Note that the uncertainty distribution is symmetric and reaches its maximum when ρ_{YZ_1} is defined by the CIA. Similar results apply to ρ_{YZ_2}.

4.3.2 Uncertainty in the multinomial case

The case of categorical variables has been already investigated in D'Orazio *et al.* (2005b). In this situation, the natural parameter space Θ defined in (4.22) is closed and convex. Hence, $m(\boldsymbol{\theta})$ is also defined in Θ. Some characteristics of the distributions in the sets (4.22) and (4.25) can be easily derived.

(i) The true, but unknown, parameter θ_{ijk}^* lies in an interval $\theta_{ijk}^{\mathrm{L}} \leq \theta_{ijk}^* \leq \theta_{ijk}^{\mathrm{U}}$. For the natural parameter space Θ defined in (4.22), the bounds are $\theta_{ijk}^{\mathrm{L}} = 0$ and $\theta_{ijk}^{\mathrm{U}} = 1$ for all (i, j, k). For the restricted parameter space Θ^{SM} defined in (4.25), the bounds may be $\theta_{ijk}^{\mathrm{L}} > 0$ and $\theta_{ijk}^{\mathrm{U}} < 1$ for some (i, j, k).

(ii) For each $(i, j, k) \in \Delta$, the uncertainty density $m(\boldsymbol{\theta})$ is equal to zero for those $\boldsymbol{\theta}$ with at least one of their components outside the bounds.

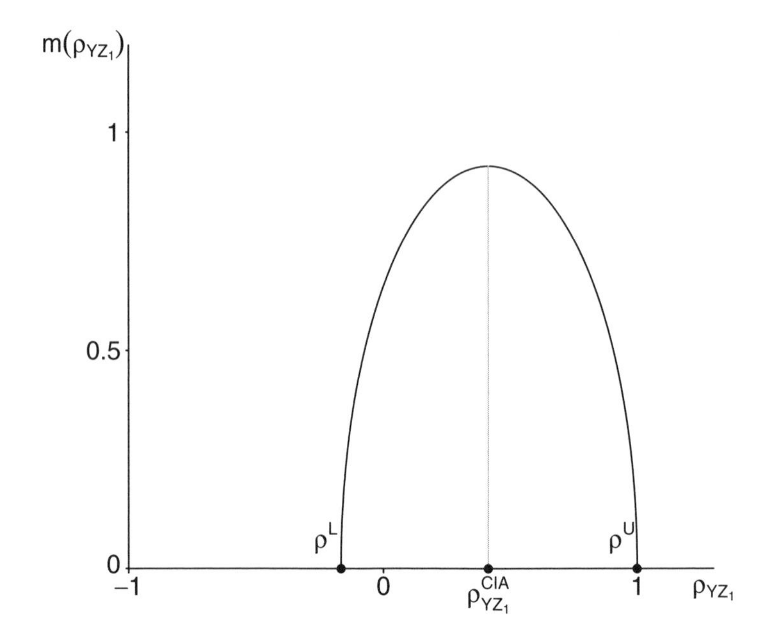

Figure 4.3 The marginal uncertainty distribution $m(\rho_{YZ_1})$ given $\boldsymbol{\rho}^*$ as in (4.44)

The first point is trivial for Θ, while for Θ^{SM} it is due to the fact that the space of solutions of the system (4.25) is a convex space as a result of a finite intersection of convex spaces. Hence, it is possible to determine in the space of solutions (4.25) the minimum and maximum θ^{L}_{ijk} and θ^{U}_{ijk} for each parameter θ_{ijk}. Note also that the bounds θ^{L}_{ijk} and θ^{U}_{ijk} are still solutions of the system (4.25). Furthermore, they can easily be derived through the system (4.27), given that the parameter space of a multinomial distribution is a normal parameter space. For the convexity of the space of solutions of the system (4.25), each θ_{ijk} in Θ^{SM} is defined by the bounds θ^{L}_{ijk} and θ^{U}_{ijk}:

$$\theta_{ijk} = \alpha\theta^{\text{L}}_{ijk} + (1-\alpha)\theta^{\text{U}}_{ijk}, \qquad \alpha \in [0, 1].$$

The corresponding parameters under the CIA,

$$\theta_{ijk} = \frac{\theta_{ij.}\theta_{i.k}}{\theta_{i..}}, \qquad (i, j, k) \in \Delta, \tag{4.45}$$

lie in the interval of plausible values, although they are not the midpoint of the intervals $[\theta^{\text{L}}_{ijk}, \theta^{\text{U}}_{ijk}]$.

Example 4.7 When $I = 2$, $J = 2$ and $K = 2$, i.e. X, Y and Z are dichotomous variables, the interval $[\theta^{\text{L}}_{ijk}, \theta^{\text{U}}_{ijk}]$ (and in particular its width) perfectly describes parameter uncertainty. For instance, consider the marginal parameter θ_{111} and its uncertainty distribution in Θ^{SM}. Knowledge of the marginal (X, Y) and (X, Z) is such that there is just one allowable distribution $\boldsymbol{\theta} \in \Theta^{SM}$, once θ_{111} is fixed in $[\theta^{L}_{ijk}, \theta^{U}_{ijk}]$. Hence

$$m(\theta_{111}) = \begin{cases} 1/(\theta^{U}_{111} - \theta^{L}_{111}) & \theta^{L}_{111} \leq \theta_{111} \leq \theta^{U}_{111}, \\ 0 & \text{otherwise.} \end{cases} \tag{4.46}$$

The same holds true for all the other parameters θ_{ijk}. The bounds for θ_{111} are:

$$\theta^{L}_{111} = \max\{0; \theta_{1.1} - \theta_{12.}\},$$
$$\theta^{U}_{111} = \min\{\theta_{11.}; \theta_{1.1}\}.$$

Hence, the mean value $\bar{\theta}_{111}$ with respect to the marginal uncertainty density (4.46),

$$\bar{\theta}_{111} = \int \theta_{111} m(\theta_{111}) \mathrm{d}\theta_{111} = \frac{\theta^{L}_{111} + \theta^{U}_{111}}{2},$$

will be generally different than (4.45).

While in the dichotomous case the width of the intervals $[\theta^{L}_{ijk}, \theta^{U}_{ijk}]$ is representative of the uncertainty of the parameters, the same does not hold true in a more complex set-up, as shown in the following example.

Example 4.8 Let X and Y be dichotomous, i.e. $I = 2$ and $J = 2$, and Z consist of three categories, i.e. $K = 3$. Hence, $\boldsymbol{\theta}$ consists of 12 parameters θ_{ijk}, $i = 1, 2$, $j = 1, 2$, $k = 1, 2, 3$. In the natural parameter space Θ, bearing in mind its definition in (4.22), these parameters should be nonnegative and sum to 1. As a result, in line with (4.29), the uncertainty density for θ_{111} is

$$m(\theta_{111}) = \begin{cases} 11\,(1 - \theta_{111})^{10} & 0 \leq \theta_{111} \leq 1, \\ 0 & \text{otherwise.} \end{cases}$$

This marginal uncertainty distribution is represented in Figure 4.4. Note that there is just one distribution with $\theta_{111} = 1$ (all the other parameters must be equal to zero), while when $\theta_{111} = 0$ the other parameters are allowed to take values in larger intervals. Consequently, $\theta = 0$ is the mode of $m(\theta_{ijk})$.

Let Θ be restricted to Θ^{SM} by means of knowledge of the distributions of (X, Y) (Table 4.1) and (X, Z) (Table 4.2). The intervals of plausible values for the different θ_{ijk} shrink to those described in Table 4.3. These intervals can be obtained by the algorithm in Capotorti and Vantaggi (2002). In order to draw the marginal uncertainty distribution connected to a single parameter, θ_{111} say, it is better to resort to the normal form of Θ^{SM}.

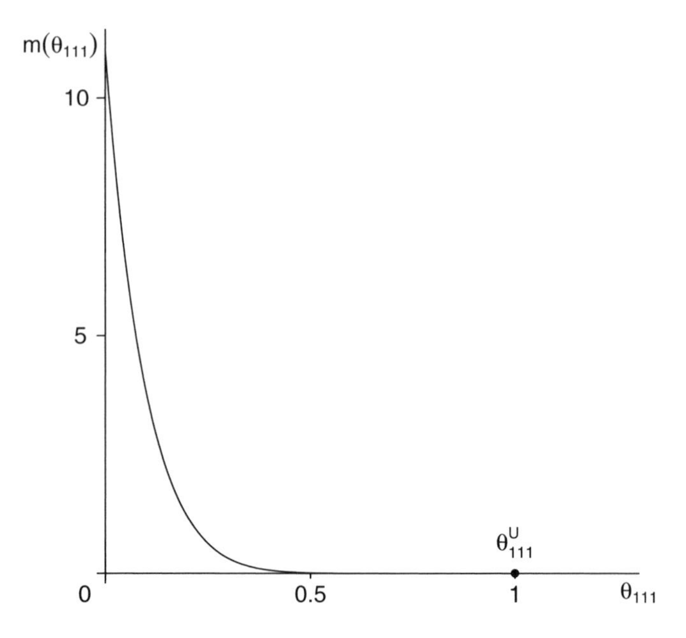

Figure 4.4 Marginal uncertainty distribution $m(\theta_{111})$ for θ_{111} in the unconstrained space Θ

Table 4.1 Parameters $\theta^*_{ij.}$ of the known (X, Y) distribution

	$Y = 1$	$Y = 2$
$X = 1$	0.43	0.16
$X = 2$	0.15	0.26

Table 4.2 Parameters $\theta^*_{i.k}$ of the known (X, Z) distribution

	$Z = 1$	$Z = 2$	$Z = 3$
$X = 1$	0.16	0.23	0.20
$X = 2$	0.14	0.16	0.11

The normal form of Θ^{SM} (i.e. the system of equations (4.27) in Remark 4.4) can be computed by the following steps. The parameter θ_{111} should satisfy the following inequalities:

$$0 \leq \theta_{111} \leq \theta_{1.1},$$

$$0 \leq \theta_{11.} - \theta_{111} \leq \theta_{12.}.$$

Table 4.3 Bounds θ^{L}_{ijk} and θ^{U}_{ijk}, $i = 1, 2,\ j = 1, 2$, and $k = 1, 2, 3$, defined by Tables 4.2 and 4.1

	θ^{L}_{ijk}	θ^{U}_{ijk}
θ_{111}	0.00	0.16
θ_{112}	0.07	0.23
θ_{113}	0.04	0.20
θ_{121}	0.00	0.16
θ_{122}	0.00	0.16
θ_{123}	0.00	0.16
θ_{211}	0.00	0.14
θ_{212}	0.00	0.15
θ_{213}	0.00	0.11
θ_{221}	0.00	0.14
θ_{222}	0.01	0.16
θ_{223}	0.00	0.11

Hence, the range of possible values for θ_{111} shrinks from [0, 1] to the interval with bounds

$$a_{111}(\boldsymbol{\theta}_{XY}, \boldsymbol{\theta}_{XZ}) = \max\{0; \theta_{1.1} - \theta_{12.}\} = 0,$$

$$b_{111}(\boldsymbol{\theta}_{XY}, \boldsymbol{\theta}_{XZ}) = \min\{\theta_{11.}; \theta_{1.1}\} = 0.16.$$

Note that, in this case, a_{111} and b_{111} are functions of the fixed tables for (X, Y) and (X, Z), and that $a_{111} = \theta^{L}_{111}$ and $b_{111} = \theta^{U}_{111}$. Given Tables 4.2 and 4.1 and a value of θ_{111}, then θ_{121} is automatically fixed, and only one other parameter θ_{1jk}, with $i = 1$ fixed and $k \geq 2$, is allowed to take values in nondegenerate intervals. Let θ_{112} be such a parameter. It should satisfy the following inequalities:

$$0 \leq \theta_{112} \leq \theta_{1.2},$$

$$0 \leq \theta_{11.} - \theta_{111} - \theta_{112} \leq \theta_{1.3}.$$

As a result, once θ_{111} is within its bounds, θ_{112} can assume values in the interval bounded by

$$a_{112}(\theta_{111}, \boldsymbol{\theta}_{XY}, \boldsymbol{\theta}_{XZ}) = \max\{0; \theta_{11.} - \theta_{111} - \theta_{1.3}\},$$

$$b_{112}(\theta_{111}, \boldsymbol{\theta}_{XY}, \boldsymbol{\theta}_{XZ}) = \min\{\theta_{1.2}; \theta_{11.} - \theta_{111}\}.$$

Note that these bounds are not the marginal ones, i.e. $a_{112} \neq \theta^{L}_{112}$ and $b_{112} \neq \theta^{U}_{112}$, but the bounds determined by holding θ_{111} fixed. Similar results hold for the parameters θ_{2jk}, with $i = 2$ fixed. Hence, the normal form of the parameter space is the following (only nondegenerate intervals are reported):

$$\begin{cases} a_{111}(\boldsymbol{\theta}_{XY}, \boldsymbol{\theta}_{XZ}) = \max\{0; \theta_{1.1} - \theta_{12.}\}, \\ b_{111}(\boldsymbol{\theta}_{XY}, \boldsymbol{\theta}_{XZ}) = \min\{\theta_{11.}; \theta_{1.1}\}, \\ a_{112}(\theta_{111}, \boldsymbol{\theta}_{XY}, \boldsymbol{\theta}_{XZ}) = \max\{0; \theta_{11.} - \theta_{111} - \theta_{1.3}\}, \\ b_{112}(\theta_{111}, \boldsymbol{\theta}_{XY}, \boldsymbol{\theta}_{XZ}) = \min\{\theta_{1.2}; \theta_{11.} - \theta_{111}\}, \\ a_{211}(\boldsymbol{\theta}_{XY}, \boldsymbol{\theta}_{XZ}) = \max\{0; \theta_{2.1} - \theta_{22.}\}, \\ b_{211}(\boldsymbol{\theta}_{XY}, \boldsymbol{\theta}_{XZ}) = \min\{\theta_{21.}; \theta_{2.1}\}, \\ a_{212}(\theta_{211}, \boldsymbol{\theta}_{XY}, \boldsymbol{\theta}_{XZ}) = \max\{0; \theta_{21.} - \theta_{211} - \theta_{2.3}\}, \\ b_{212}(\theta_{211}, \boldsymbol{\theta}_{XY}, \boldsymbol{\theta}_{XZ}) = \min\{\theta_{2.2}; \theta_{21.} - \theta_{211}\}. \end{cases} \tag{4.47}$$

Hence, having fixed θ_{111} between $\theta_{111}^{\mathrm{L}}$ and $\theta_{111}^{\mathrm{U}}$, by (4.29) and inequalities (4.47), the marginal uncertainty density $m(\theta_{111})$ assumes the form

$$m(\theta_{111}) = \begin{cases} cd & \theta_{111} = 0, \\ c\int_{0.23-\theta_{111}}^{\theta_{111}} \mathrm{d}\theta_{111} = c\theta_{111} & 0 < \theta_{111} \leq 0.16, \\ 0 & \text{otherwise}, \end{cases}$$

where

$$d = \int_0^{0.14} \mathrm{d}\theta_{211} \int_{\max\{0; 0.4-\theta_{211}\}}^{0.16} \mathrm{d}\theta_{212} = 0.0216$$

and c is the normalizing constant

$$c = \frac{2}{\left(\theta_{111}^{\mathrm{U}}\right)^2 - \left(\theta_{111}^{\mathrm{L}}\right)^2} = \frac{2}{0.16^2}.$$

This density is shown in Figure 4.5. The parameter under the CIA is still in the space Θ^{SM}. Its value is

$$\theta_{111}^{\mathrm{CIA}} = \frac{\theta_{11.}\theta_{1.1}}{\theta_{1..}} = 0.1166.$$

Again, $\theta_{ijk}^{\mathrm{CIA}}$ is not the centre, mean or median, of the marginal uncertainty distribution $m(\theta_{ijk})$.

Generally speaking, the marginal uncertainty distributions for θ_{ijk} in the natural parameter space coincide for every (i, j, k) and are equal to

$$m(\theta_{ijk}) = \begin{cases} (IJK - 1)\left(1 - \theta_{ijk}\right)^{IJK-2} & 0 \leq \theta_{ijk} \leq 1, \\ 0 & \text{otherwise}. \end{cases}$$

Remark 4.6 Although the mean $\bar{\theta}_{ijk}$ of θ_{ijk} with respect to its marginal uncertainty distribution,

$$\bar{\theta}_{ijk} = \int_{\theta_{ijk}^{\mathrm{L}}}^{\theta_{ijk}^{\mathrm{U}}} \theta_{ijk} m(\theta_{ijk}) \mathrm{d}\theta_{ijk},$$

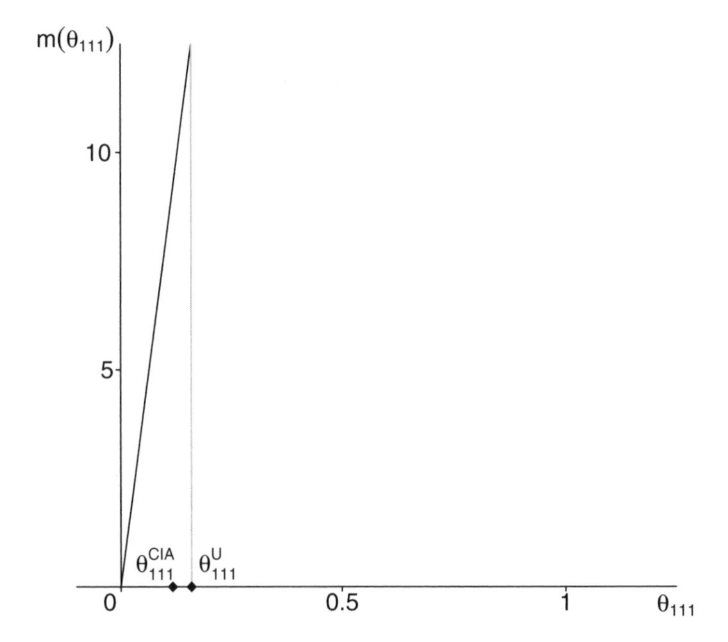

Figure 4.5 Marginal uncertainty distribution $m(\theta_{111})$ for θ_{111} in Θ^{SM} under complete knowledge of the marginal distributions $\theta_{ij.}$ and $\theta_{i.k}$, as given in Tables 4.1 and 4.2. Note that $\theta_{111} = \theta_{111}^{\mathrm{L}} = 0$ is a discontinuity point with $m(0) = 2/0.16^2$. Furthermore, note that the mode of this density is at $\theta_{111} = \theta_{111}^{\mathrm{U}}$

does not have an interpretation in terms of the CIA, it is a representative value of the uncertainty space. Furthermore, $\{\bar{\theta}_{ijk}\}$ is still a distribution. In fact, these parameters are all nonnegative, and

$$\sum_{ijk} \bar{\theta}_{ijk} = \sum_{ijk} \int_{\Theta^{\mathrm{SM}}} \theta_{ijk} m(\boldsymbol{\theta}) \mathrm{d}\boldsymbol{\theta}$$

$$= \int_{\Theta^{\mathrm{SM}}} \left[\sum_{ijk} \theta_{ijk} \right] m(\boldsymbol{\theta}) \mathrm{d}\boldsymbol{\theta} = \int_{\Theta^{\mathrm{SM}}} m(\boldsymbol{\theta}) \mathrm{d}\boldsymbol{\theta} = 1.$$

4.4 Estimation of Uncertainty

Let $(\mathbf{x}_a, \mathbf{y}_a)$, $a = 1, \ldots, n_A$, and $(\mathbf{x}_b, \mathbf{z}_b)$, $b = 1, \ldots, n_B$, be $n_A + n_B$ records generated independently by the r.v. $(\mathbf{X}, \mathbf{Y}, \mathbf{Z})$ with distribution $f(\mathbf{x}, \mathbf{y}, \mathbf{z}; \boldsymbol{\theta})$, $\boldsymbol{\theta} \in \Theta$, where the units in A have $\mathbf{Z}$ missing and the units in B have $\mathbf{Y}$ missing.

When no assumptions on the statistical relationship between $\mathbf{Y}$ and $\mathbf{Z}$ given $\mathbf{X}$ are made and no auxiliary information is available, Proposition 3.1 holds: there is a set of maximum likelihood estimates, the likelihood ridge. It is not possible to distinguish or privilege one parameter in the likelihood ridge with respect to another, given information in the available sample $A \cup B$. It is only possible to assert that parameters in the likelihood ridge are privileged by the available sample if compared to those outside the likelihood ridge.

Proposition 4.1 *The likelihood ridge is the ML estimate of* Θ^{SM}.

Proof. This is a consequence of Proposition 3.1. The likelihood ridge is composed of those, and only those, $\boldsymbol{\theta} \in \Theta$ satisfying equations (3.6), (3.7), and (3.8). Equations (3.6)–(3.8) define the ML estimate of $\boldsymbol{\theta}_{\mathbf{XY}}$ and $\boldsymbol{\theta}_{\mathbf{XZ}}$, i.e. of the constraints (4.1) and (4.2) for Θ^{SM}.

In order better to understand the likelihood ridge and its characteristics, it is necessary to study the likelihood function in the statistical matching problem:

$$L(\boldsymbol{\theta}|A \cup B) = L(\boldsymbol{\theta}_{\mathbf{X}}|A \cup B)L(\boldsymbol{\theta}_{\mathbf{Y}|\mathbf{X}}|A)L(\boldsymbol{\theta}_{\mathbf{Z}|\mathbf{X}}|B), \qquad \boldsymbol{\theta} \in \Theta,$$

or, equivalently, the loglikelihood function, given that the logarithm preserves the extrema (maxima and minima) of the likelihood function:

$$\ell(\boldsymbol{\theta}|A \cup B) = \ell(\boldsymbol{\theta}_{\mathbf{X}}|A \cup B) + \ell(\boldsymbol{\theta}_{\mathbf{Y}|\mathbf{X}}|A) + \ell(\boldsymbol{\theta}_{\mathbf{Z}|\mathbf{X}}|B), \qquad \boldsymbol{\theta} \in \Theta, \qquad (4.48)$$

where $\ell(\boldsymbol{\theta}_{\mathbf{X}}|A \cup B)$ is the loglikelihood of $\boldsymbol{\theta}_{\mathbf{X}}$ on the completely observed sample $A \cup B$, $\ell(\boldsymbol{\theta}_{\mathbf{Y}|\mathbf{X}}|A)$ is the loglikelihood of $\boldsymbol{\theta}_{\mathbf{Y}|\mathbf{X}}$ on the complete sample A, and $\ell(\boldsymbol{\theta}_{\mathbf{Z}|\mathbf{X}}|B)$ is the loglikelihood of $\boldsymbol{\theta}_{\mathbf{Z}|\mathbf{X}}$ on the complete sample B.

The loglikelihood function (4.48) has the following characteristics.

(a) It is a function with many ridges. These ridges are obtained by partitioning the parameter space Θ into parameter subsets $\mathcal{P}_{\boldsymbol{\theta}_{\mathbf{X}}\boldsymbol{\theta}_{\mathbf{Y}|\mathbf{X}}\boldsymbol{\theta}_{\mathbf{Z}|\mathbf{X}}}$ determined by those $\boldsymbol{\theta}$ with the same $\boldsymbol{\theta}_{\mathbf{X}}$, $\boldsymbol{\theta}_{\mathbf{Y}|\mathbf{X}}$ and $\boldsymbol{\theta}_{\mathbf{Z}|\mathbf{X}}$, for all $\boldsymbol{\theta}_{\mathbf{X}} \in \Theta_{\mathbf{X}}$, $\boldsymbol{\theta}_{\mathbf{Y}|\mathbf{X}} \in \Theta_{\mathbf{Y}|\mathbf{X}}$ and $\boldsymbol{\theta}_{\mathbf{Z}|\mathbf{X}} \in \Theta_{\mathbf{Z}|\mathbf{X}}$, such that

$$\bigcup_{\boldsymbol{\theta}_{\mathbf{X}}\boldsymbol{\theta}_{\mathbf{Y}|\mathbf{X}}\boldsymbol{\theta}_{\mathbf{Z}|\mathbf{X}}} \mathcal{P}_{\boldsymbol{\theta}_{\mathbf{X}}\boldsymbol{\theta}_{\mathbf{Y}|\mathbf{X}}\boldsymbol{\theta}_{\mathbf{Z}|\mathbf{X}}} = \Theta$$

and

$$\mathcal{P}_{\boldsymbol{\theta}_{\mathbf{X}}\boldsymbol{\theta}_{\mathbf{Y}|\mathbf{X}}\boldsymbol{\theta}_{\mathbf{Z}|\mathbf{X}}} \bigcap \mathcal{P}_{\boldsymbol{\eta}_{\mathbf{X}}\boldsymbol{\eta}_{\mathbf{Y}|\mathbf{X}}\boldsymbol{\eta}_{\mathbf{Z}|\mathbf{X}}} = \emptyset$$

if any of the following inequalities occur:

$$\boldsymbol{\theta}_{\mathbf{X}} \neq \boldsymbol{\eta}_{\mathbf{X}}, \quad \boldsymbol{\theta}_{\mathbf{Y}|\mathbf{X}} \neq \boldsymbol{\eta}_{\mathbf{Y}|\mathbf{X}}, \quad \boldsymbol{\theta}_{\mathbf{Z}|\mathbf{X}} \neq \boldsymbol{\eta}_{\mathbf{Z}|\mathbf{X}}.$$

The loglikelihood function (4.48) is constant for $\boldsymbol{\theta} \in \mathcal{P}_{\boldsymbol{\theta}_{\mathbf{X}}\boldsymbol{\theta}_{\mathbf{Y}|\mathbf{X}}\boldsymbol{\theta}_{\mathbf{Z}|\mathbf{X}}}$, $\boldsymbol{\theta}_{\mathbf{X}} \in \Theta_{\mathbf{X}}$, $\boldsymbol{\theta}_{\mathbf{Y}|\mathbf{X}} \in \Theta_{\mathbf{Y}|\mathbf{X}}$, $\boldsymbol{\theta}_{\mathbf{Z}|\mathbf{X}} \in \Theta_{\mathbf{Z}|\mathbf{X}}$.

(b) The extrema of (4.48) for $\boldsymbol{\theta} \in \Theta$ can equivalently be found from the extrema of

$$\ell(\boldsymbol{\theta}_{\mathbf{X}}, \boldsymbol{\theta}_{\mathbf{Y}|\mathbf{X}}, \boldsymbol{\theta}_{\mathbf{Z}|\mathbf{X}}) = \ell(\boldsymbol{\theta}_{\mathbf{X}}|A \cup B) + \ell(\boldsymbol{\theta}_{\mathbf{Y}|\mathbf{X}}|A) + \ell(\boldsymbol{\theta}_{\mathbf{Z}|\mathbf{X}}|B) \tag{4.49}$$

as a function of $\boldsymbol{\theta}_{\mathbf{X}} \in \Theta_{\mathbf{X}}$, $\boldsymbol{\theta}_{\mathbf{Y}|\mathbf{X}} \in \Theta_{\mathbf{Y}|\mathbf{X}}$ and $\boldsymbol{\theta}_{\mathbf{Z}|\mathbf{X}} \in \Theta_{\mathbf{Z}|\mathbf{X}}$. Then, $\tilde{\boldsymbol{\theta}}$ is an extremum (minimum or maximum) of (4.48) if and only if $\tilde{\boldsymbol{\theta}} \in \mathcal{P}_{\tilde{\boldsymbol{\theta}}_{\mathbf{X}}\tilde{\boldsymbol{\theta}}_{\mathbf{Y}|\mathbf{X}}\tilde{\boldsymbol{\theta}}_{\mathbf{Z}|\mathbf{X}}}$ and $(\tilde{\boldsymbol{\theta}}_{\mathbf{X}}, \tilde{\boldsymbol{\theta}}_{\mathbf{Y}|\mathbf{X}}, \tilde{\boldsymbol{\theta}}_{\mathbf{Z}|\mathbf{X}})$ is an extremum of (4.49).

Part (b) throws extra light on Proposition 3.1. Assume that the ML estimators of $\boldsymbol{\theta}_{\mathbf{X}}$, $\boldsymbol{\theta}_{\mathbf{Y}|\mathbf{X}}$ and $\boldsymbol{\theta}_{\mathbf{Z}|\mathbf{X}}$ exist and are unique. Then:

(i) $(\hat{\boldsymbol{\theta}}_{\mathbf{X}}, \hat{\boldsymbol{\theta}}_{\mathbf{Y}|\mathbf{X}}, \hat{\boldsymbol{\theta}}_{\mathbf{Z}|\mathbf{X}})$ is a maximum of (4.49);

(ii) the set of parameters $\boldsymbol{\theta} \in \Theta$ maximizing (4.48) is

$$\left\{\boldsymbol{\theta} \in \mathcal{P}_{\hat{\boldsymbol{\theta}}_{\mathbf{X}}\hat{\boldsymbol{\theta}}_{\mathbf{Y}|\mathbf{X}}\hat{\boldsymbol{\theta}}_{\mathbf{Z}|\mathbf{X}}}\right\}$$

and $\mathcal{P}_{\hat{\boldsymbol{\theta}}_{\mathbf{X}}\hat{\boldsymbol{\theta}}_{\mathbf{Y}|\mathbf{X}}\hat{\boldsymbol{\theta}}_{\mathbf{Z}|\mathbf{X}}}$ is the likelihood ridge, i.e. the ML estimate $\hat{\Theta}^{\mathrm{SM}}$ of Θ^{SM}.

When the loglikelihood function (4.48) is differentiable with respect to $\boldsymbol{\theta}$, the following result holds.

Proposition 4.2 *Let $\ell(\boldsymbol{\theta}|A \cup B)$ be a differentiable function of $\boldsymbol{\theta} \in \Theta$. Then, a necessary condition for $\tilde{\boldsymbol{\theta}}$ to be an extremum of $\ell(\boldsymbol{\theta}|A \cup B)$ is that $\tilde{\boldsymbol{\theta}}$ belongs to the following set:*

$$\left\{\boldsymbol{\theta} \in \Theta : \frac{\mathrm{d}\ell(\boldsymbol{\theta}_{\mathbf{X}}|A \cup B)}{\mathrm{d}\boldsymbol{\theta}_{\mathbf{X}}} = 0; \frac{\mathrm{d}\ell(\boldsymbol{\theta}_{\mathbf{Y}|\mathbf{X}}|A)}{\mathrm{d}\boldsymbol{\theta}_{\mathbf{Y}|\mathbf{X}}} = 0; \frac{\mathrm{d}\ell(\boldsymbol{\theta}_{\mathbf{Z}|\mathbf{X}}|B)}{\mathrm{d}\boldsymbol{\theta}_{\mathbf{Z}|\mathbf{X}}} = 0\right\}. \tag{4.50}$$

Proof. As stated in (a), the vector space Θ is partitioned into subspaces $\mathcal{P}_{\hat{\boldsymbol{\theta}}_{\mathbf{X}}\hat{\boldsymbol{\theta}}_{\mathbf{Y}|\mathbf{X}}\hat{\boldsymbol{\theta}}_{\mathbf{Z}|\mathbf{X}}}$, for $\boldsymbol{\theta}_{\mathbf{X}} \in \Theta_{\mathbf{X}}$, $\boldsymbol{\theta}_{\mathbf{Y}|\mathbf{X}} \in \Theta_{\mathbf{Y}|\mathbf{X}}$ and $\boldsymbol{\theta}_{\mathbf{Z}|\mathbf{X}} \in \Theta_{\mathbf{Z}|\mathbf{X}}$. Hence, extrema of $\ell(\boldsymbol{\theta}|A \cup B)$ can be found as in (b). Note that the subspaces $\boldsymbol{\theta}_{\mathbf{X}} \in \Theta_{\mathbf{X}}$, $\boldsymbol{\theta}_{\mathbf{Y}|\mathbf{X}} \in \Theta_{\mathbf{Y}|\mathbf{X}}$ and $\boldsymbol{\theta}_{\mathbf{Z}|\mathbf{X}} \in \Theta_{\mathbf{Z}|\mathbf{X}}$ are orthogonal. A necessary condition for $\boldsymbol{\theta}$ to be a (local or global) extremum of (4.49) is that its gradient with respect to $(\boldsymbol{\theta}_{\mathbf{X}}, \boldsymbol{\theta}_{\mathbf{Y}|\mathbf{X}}, \boldsymbol{\theta}_{\mathbf{Z}|\mathbf{X}})$ is null. Hence,

$$\frac{\mathrm{d}\ell(\boldsymbol{\theta}_{\mathbf{X}}, \boldsymbol{\theta}_{\mathbf{Y}|\mathbf{X}}, \boldsymbol{\theta}_{\mathbf{Z}|\mathbf{X}})}{\mathrm{d}\boldsymbol{\theta}_{\mathbf{X}}} = \frac{\mathrm{d}\ell(\boldsymbol{\theta}_{\mathbf{X}}|A \cup B)}{\mathrm{d}\boldsymbol{\theta}_{\mathbf{X}}} = 0, \tag{4.51}$$

$$\frac{\mathrm{d}\ell(\boldsymbol{\theta}_{\mathbf{X}}, \boldsymbol{\theta}_{\mathbf{Y}|\mathbf{X}}, \boldsymbol{\theta}_{\mathbf{Z}|\mathbf{X}})}{\mathrm{d}\boldsymbol{\theta}_{\mathbf{Y}|\mathbf{X}}} = \frac{\mathrm{d}\ell(\boldsymbol{\theta}_{\mathbf{Y}|\mathbf{X}}|A)}{\mathrm{d}\boldsymbol{\theta}_{\mathbf{Y}|\mathbf{X}}} = 0, \tag{4.52}$$

$$\frac{\mathrm{d}\ell(\boldsymbol{\theta}_{\mathbf{X}}, \boldsymbol{\theta}_{\mathbf{Y}|\mathbf{X}}, \boldsymbol{\theta}_{\mathbf{Z}|\mathbf{X}})}{\mathrm{d}\boldsymbol{\theta}_{\mathbf{Z}|\mathbf{X}}} = \frac{\mathrm{d}\ell(\boldsymbol{\theta}_{\mathbf{Z}|\mathbf{X}}|B)}{\mathrm{d}\boldsymbol{\theta}_{\mathbf{Z}|\mathbf{X}}} = 0. \tag{4.53}$$

Remark 4.7 Note that, for some parametric distributions $f(\mathbf{x}, \mathbf{y}, \mathbf{z}|\boldsymbol{\theta})$, conditions (4.51)–(4.53) ensure a unique (global) maximum. This is the case for the multinormal and multinomial distributions. For these distributions, it can be stated that:

(i) (4.49) is maximized at a unique point $(\hat{\boldsymbol{\theta}}_{\mathbf{X}}, \hat{\boldsymbol{\theta}}_{\mathbf{Y}|\mathbf{X}}, \hat{\boldsymbol{\theta}}_{\mathbf{Z}|\mathbf{X}})$;

(ii) $(\hat{\boldsymbol{\theta}}_{\mathbf{X}}, \hat{\boldsymbol{\theta}}_{\mathbf{Y}|\mathbf{X}}, \hat{\boldsymbol{\theta}}_{\mathbf{Z}|\mathbf{X}})$ is the ML solution of (4.48) under the CIA;

(iii) the likelihood ridge is the set (4.50);

(iv) there do not exist local maxima and global or local minima.

In order to compute marginal uncertainty densities, in complex cases it is possible to resort to approximate solutions. D'Orazio *et al.* (2005b) suggest the following exploration of the likelihood ridge.

Algorithm 4.1 *The exploration is composed of the following steps.*

(a) Randomly generate a starting value $\boldsymbol{\theta}^0$ *in* Θ.

(b) Apply the EM algorithm on the overall sample $A \cup B$, *and register the final solution (once the EM algorithm stops because it reaches the maximum number of iterations or because two successive estimates* $\boldsymbol{\theta}^{\iota-1}$ *and* $\boldsymbol{\theta}^{\iota}$ *are sufficiently close).*

(c) Carry out the previous two steps a sufficient number of times. Select only those distinct final estimates with maximum likelihood. Let $\hat{\boldsymbol{\theta}}^{(v)}$, $v = 1, \ldots, V$, *denote these parameters. These* V *parameters are a sketch of the likelihood ridge.*

(d) Estimate the uncertainty distribution $m(\theta_{ijk})$ *by means of the values assumed by* θ_{ijk} *among the* V *maximum likelihood parameters* $\hat{\boldsymbol{\theta}}^{(v)}$, $v = 1, \ldots, V$.

4.4.1 Maximum likelihood estimation of uncertainty in the multinormal case

Let $(\mathbf{X}, \mathbf{Y}, \mathbf{Z})$ be multinormal r.v.s with parameter $\boldsymbol{\theta} = (\boldsymbol{\mu}, \boldsymbol{\Sigma})$, and let $A \cup B$ be a sample of $n_A + n_B$ units whose n_A units in A have $\mathbf{Z}$ missing and n_B units in B have $\mathbf{Y}$ missing. Propositions 4.1 and 4.2 claim that the ML estimate of $\boldsymbol{\theta}$ under the CIA is essential for determining an ML estimate of Θ^{SM}. Section 2.1.2 shows how to estimate $\boldsymbol{\theta}$ under the CIA.

(a) First, consider ML estimates of:

- $\boldsymbol{\theta}_{\mathbf{X}}$, i.e. $\hat{\boldsymbol{\mu}}_{\mathbf{X}}$ and $\hat{\boldsymbol{\Sigma}}_{\mathbf{X}}$;
- $\boldsymbol{\theta}_{\mathbf{Y}|\mathbf{X}}$, i.e. $\hat{\boldsymbol{\beta}}_{\mathbf{Y}|\mathbf{X}}$, $\hat{\boldsymbol{\alpha}}_{\mathbf{Y}}$, and $\hat{\boldsymbol{\Sigma}}_{\mathbf{Y}|\mathbf{X}}$;
- $\boldsymbol{\theta}_{\mathbf{Z}|\mathbf{X}}$, i.e. $\hat{\boldsymbol{\beta}}_{\mathbf{Z}|\mathbf{X}}$, $\hat{\boldsymbol{\alpha}}_{\mathbf{Z}}$, and $\hat{\boldsymbol{\Sigma}}_{\mathbf{Z}|\mathbf{X}}$.

(b) Then, obtain ML estimates of $\boldsymbol{\theta}$ coherently with the previous estimates, as in steps (i), (ii) and (iii) of Section 2.1.2. Uncertainty on the estimable parameters is degenerate, and is concentrated on the ML estimates.

(c) The likelihood ridge $\hat{\Theta}^{\text{SM}}$ is composed of all those $\boldsymbol{\theta} \in \Theta$ whose estimable parameters are set at their ML estimates, and $\boldsymbol{\Sigma}_{\mathbf{YZ}}$ in the set

$$\hat{\Theta}^{\text{SM}}_{\boldsymbol{\Sigma}_{\mathbf{YZ}}} = \left\{ \boldsymbol{\Sigma}_{\mathbf{YZ}} : \begin{pmatrix} \hat{\boldsymbol{\Sigma}}_{\mathbf{XX}} & \hat{\boldsymbol{\Sigma}}_{\mathbf{XY}} & \hat{\boldsymbol{\Sigma}}_{\mathbf{XZ}} \\ \hat{\boldsymbol{\Sigma}}_{\mathbf{YX}} & \hat{\boldsymbol{\Sigma}}_{\mathbf{YY}} & \boldsymbol{\Sigma}_{\mathbf{YZ}} \\ \hat{\boldsymbol{\Sigma}}_{\mathbf{ZX}} & \boldsymbol{\Sigma}_{\mathbf{ZY}} & \hat{\boldsymbol{\Sigma}}_{\mathbf{ZZ}} \end{pmatrix} \text{ positive semidefinite} \right\}. \tag{4.54}$$

The set (4.54) can be used to estimate the uncertainty of the covariance matrix $\boldsymbol{\Sigma}_{\mathbf{YZ}}$ and each of its components. Estimates of the marginal uncertainty densities, say $\hat{m}(\sigma_{Y_q Z_r})$, $q = 1, \ldots, Q$, $r = 1, \ldots, R$, can be determined as in Section 4.3.1, and substituting perfect knowledge of the marginal parameters $\boldsymbol{\theta}_{\mathbf{XY}}$ and $\boldsymbol{\theta}_{\mathbf{XZ}}$ with their ML estimate on $A \cup B$.

Note that the set $\hat{\Theta}^{\text{SM}}_{\boldsymbol{\Sigma}_{\mathbf{YZ}}}$, together with the estimated parameters in step (a), allows the computation of the uncertainty sets and of the corresponding uncertainty densities for other parameters of interest in the multinormal case. For instance, suppose it is necessary to assess how uncertain the regression parameters of $\mathbf{Z}$ on $\mathbf{Y}$ and $\mathbf{X}$ (i.e. $\boldsymbol{\beta}_{\mathbf{ZY.X}}$, $\boldsymbol{\beta}_{\mathbf{ZX.Y}}$, and $\boldsymbol{\Sigma}_{\mathbf{ZZ|XY}}$) are. First of all, uncertainty on $\boldsymbol{\Sigma}_{\mathbf{YZ|X}}$ is defined by the set

$$\hat{\Theta}^{\text{SM}}_{\boldsymbol{\Sigma}_{\mathbf{YZ|X}}} = \left\{ \boldsymbol{\Sigma}_{\mathbf{YZ}} - \hat{\boldsymbol{\Sigma}}_{\mathbf{YX}} \hat{\boldsymbol{\Sigma}}^{-1}_{\mathbf{XX}} \hat{\boldsymbol{\Sigma}}_{\mathbf{XZ}},\ \boldsymbol{\Sigma}_{\mathbf{YZ}} \in \hat{\Theta}^{\text{SM}}_{\boldsymbol{\Sigma}_{\mathbf{YZ}}} \right\}.$$

Hence, through step (f) of Section 3.2.4 in the known partial correlation coefficient case, the following sets can be found:

$$\begin{aligned}
\hat{\Theta}^{\text{SM}}_{\boldsymbol{\beta}_{\mathbf{ZY.X}}} &= \left\{ \boldsymbol{\Sigma}_{\mathbf{YZ|X}} \hat{\boldsymbol{\Sigma}}^{-1}_{\mathbf{YY|X}},\ \boldsymbol{\Sigma}_{\mathbf{YZ|X}} \in \hat{\Theta}^{\text{SM}}_{\boldsymbol{\Sigma}_{\mathbf{YZ|X}}} \right\}, \\
\hat{\Theta}^{\text{SM}}_{\boldsymbol{\beta}_{\mathbf{ZX.Y}}} &= \left\{ \hat{\boldsymbol{\beta}}_{\mathbf{ZX}} - \hat{\boldsymbol{\beta}}_{\mathbf{YX}} \boldsymbol{\beta}_{\mathbf{ZY.X}},\ \boldsymbol{\beta}_{\mathbf{ZY.X}} \in \hat{\Theta}^{\text{SM}}_{\boldsymbol{\beta}_{\mathbf{ZY.X}}} \right\}, \\
\hat{\Theta}^{\text{SM}}_{\boldsymbol{\Sigma}_{\mathbf{ZZ|XY}}} &= \left\{ \boldsymbol{\Sigma}_{\mathbf{ZZ|X}} - \hat{\boldsymbol{\Sigma}}_{\mathbf{ZY|X}} \hat{\boldsymbol{\Sigma}}^{-1}_{\mathbf{YY|X}} \hat{\boldsymbol{\Sigma}}_{\mathbf{YZ|X}},\ \boldsymbol{\Sigma}_{\mathbf{YZ|X}} \in \hat{\Theta}^{\text{SM}}_{\boldsymbol{\Sigma}_{\mathbf{YZ|X}}} \right\}.
\end{aligned}$$

4.4.2 Maximum likelihood estimation of uncertainty in the multinomial case

Let $A \cup B$ be $n_A + n_B$ i.i.d. realizations of (X, Y, Z), a multinomial r.v. as in Section 2.1.3. By equations (2.33)–(2.35), the likelihood ridge is given by every distribution $\boldsymbol{\theta} = \{\theta_{ijk}\}$ that satisfies the following set of equations:

$$\begin{cases}
\sum_k \theta_{ijk} = \hat{\theta}_{ij.} = \frac{n^A_{ij.}}{n^A_{i..}} \left(\frac{n^A_{i..} + n^B_{i..}}{n} \right), \\
\sum_j \theta_{ijk} = \hat{\theta}_{i.k} = \frac{n^B_{i.k}}{n^B_{i..}} \left(\frac{n^A_{i..} + n^B_{i..}}{n} \right), \\
\theta_{ijk} \geq 0.
\end{cases} \tag{4.55}$$

From the definition of the multinomial parameters under the CIA, equation (4.45), the ML estimator of θ_{ijk}^{CIA} is

$$\hat{\theta}_{ijk}^{\mathrm{CIA}} = \frac{n_{ij.}^{A}}{n_{i..}^{A}} \frac{n_{i.k}^{B}}{n_{i..}^{B}} \frac{n_{i..}^{A} + n_{i..}^{B}}{n}, \qquad \forall\, i, j, k,$$

and this distribution is clearly inside the likelihood ridge (4.55).

The bounds of the intervals of values defined by (4.55) can be found using the algorithm in Capotorti and Vantaggi (2002).

Example 4.9 Let A and B be two samples of $n_A = 140$ and $n_B = 220$ units, and let Tables 2.3 and 2.4 be the corresponding contingency tables. The unrestricted ML estimates of the estimable parameters $\theta_{i..}$, $\theta_{j|i}$ and $\theta_{k|i}$ have already been computed in Tables 2.5, 2.6 and 2.7 respectively.

The likelihood ridge allows θ_{111} to take values in the interval bounded by

$$\hat{\theta}_{111}^{\mathrm{L}} = \max\left\{0; \hat{\theta}_{1.1} - \hat{\theta}_{12.}\right\},$$
$$\hat{\theta}_{111}^{\mathrm{U}} = \min\left\{\hat{\theta}_{11.}; \hat{\theta}_{1.1}\right\}.$$

The bounds described by the likelihood ridge for all the parameters are given in Table 4.4.

Having fixed θ_{111} in the above interval, according to the normal form of the parameter space (see inequalities (4.47)), $\hat{\theta}_{112}$ can assume values in the interval

Table 4.4 Bounds for θ_{ijk} estimated by the likelihood ridge, and those found through the exploration of the likelihood ridge with the EM algorithm, mean and variance of the marginal uncertainty distributions

θ_{ijk}	$\hat{\theta}_{ijk}^{\mathrm{L}}$	$\hat{\theta}_{ijk}^{\mathrm{U}}$	$\hat{\theta}_{ijk}^{\mathrm{L\|EM}}$	$\hat{\theta}_{ijk}^{\mathrm{U\|EM}}$	$\hat{\hat{\theta}}_{ijk}$	$\mathrm{Var}(\theta_{ijk})$
$\hat{\theta}_{111}$	0.00	0.20	8.39E-06	0.199 990	0.099 933	0.001 699
$\hat{\theta}_{112}$	0.00	0.10	1.28E-07	0.100 000	0.049 936	0.000 622
$\hat{\theta}_{113}$	0.00	0.20	2.78E-07	0.199 985	0.100 131	0.001 690
$\hat{\theta}_{121}$	0.00	0.20	8.10E-06	0.199 993	0.100 067	0.001 699
$\hat{\theta}_{122}$	0.00	0.10	6.73E-07	0.100 001	0.050 064	0.000 622
$\hat{\theta}_{123}$	0.00	0.20	1.72E-05	0.200 001	0.099 869	0.001 690
$\hat{\theta}_{211}$	0.05	0.25	0.049998	0.250 001	0.149 897	0.001 603
$\hat{\theta}_{212}$	0.00	0.10	1.56E-06	0.099 999	0.050 066	0.000 691
$\hat{\theta}_{213}$	0.00	0.10	1.83E-07	0.099 999	0.050 037	0.000 692
$\hat{\theta}_{221}$	0.05	0.25	0.049997	0.250 001	0.150 103	0.001 603
$\hat{\theta}_{222}$	0.00	0.10	4.15E-07	0.099 999	0.049 934	0.000 691
$\hat{\theta}_{223}$	0.00	0.10	1.10E-06	0.100 000	0.049 963	0.000 692

bounded by

$$\hat{b}_{112}^{\mathrm{L}} = \max\left\{0; \hat{\theta}_{11.} - \hat{\theta}_{1.3} - \theta_{111}\right\} = 0,$$
$$\hat{b}_{112}^{\mathrm{U}} = \min\left\{\hat{\theta}_{1.2}; \hat{\theta}_{11.} - \theta_{111}\right\} = 0.20.$$

Noting that, by the estimates in Tables 2.5–2.7,

$$\hat{\theta}_{11.} - \hat{\theta}_{1.3} < \hat{\theta}_{11.} - \hat{\theta}_{1.2},$$

the estimate of the marginal uncertainty density for θ_{111} is (see Figure 4.6):

$$\hat{m}(\theta_{111}) = \begin{cases} c(0.05 + \theta_{111}) & 0 \le \theta_{111} < 0.05, \\ 0.1c & 0.05 \le \theta_{111} < 0.15, \\ c(0.25 - \theta_{111}) & 0.15 \le \theta_{111} \le 0.20, \end{cases}$$

where c is the normalizing constant

$$c = \frac{1}{0.0175}.$$

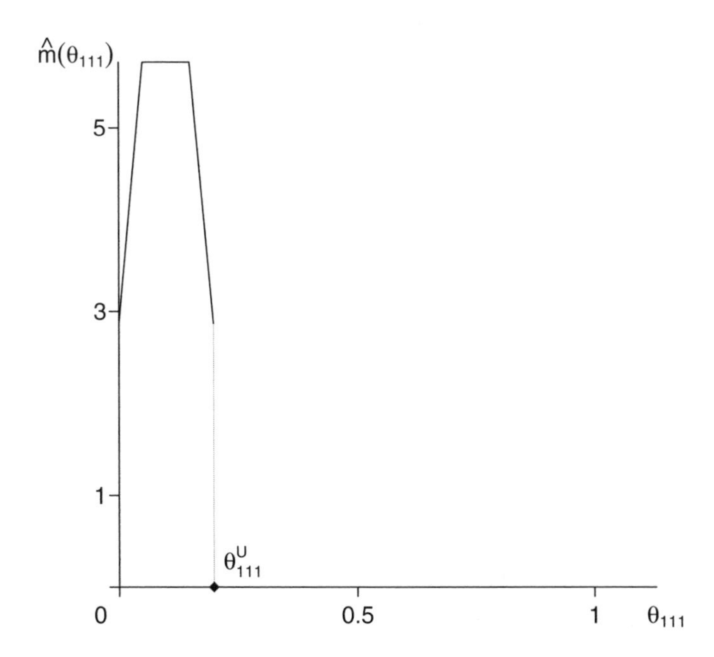

Figure 4.6 Estimated marginal uncertainty density $\hat{m}(\theta_{111})$

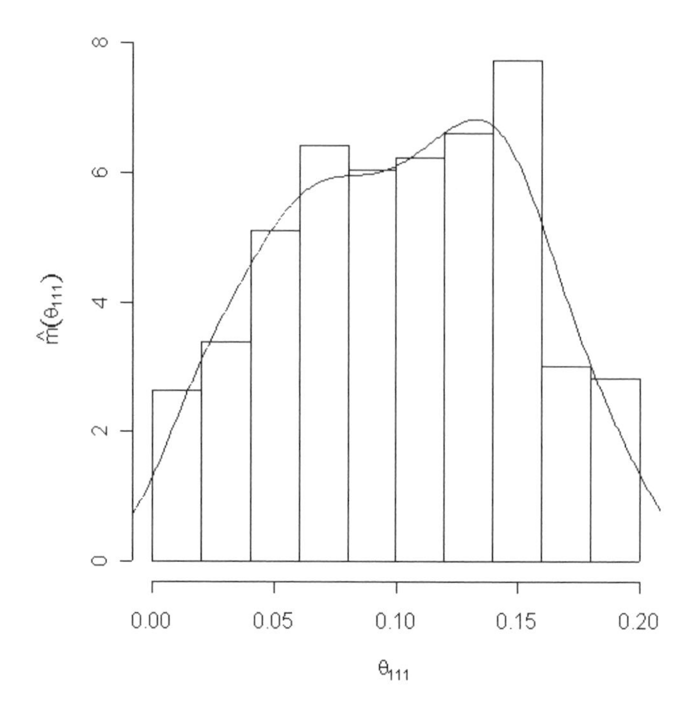

Figure 4.7 Marginal uncertainty density $\hat{m}(\theta_{111})$ of Figure 4.6 obtained by exploration of the likelihood ridge with the EM algorithm, with a histogram and a nonparametric approximation

In this case, the marginal uncertainty distribution $\hat{m}(\theta_{111})$ is a symmetric function between $\hat{\theta}^{\mathrm{L}}_{111}$ and $\hat{\theta}^{\mathrm{U}}_{111}$. Furthermore,

$$\hat{\theta}^{\mathrm{CIA}}_{111} = \frac{\hat{\theta}_{11.}\hat{\theta}_{1.1}}{\hat{\theta}_{1..}} = \frac{0.25 \times 0.2}{0.5} = 0.1$$

coincides with $\hat{\hat{\theta}}_{111}$, although this is not generally true.

Exploration of the likelihood ridge with 100 000 runs of the EM algorithm, as explained in Algorithm 4.1, has given an estimate of $\hat{m}(\theta_{111})$ as in Figure 4.7.

4.5 Reduction of Uncertainty: Use of Parameter Constraints

How can the uncertainty which characterizes statistical matching be reduced? In other words, for those components θ_t of $\boldsymbol{\theta}$ with a marginal uncertainty density $m(\theta_t)$ not concentrated on just one point, is it possible to reduce the variability of $m(\theta_t)$? Again, external knowledge, i.e. knowledge not contained in the data set itself, plays a very important role.

Example 4.10 Assume the variables 'Age' and 'Marital Status' are examined. These variables admit a very straightforward kind of external knowledge: for example, it is not possible for a unit in a population to be both younger than marriageable age and married. This restriction corresponds to a parameter constraint: the joint probability of the categories 'Age = younger than marriageable age' and 'Marital Status = married' *must* be zero.

Generally speaking, parameter constraints are all those rules that make some of the distributions in Θ illogical for the phenomenon being investigated. Their introduction is useful for eliminating impossible worlds, thus decreasing uncertainty. There are many examples of logical constraints. One area where they are frequently used is statistical data editing, i.e. the data cleaning phase generally performed by the national statistical institutes; see Fellegi and Holt (1976) and De Waal (2003).

Let the imposed parameter constraints restrict Θ to a subspace $\Omega \subset \Theta$ which is closed and convex. The goal becomes the maximization of the likelihood function subject to constraints, i.e. to find that set of $\boldsymbol{\theta} \in \Omega$ such that the likelihood function $L(\boldsymbol{\theta}|A \cup B)$ is maximized. There are two possibilities.

(i) Ω has a nonempty intersection with the unconstrained likelihood ridge,

$$\Omega \bigcap \mathcal{P}_{\hat{\boldsymbol{\theta}}_{\mathbf{X}}\hat{\boldsymbol{\theta}}_{\mathbf{Y}|\mathbf{X}}\hat{\boldsymbol{\theta}}_{\mathbf{Z}|\mathbf{X}}} \neq \emptyset; \tag{4.56}$$

(ii) Ω has an empty intersection with the unconstrained likelihood ridge,

$$\Omega \bigcap \mathcal{P}_{\hat{\boldsymbol{\theta}}_{\mathbf{X}}\hat{\boldsymbol{\theta}}_{\mathbf{Y}|\mathbf{X}}\hat{\boldsymbol{\theta}}_{\mathbf{Z}|\mathbf{X}}} = \emptyset. \tag{4.57}$$

In the first case, the ML estimate of Θ^{SM} is obviously

$$\hat{\Theta}^{\mathrm{SM}} = \Omega \bigcap \mathcal{P}_{\hat{\boldsymbol{\theta}}_{\mathbf{X}}\hat{\boldsymbol{\theta}}_{\mathbf{Y}|\mathbf{X}}\hat{\boldsymbol{\theta}}_{\mathbf{Z}|\mathbf{X}}}.$$

In fact, the imposed parameter constraints restrict Θ to Ω, but leave the values of the likelihood function unchanged. Hence, if at least one $\boldsymbol{\theta}$ maximizes the likelihood function in Θ and satisfies the constraints, then it is still an ML solution, while all those $\boldsymbol{\theta} \notin \mathcal{P}_{\hat{\boldsymbol{\theta}}_{\mathbf{X}}\hat{\boldsymbol{\theta}}_{\mathbf{Y}|\mathbf{X}}\hat{\boldsymbol{\theta}}_{\mathbf{Z}|\mathbf{X}}}$ cannot maximize the likelihood function. Furthermore, ML estimates of the estimable parameters $\boldsymbol{\theta}_{\mathbf{X}}$, $\boldsymbol{\theta}_{\mathbf{Y}|\mathbf{X}}$ and $\boldsymbol{\theta}_{\mathbf{Z}|\mathbf{X}}$ are also unchanged. In the second case, constrained ML estimators should be considered. When, as in the multinormal and multinomial cases, the loglikelihood function is differentiable with respect to $\boldsymbol{\theta}$, and the solutions of (4.51)–(4.53) define a unique ML estimate of $\boldsymbol{\theta}_{\mathbf{X}}$, $\boldsymbol{\theta}_{\mathbf{Y}|\mathbf{X}}$ and $\boldsymbol{\theta}_{\mathbf{Z}|\mathbf{X}}$, a general result can be derived. According to the equality in (4.57), the intersection between the set (4.50) and Ω is the empty set, i.e. Ω does not contain any derivative of the observed likelihood $L(\boldsymbol{\theta}|A \cup B)$ equal to zero in $\boldsymbol{\theta}$. Hence, the likelihood function $L(\boldsymbol{\theta}|A \cup B)$ may admit local maxima only on the border of the subspace Ω. In this case an iterative algorithm to find constrained maxima of the likelihood function can be used (see Judge *et al.*, 1980, Chapter 17).

Remark 4.8 Great caution should be exercised in the definition of the set of parameter constraints. It may happen that the constraints are not compatible each other, i.e. Θ is restricted to the empty set; see Bergsma and Rudas (2002) and Hansen and Jaumard (1990). From now on, we suppose that the chosen logical constraints are compatible.

Remark 4.9 Section 4.3 showed that even complete knowledge of the bivariate r.v.s $(\mathbf{X}, \mathbf{Y})$ and $(\mathbf{X}, \mathbf{Z})$ is always compatible with parameters defined by the CIA. The introduction of parameter constraints may be such that the possibly restricted likelihood ridge $\hat{\Theta}^{\mathrm{SM}}$ does not admit these solutions.

4.5.1 The multinomial case

Two frequently used parameter constraints when the r.v.s are categorical are:

- *existence of some quantities*, for example it is not acceptable for a unit in the population to be both 10 years old and married;
- *inequality constraints*, e.g. a person with a degree has higher probability of being a manager than a worker.

Formally, they are defined by the equations:

$$\theta_{ijk} = 0, \qquad \text{for some } (i, j, k), \tag{4.58}$$

$$\theta_{ijk} \leq \theta_{i'j'k'}, \qquad \text{for some } (i, j, k),\ (i', j', k'). \tag{4.59}$$

Constraint (4.58) is usually called a *structural zero*; see Agresti (1990). More precisely, a structural zero occurs when:

(i) at least one pair of categories in (i, j, k) is incompatible;

(ii) each pair in (i, j, k) is plausible but the triplet is incompatible.

The main effect of these constraints is the possible reduction of the likelihood ridge. Parameter constraints can be so informative that the likelihood ridge reduces to a unique distribution, e.g.

$$\Omega \bigcap \mathcal{P}_{\hat{\boldsymbol{\theta}}_{\mathbf{X}} \hat{\boldsymbol{\theta}}_{\mathbf{Y}|\mathbf{X}} \hat{\boldsymbol{\theta}}_{\mathbf{Z}|\mathbf{X}}} = \hat{\boldsymbol{\theta}}.$$

For instance, a sufficient condition for a unique ML estimate is the definition of $(J-1)(K-1)$ independent structural zero constraints for each $X = i$, $i = 1, \ldots, I$ (i.e. maximum dependence among Y and Z conditional to X; see D'Orazio *et al.*, 2005b).

Remark 4.10 Among parameter constraints, structural zeros are also very effective because, with the exception of limit cases, distributions $\{\theta_{ijk}\} \in \Theta$ satisfying the CIA are not in Ω. In fact, under the CIA,

$$\theta_{ijk}^{\text{CIA}} = \frac{\theta_{ij.}\theta_{i.k}}{\theta_{i..}}$$

and when the structural zero $\theta_{ijk} = 0$ is imposed, $\theta_{ijk}^{\text{CIA}}$ can be computed only when $\hat{\theta}_{ij.} = 0$ and/or $\hat{\theta}_{i.k} = 0$. In other words, the use of structural zeros most of the time implies that the CIA cannot be assumed.

When (4.56) holds, the constrained likelihood ridge reduces to the set of solutions of:

$$\begin{cases} \sum_{\mathrm{k}} \theta_{ijk} = \hat{\theta}_{ij.} = \frac{n_{ij.}^A}{n_{i..}^A} \frac{n_{i..}^A + n_{i..}^B}{n}, \\ \sum_{\mathrm{j}} \theta_{ijk} = \hat{\theta}_{i.k} = \frac{n_{i.k}^B}{n_{i..}^B} \frac{n_{i..}^A + n_{i..}^B}{n}, \\ \boldsymbol{\theta} \in \Omega. \end{cases} \tag{4.60}$$

When (4.57) holds, the constrained likelihood ridge can be determined by (usually iterative) constrained ML algorithms.

D'Orazio *et al.* (2005b) use an algorithm in the projection method family (Judge *et al.* 1980, p. 749), described in Winkler (1993), which is an adaptation of Algorithm 4.1. This algorithm modifies the usual EM algorithm (Dempster *et al.*, 1977) according to the following rules.

(a) Initialize the algorithm with a parameter satisfying the imposed parameter constraints: $\hat{\boldsymbol{\theta}}^0 \in \Omega$.

(b) At iteration ι, $\iota \geq 1$, check whether the EM unconstrained estimate $\hat{\boldsymbol{\theta}}^\iota$ is in the constrained space Ω. If it is, then the solution is left unchanged. If it is not, then $\hat{\boldsymbol{\theta}}^\iota$ is projected to the boundary of the closed and convex subspace Ω. This constrained EM algorithm is iterated until convergence.

(c) Carry out the previous two steps a sufficient number of times. Select only those distinct final estimates with maximum likelihood. Let $\hat{\boldsymbol{\theta}}^{(v)}$, $v = 1, \ldots, V$, denote these parameters. These V parameters are a sketch of the likelihood ridge.

(d) Estimate the distribution $m(\theta_{ijk})$ by means of the values assumed by θ_{ijk} among the V maximum likelihood parameters $\hat{\boldsymbol{\theta}}^{(v)}$, $v = 1, \ldots, V$.

In the multinomial case, Haberman's (1977) Theorem 4 (see also Winkler, 1993) suggests that, if $\hat{\boldsymbol{\theta}}^{\iota-1}$ and $\hat{\boldsymbol{\theta}}^\iota$, $\iota \geq 1$, are successive estimates, and $\hat{\boldsymbol{\theta}}^{\iota-1} \in \Omega$ while $\hat{\boldsymbol{\theta}}^\iota \notin \Omega$, then $\hat{\boldsymbol{\theta}}^\iota$ should be substituted by $\alpha\hat{\boldsymbol{\theta}}^{\iota-1} + (1-\alpha)\hat{\boldsymbol{\theta}}^\iota$, with $0 \leq \alpha \leq 1$ chosen so that it lies on the boundary of Ω. Given that Ω is closed and convex, α exists and is unique. Furthermore, Haberman's theorem states that the likelihood of the successive M-step solutions of this modified EM algorithm (also called EMH) is nondecreasing. Nevertheless, great caution should be exercised with this approach. In fact, this algorithm may become stuck at a solution on the boundary of Ω which is not a local maximum (see Judge *et al.* 1980; Winkler, 1993). For this reason, it is worthwhile to preserve the value of the likelihood of the solution of this algorithm, and compare it with other results obtained from different starting points. Note that it is not difficult to determine α when structural zeros and inequality constraints are imposed.

- Structural zero constraints (4.58) may easily be satisfied by setting the corresponding $\hat{\theta}^0_{ijk}$ to zero in the initialization step of the EM algorithm; for details see Schafer (1997, pp. 52–53).

- Inequality constraints (4.59) are satisfied by setting

$$\alpha = \frac{\hat{\theta}^{\iota}_{i'j'k'} - \hat{\theta}^{\iota}_{ijk}}{\hat{\theta}^{\iota-1}_{ijk} - \hat{\theta}^{\iota-1}_{i'j'k'} - \hat{\theta}^{\iota}_{ijk} + \hat{\theta}^{\iota}_{i'j'k'}}.$$

  If more than one inequality constraint is imposed, it is enough to compute the corresponding α for each inequality and choose the smallest one.

The algorithm has been implemented in an R function based on the CAT library written by Schafer, which implements the EM algorithm for categorical data. The code is reported in Section E.4.

Remark 4.11 If a loglinear model different from the saturated one is considered for (X, Y, Z), and if this model does not admit a closed-form ML estimate of the corresponding parameters, the above approach can be adapted by substituting the EM algorithm with the ECM algorithm (Meng and Rubin, 1993), as in Winkler (1993).

Remark 4.12 All the parameters $\hat{\boldsymbol{\theta}}$ in the likelihood ridge $\hat{\Theta}^{\mathrm{SM}}$ are consistent with the unique ML estimate of $\boldsymbol{\theta}_{XY}$ and $\boldsymbol{\theta}_{XZ}$. Hence, denoting by $\hat{\hat{\boldsymbol{\theta}}}$ the average of the distributions in $\hat{\Theta}^{\mathrm{SM}}$ (see Remark 4.6), it is possible to find these estimates through the marginalization of $\hat{\hat{\boldsymbol{\theta}}}$. Note that this is also possible when the likelihood ridge is explored through the EMH algorithm.

Example 4.11 Let X, Y and Z be categorical r.v.s with respectively $I = 2$, $J = 2$ and $K = 3$ categories. The overall uncertainty space is such that each parameter θ_{ijk} has a marginal uncertainty density like that in Figure 4.4. Let us see what happens when a sample $A \cup B$ is observed. Let A and B be represented by Tables 4.5 and 4.6 respectively.

The unrestricted ML estimators of the parameters $\theta_{ij.}$ and $\theta_{i.k}$ are exactly those represented in Tables 4.2 and 4.1. As a result, the estimated uncertainty distributions

Table 4.5 Contingency table for (X, Y) computed on sample A, $n_A = 100$

	$Y = 1$	$Y = 2$	Total
$X = 1$	43	16	59
$X = 2$	15	26	41
Total	58	42	100

Table 4.6 Contingency table for (X, Z) computed on sample B, $n_B = 100$

	$Z = 1$	$Z = 2$	$Z = 3$	Total
$X = 1$	16	23	20	59
$X = 2$	14	16	11	41
Total	30	39	31	100

are those found in Example 4.7, and Figure 4.5 shows the estimate of $m(\theta_{111})$. This is quite a general result: the estimated uncertainty space $\hat{\Theta}^{\mathrm{SM}}$ depends only on the values assumed by $\hat{\boldsymbol{\theta}}_{\mathbf{XY}}$ and $\hat{\boldsymbol{\theta}}_{\mathbf{XZ}}$. If these values are estimated independently of the sample sizes, or are the exact true values, the corresponding uncertainty space does not change. As far as this example is concerned, the likelihood ridge implies that each parameter has bounds as in the columns $\hat{\theta}^{\mathrm{L}}_{ijk}$ and $\hat{\theta}^{\mathrm{U}}_{ijk}$ of the unrestricted case in Table 4.7. Given that, for each parameter, upper and lower bounds are distinct, each parameter is uncertain. The imposition of a structural zero has the effect of modifying the likelihood ridge $\hat{\Theta}^{\mathrm{SM}}$ dramatically. In the following, two possible situations are illustrated.

(R1) Restricted 1. Let $\theta_{111} = 0$ be an imposed structural zero. This constraint has different impact on the parameters θ_{ijk}, as shown in the $\hat{\theta}^{\mathrm{L}}_{ijk}$ and $\hat{\theta}^{\mathrm{U}}_{ijk}$ columns of the restricted 1 case in Table 4.7. The following are consequences of inequalities (4.47).

- Parameter estimates for θ_{1jk}, $j = 1, 2$, $k = 1, 2, 3$, are no longer uncertain. Their lower and upper bounds coincide.
- Parameter estimates for θ_{2jk}, $j = 1, 2$, $k = 1, 2, 3$, are as uncertain as in the unrestricted case.

Note that the not uncertain parameters $\hat{\theta}_{1jk}$ are able to reproduce the unrestricted ML frequencies $\hat{\theta}_{1..}$, $\hat{\theta}_{1j.}$ and $\hat{\theta}_{1.k}$.

Table 4.7 Bounds $\hat{\theta}^{L}_{ijk}$ and $\hat{\theta}^{U}_{ijk}$ of the parameters in the likelihood ridge, given sample A in Table 4.5 and sample B in Table 4.6, in the unrestricted, restricted 1, and restricted 2 case

	Unrestricted				Restricted 1		Restricted 2	
θ_{ijk}	$\hat{\theta}^{L}_{ijk}$	$\hat{\theta}^{U}_{ijk}$	$\hat{\hat{\theta}}_{ijk}$	$\mathrm{Var}(\theta_{ijk})$	$\hat{\theta}^{L}_{ijk}$	$\hat{\theta}^{U}_{ijk}$	$\hat{\theta}^{L}_{ijk}$	$\hat{\theta}^{U}_{ijk}$
$\hat{\theta}_{111}$	0	0.16	0.117	0.0010	0	0.00	0.176	0.176
$\hat{\theta}_{112}$	0.07	0.23	0.167	0.0014	0.23	0.23	0	0
$\hat{\theta}_{113}$	0.04	0.20	0.146	0.0013	0.20	0.20	0.219	0.219
$\hat{\theta}_{121}$	0	0.16	0.043	0.0010	0.16	0.16	0	0
$\hat{\theta}_{122}$	0	0.16	0.063	0.0014	0	0	0.195	0.195
$\hat{\theta}_{123}$	0	0.16	0.054	0.0013	0	0	0	0
$\hat{\theta}_{211}$	0	0.14	0.051	0.0008	0	0.14	0	0.14
$\hat{\theta}_{212}$	0	0.15	0.059	0.0010	0	0.15	0	0.15
$\hat{\theta}_{213}$	0	0.11	0.040	0.0006	0	0.11	0	0.11
$\hat{\theta}_{221}$	0	0.14	0.089	0.0008	0	0.14	0	0.14
$\hat{\theta}_{222}$	0.01	0.16	0.101	0.0009	0.01	0.16	0.01	0.16
$\hat{\theta}_{223}$	0	0.11	0.070	0.0006	0	0.11	0	0.11

(R2) Restricted 2. Let us now consider a different structural zero: $\theta_{112} = 0$.

- As before, parameters θ_{1jk} are no longer uncertain. However, they change dramatically. In fact, the imposed structural zero is outside the unrestricted likelihood ridge (i.e. it is not in the unrestricted bounds described in the columns $\hat{\theta}^{L}_{ijk}$ and $\hat{\theta}^{U}_{ijk}$ of the unrestricted case in Table 4.7). As a result, almost all the estimates $\hat{\theta}_{1jk}$ are outside the likelihood ridge. Now, these estimates are able to reproduce only the unrestricted ML estimate of $\theta_{1..}$.
- Again, the bounds for θ_{2jk} are unaffected by this structural zero.

As far as the variability of the marginal uncertainty densities is concerned (column $\mathrm{Var}(\theta_{ijk})$ in Table 4.7), it remains unchanged for the parameters θ_{2jk} under the restricted cases. However, these distributions do not show any variability for the parameters θ_{1jk} in the restricted cases.

As a matter of fact, structural zeros are a very powerful tool for making uncertainty decrease, at least on the relevant parameters. In this case, the effect of inequality constraints is still important, but less dramatic. Let us consider the following three cases (see Table 4.8):

(R3) $\theta_{111} \leq \theta_{112}$;

(R4) $\theta_{111} \geq \theta_{112}$;

(R5) $\theta_{111} \geq \theta_{211}$.

Table 4.8 Estimated bounds θ^{L}_{ijk} and θ^{U}_{ijk} when inequality constraints are imposed

	Restricted 3		Restricted 4		Restricted 5	
	$\hat{\theta}^{L}_{ijk}$	$\hat{\theta}^{U}_{ijk}$	$\hat{\theta}^{L}_{ijk}$	$\hat{\theta}^{U}_{ijk}$	$\hat{\theta}^{L}_{ijk}$	$\hat{\theta}^{U}_{ijk}$
θ_{111}	0	0.16	0.115	0.16	0.007	0.16
θ_{112}	0.116	0.230	0.070	0.160	0.07	0.230
θ_{113}	0.040	0.2	0.11	0.2	0.04	0.2
θ_{121}	0	0.159	0	0.045	0	0.153
θ_{122}	0	0.114	0.07	0.16	0	0.16
θ_{123}	0	0.16	0	0.09	0	0.16
θ_{211}	0	0.14	0	0.14	0	0.14
θ_{212}	0	0.15	0	0.15	0	0.15
θ_{213}	0	0.11	0	0.11	0	0.11
θ_{221}	0	0.14	0	0.14	0	0.14
θ_{222}	0.01	0.16	0.01	0.16	0.01	0.16
θ_{223}	0	0.11	0	0.11	0	0.11

The first inequality recalls the relation between θ_{111} and θ_{112} that can be observed in the unrestricted case of Table 4.7. The effect is actually rather small. The second inequality constraint (R4) has a very important effect. Almost all the parameters θ_{1jk}, $j = 1, 2$, $k = 1, 2, 3$, are associated with completely different intervals. Neither (R3) nor (R4) affects the other parameters θ_{2jk}. Finally, the last inequality (R5) does not show any significant change between the estimated upper and lower bounds. A more precise analysis considers the dispersion of each single parameter in its estimated uncertainty space. Figure 4.8 shows the estimated marginal uncertainty densities in the unrestricted and (R3), (R4), and (R5) cases. These pictures show that all the imposed inequality constraints have an effect, which is actually larger for (R4) and milder for (R5). This aspect is reinforced by the computation of the variance of θ_{111} in its uncertainty space with respect to the estimated uncertainty density $\hat{m}(\theta_{111})$ (Table 4.9). Although θ_{111} seems to belong to the same interval, the inequality in (R3) induces a reduction of the variability. This reduction is larger for case (R5), and dramatic for (R4).

Table 4.9 Estimated dispersion of θ_{111} in its uncertainty space, in the unrestricted case and when inequality constraints are imposed

	Unrestricted	Restricted 3	Restricted 4	Restricted 5
$\mathrm{Var}(\theta_{111})$	0.001 45	0.001 31	0.000 09	0.000 54

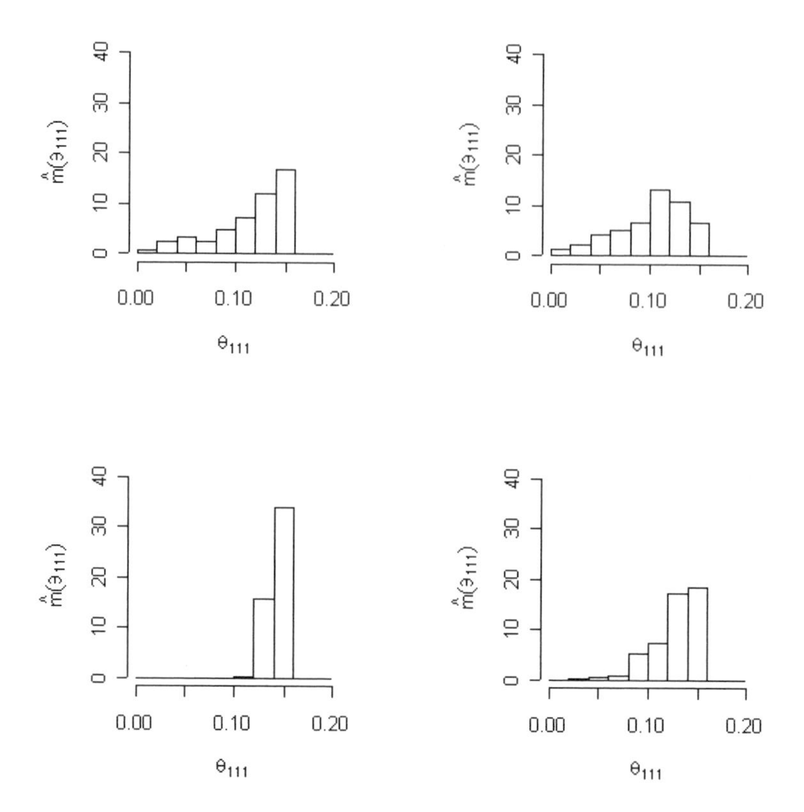

Figure 4.8 Estimated uncertainty densities for $m(\theta_{111})$ in the unconstrained (top left), restricted 3 (top right), restricted 4 (bottom left), restricted 5 (bottom right) cases. Θ^{SM} has been explored via the (constrained, whenever necessary) EM algorithm

4.6 Further Aspects of Maximum Likelihood Estimation of Uncertainty

Section 4.2 defines uncertainty as all the set of distributions Θ^{SM} compatible with complete knowledge of the bivariate r.v.s $(\mathbf{X}, \mathbf{Y})$ and $(\mathbf{X}, \mathbf{Z})$. Section 4.4 describes how to estimate this set by maximum likelihood given the observed data $A \cup B$. This estimate is the likelihood ridge $\hat{\Theta}^{\mathrm{SM}}$. The statistical properties of the likelihood ridge as an estimator of Θ^{SM} are of interest. However, such properties are nonstandard, given that $\hat{\Theta}^{\mathrm{SM}}$ is a random set.

One important characteristic of the likelihood ridge is that it is defined by the ML estimator of the estimable parameters:

$$\hat{\Theta}^{\mathrm{SM}} = \mathcal{P}_{\hat{\boldsymbol{\theta}}_{\mathbf{X}} \hat{\boldsymbol{\theta}}_{\mathbf{Y}|\mathbf{X}} \hat{\boldsymbol{\theta}}_{\mathbf{Z}|\mathbf{X}}}.$$

Let $\hat{\boldsymbol{\theta}}_{\mathbf{X}}$, $\hat{\boldsymbol{\theta}}_{\mathbf{Y}|\mathbf{X}}$, and $\hat{\boldsymbol{\theta}}_{\mathbf{Z}|\mathbf{X}}$ be consistent estimators of $\boldsymbol{\theta}_{\mathbf{X}}$, $\boldsymbol{\theta}_{\mathbf{Y}|\mathbf{X}}$, and $\boldsymbol{\theta}_{\mathbf{Z}|\mathbf{X}}$. Then, almost surely,

$$\mathcal{P}_{\lim \hat{\boldsymbol{\theta}}_{\mathbf{X}} \hat{\boldsymbol{\theta}}_{\mathbf{Y}|\mathbf{X}} \hat{\boldsymbol{\theta}}_{\mathbf{Z}|\mathbf{X}}} = \Theta^{\mathrm{SM}},$$

where the limit is with respect to n_A and n_B to $+\infty$. The previous property is not sufficient for establishing consistency of the random set $\mathcal{P}_{\hat{\boldsymbol{\theta}}_{\mathbf{X}} \hat{\boldsymbol{\theta}}_{\mathbf{Y}|\mathbf{X}} \hat{\boldsymbol{\theta}}_{\mathbf{Z}|\mathbf{X}}}$. A sufficient condition for $\hat{\Theta}^{\mathrm{SM}}$ to be a consistent estimator of Θ^{SM}, i.e.

$$\mathcal{P}_{\hat{\boldsymbol{\theta}}_{\mathbf{X}} \hat{\boldsymbol{\theta}}_{\mathbf{Y}|\mathbf{X}} \hat{\boldsymbol{\theta}}_{\mathbf{Z}|\mathbf{X}}} \xrightarrow{\text{a.s.}} \Theta^{\mathrm{SM}},$$

is that the bounds θ_t^{L} and θ_t^{U}, $t = 1, \ldots, T$, are continuous functions of the estimable parameters $\boldsymbol{\theta}_{\mathbf{X}}$, $\boldsymbol{\theta}_{\mathbf{Y}|\mathbf{X}}$, and $\boldsymbol{\theta}_{\mathbf{Z}|\mathbf{X}}$.

Remark 4.13 Remark 4.4 claims that the parameter spaces of both the multinormal and multinomial distributions are normal spaces. Furthermore, it can be proved that the functions $a_t(.)$ and $b_t(.)$ in (4.27) are continuous functions of their arguments. Hence, the estimator $\hat{\Theta}^{\mathrm{SM}}$ is a consistent estimator of Θ^{SM} for both the multinormal and the multinomial cases.

What can be said about finite sample properties? In other words, what are the sampling properties of the estimator likelihood ridge? Variability of the estimator of uncertainty is a very important aspect. In fact, variability indicates how much can we trust the estimates of the uncertainty measures defined in Section 4.4. The notion of variability of the estimator of an uncertainty measure can be defined as in Kenward *et al.* (2001). In that case, uncertainty estimates are inflated in order to take into account sample variability by computing confidence intervals of the extremes of the estimate of what they call the 'uncertain region'. This procedure may be appropriate when uncertainty is defined in terms of the width of the intervals $[\theta_t^{\mathrm{L}}, \theta_t^{\mathrm{U}}]$, $t = 1, \ldots, T$. Further research should be devoted to these aspects.

Example 4.12 Let A and B be two samples drawn from an r.v. (X, Y, Z) distributed as a multinormal r.v. with mean vector

$$\boldsymbol{\mu} = \begin{pmatrix} 1 \\ 1 \\ 1 \end{pmatrix}$$

and covariance matrix

$$\boldsymbol{\Sigma} = \boldsymbol{\rho} = \begin{pmatrix} 1 & 0.66 & 0.63 \\ 0.66 & 1 & -0.10 \\ 0.63 & -0.10 & 1 \end{pmatrix}.$$

If all the parameters but ρ_{YZ} were known, as in Section 4.3.1, then ρ_{YZ} would assume values only in the interval bounded by (4.41) and (4.42):

$$\rho_{YZ}^{\mathrm{L}} = -0.1676, \qquad \rho_{YZ}^{\mathrm{U}} = 0.9992.$$

Figures 4.9–4.12 show the estimates $\hat{\rho}^{\mathrm{L}}_{YZ}$ and $\hat{\rho}^{\mathrm{U}}_{YZ}$ for 100 samples drawn independently in four distinct cases: $n_A = n_B = 10$; $n_A = n_B = 50$; $n_A = n_B = 200$; $n_A = n_B = 1000$. As expected, the larger the sample size, the more stable are the estimates of the estimable parameters. Furthermore, these estimates tend to converge to the highlighted points ρ^{L}_{YZ} and ρ^{U}_{YZ}. This is an effect of the consistency of $\hat{\Theta}^{\mathrm{SM}}$.

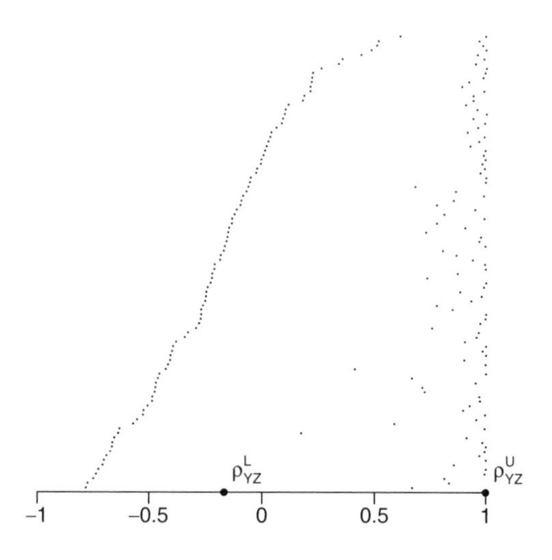

Figure 4.9 True (ρ^{L}_{YZ} and ρ^{U}_{YZ}) and estimated ($\hat{\rho}^{\mathrm{L}}_{YZ}$ and $\hat{\rho}^{\mathrm{U}}_{YZ}$) bounds for 100 samples A and B with $n_A = n_B = 10$

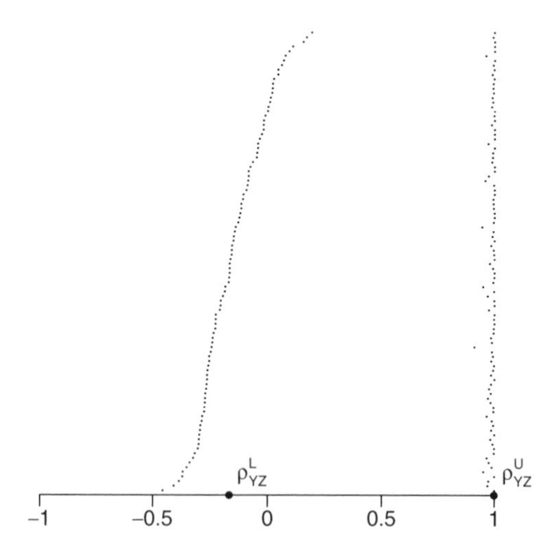

Figure 4.10 True (ρ^{L}_{YZ} and ρ^{U}_{YZ}) and estimated ($\hat{\rho}^{\mathrm{L}}_{YZ}$ and $\hat{\rho}^{\mathrm{U}}_{YZ}$) bounds for 100 samples A and B with $n_A = n_B = 50$

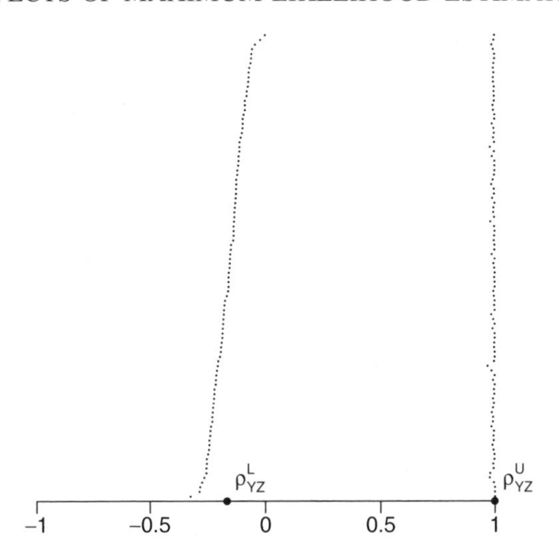

Figure 4.11 True (ρ^{L}_{YZ} and ρ^{U}_{YZ}) and estimated ($\hat{\rho}^{L}_{YZ}$ and $\hat{\rho}^{U}_{YZ}$) bounds for 100 samples A and B with $n_A = n_B = 200$

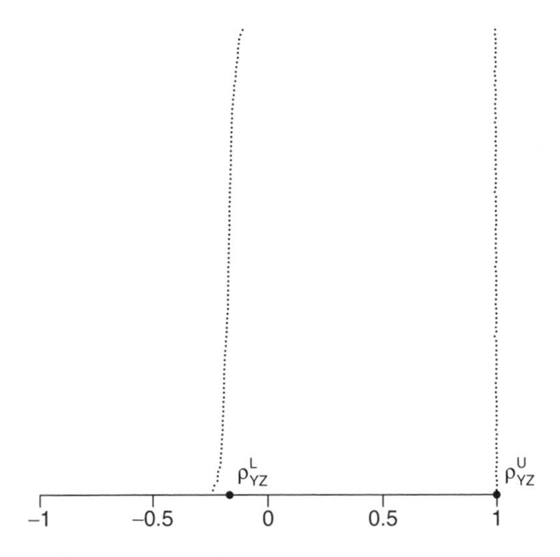

Figure 4.12 True (ρ^{L}_{YZ} and ρ^{U}_{YZ}) and estimated ($\hat{\rho}^{L}_{YZ}$ and $\hat{\rho}^{U}_{YZ}$) bounds for 100 samples A and B with $n_A = n_B = 1000$

One might wonder whether the study of uncertainty by means of the likelihood ridge can also be interpreted under the Bayesian paradigm. Actually, the likelihood ridge coincides with the maximum posterior density space when a flat (whenever appropriate, improper) prior distribution on $\boldsymbol{\theta}$ is imposed. However, there are no other similarities. In particular, the uncertainty distribution $\hat{m}(\boldsymbol{\theta})$ cannot be seen as a posterior distribution of $\boldsymbol{\theta}$, given the sample $A \cup B$, in a rigorous Bayesian framework. In fact, for $\hat{m}(\boldsymbol{\theta})$ to be a posterior the prior distribution on $\boldsymbol{\theta}$ should be flat for those parameters in the likelihood ridge and null outside, i.e.

$$\pi(\boldsymbol{\theta}) = \begin{cases} c & \boldsymbol{\theta} \in \hat{\Theta}^{\mathrm{SM}}, \\ 0 & \text{otherwise.} \end{cases}$$

This prior distribution is *data dependent*: it should assume with probability one the ML estimates of the estimable parameters, and should be flat for the inestimable ones. A data dependent prior is difficult to justify in a rigorous Bayesian framework. This concept belongs to the *objective Bayesian* (Williamson, 2004) framework, and it has not yet been analysed in the context of statistical matching.

4.7 An Example with Real Data

In D'Orazio *et al.* (2005b) an example involving real socio-economic variables is discussed in detail. In this example the variables 'Age' (AGE), 'Educational Level' (EDU) and 'Professional Status' (PRO) observed on 2313 employees (aged at least 15 years) were extracted from the 2000 pilot survey of the Italian Population and Households Census. To reproduce the statistical matching problem, the original file was randomly split into two almost equal subsets. The variable 'Educational Level' was removed from the second subset (file B), containing 1148 units, and the variable 'Professional Status' was removed from the first subset (file A), consisting of the remaining 1165 observations. Hence, 'Age' plays the role of X, 'Educational Level' plays the role of Y, and 'Professional Status' plays the role of Z. The categories of the variables are shown in Table 4.10.

A characteristic of the previous three variables is that they admit some parameter constraints.

(a) Structural zeros. Some structural zeros are induced by the observed tables: for example, in Italy a 17-year-old person cannot have a university degree:

$$\theta_{1\mathrm{D}k} = 0, \qquad k = \mathrm{M, E, W}.$$

Table 4.10 Response categories for the variables considered in the example

Variables	Transformed response categories
AGE	1 = 15–17 years old; 2 = 18–22; 3 = 23–64; 4 = 65 and older
EDU	C = None or compulsory school; V = Vocational school;
	S = Secondary school; D = Degree
PRO	M = Manager; E = Clerk; W = Worker

Some other structural zeros are induced by the unobserved table (Y, Z): for example, the categories corresponding to managers (PRO = M) with at most a compulsory school educational level (EDU = C) should all be set to zero:

$$\theta_{i\mathrm{CM}} = 0, \qquad i = 1, 2, 3.$$

(b) Inequality constraints. For instance, in the reference population, managers aged 23–64 with a degree are more frequent than clerks of the same age and educational level:

$$\theta_{3\mathrm{DM}} \geq \theta_{3\mathrm{DE}}.$$

As a general result, the experiments in D'Orazio *et al.* (2005b) show that estimates under the CIA fail to respect some (but not all) structural zeros. In particular, the structural zeros not respected are those related to the variables not jointly observed, i.e. structural zeros for (PRO, EDU). On the other hand, all the other structural zeros are inferred by the zeros appearing in the observed tables A and B.

This example shows also how uncertainty is reduced through the use of the available constraints. In particular, the likelihood ridge is studied under the following three scenarios:

S0 unrestricted likelihood ridge;

S1 restricted likelihood ridge imposing only structural zeros in (a);

S2 restricted likelihood ridge imposing both structural zeros in (a) and the inequality constraint in (b).

As expected, when no restrictions are imposed on the starting point (S0), the results of the EM algorithm produce a variety of estimates corresponding to the structural zeros: they represent the unconstrained likelihood ridge. In this case, the (nonnull) ML estimates of the parameters under the CIA are always included in the interval found through EM. On the other hand, when structural zeros are introduced as starting points of the EM algorithm (S1), ML estimates are forced to be null corresponding to the structural zeros.

In general, the different strategies produce the following result.

- The widths of the uncertainty intervals of all the parameters under S0 reduce when strategy S2 is adopted.

- About a half of the parameters have the width of the uncertainty intervals determined under S1 reduced when strategy S2 is adopted. The other parameters are unaffected.

Since the width is not the only element to consider as an evaluation measure of the uncertainty on the parameters, the marginal uncertainty densities of each single parameter were computed. These densities were computed with the algorithms described respectively in Algorithm 4.1 (for S0) and in Section 4.5 (for S1 and S2). These algorithms were run 100 000 times each. Estimates of $m(\theta_{ijk})$ were computed on the subset of the distinct results (replications of the same $\boldsymbol{\theta}$ among the 100 000 results are discarded).

D'Orazio *et al.* (2005b) already showed for two particular cells (respectively AGE = 3, EDU = D, PRO = M, and AGE = 3, EDU = S, PRO = E) that the marginal uncertainty distributions are less dispersed when passing from strategy S0 to S1 and S2. In particular, this result also holds for those cells where the width of the interval did not change from S1 to S2, i.e. the reduction of uncertainty is due to the reduction of the variability of the corresponding uncertainty distribution. Here, other estimated marginal uncertainty densities are reported.

Figure 4.13 describes the estimated marginal uncertainty densities for $\theta_{3\mathrm{SW}}$ for the three different strategies. It clearly shows that the imposition of logical constraints induces a dramatic decrease in the width of the uncertainty space and also in the overall dispersion of the marginal uncertainty distribution. Figure 4.14, which refers to $\theta_{2\mathrm{SW}}$, shows how the constraints move part of the uncertainty density mass towards the true value (the dashed line). Finally, Figure 4.15, which refers to $\theta_{3\mathrm{DE}}$, shows that there can be cases where we need something more than structural zeros in order to reduce uncertainty. Actually, the ML estimates of the estimable parameters in S1 move the mass of the estimated uncertainty density in the wrong direction. This situation is recovered when inequality constraints are also imposed.

A further reduction of uncertainty can be obtained by trimming the extremes of the marginal uncertainty densities, i.e. those ML estimates which are quite rare in

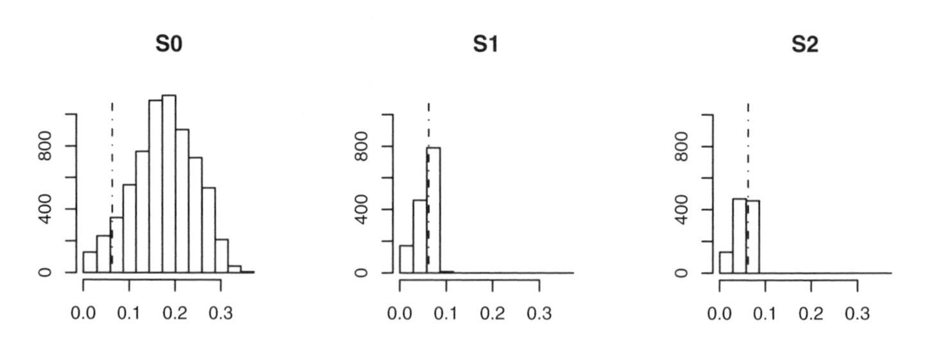

Figure 4.13 Marginal uncertainty densities under the S0 (left), S1 (centre) and S2 (right) strategies for the parameter $\theta_{3\mathrm{SW}}$

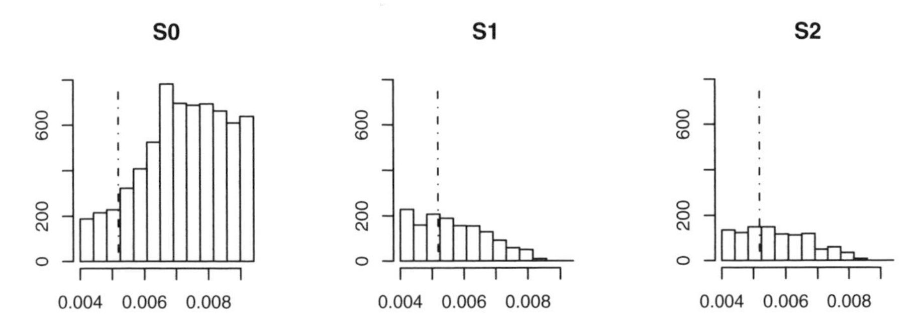

Figure 4.14 Marginal uncertainty densities under the S0 (left), S1 (centre) and S2 (right) strategies for the parameter θ_{2SW}

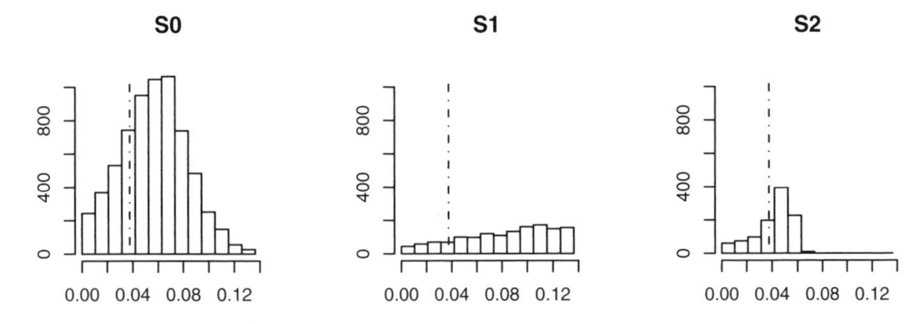

Figure 4.15 Marginal uncertainty densities under the S0 (left), S1 (centre) and S2 (right) strategies for the parameter θ_{3DE}

the likelihood ridge, as already introduced in Sections 4.3 and 4.5.1. For example, when 5% of the ML estimates in the tails of the uncertainty density of θ_{3SE} are trimmed, the resulting range is reduced by about 50% (Table 4.11). Obviously, this operation is performed on each single parameter without regard to what happens to other cells. In fact, a distribution with an extreme parameter in one cell does not necessarily have extreme values for other cells.

Note that logical constraints can be also useful for improving the ML estimators of estimable parameters, in this example the parameters $\theta_{ij.}$ for (AGE, PRO) and $\theta_{i.k}$ for (AGE, EDU). In fact, the frequencies of the marginal tables A and B that are not compatible with the restricted space induced by the constraints, are necessarily changed. For instance, D'Orazio *et al.* (2005b) show how the incompatible estimates move towards the true parameters when passing from S0 to S1.

Table 4.11 True probabilities (θ), extremes of the likelihood ridge estimates (min and max), extremes of the 95% trimmed interval (95% lv and 95% uv), gain of the interval (95% lv, 95% uv) with respect to the interval (min, max), average values in the ridge ($\bar{\theta}$) for the S2 case. All the values are computed over the 100 000 runs of EM

ijk	θ_{ijk}	Min	Max	95% lv	95% uv	Gain	$\bar{\theta}$
3SM	0.0614	0.0186	0.1290	0.0446	0.1174	0.3412	0.0850
4SM	0.0017	0.0024	0.0031	0.0024	0.0031	0.0729	0.0027
3DM	0.0951	0.0260	0.1364	0.0376	0.1104	0.3412	0.0700
4DM	0.0022	0.0013	0.0021	0.0014	0.0021	0.0729	0.0018
2VE	0.0004	0.0000	0.0043	0.0000	0.0037	0.1486	0.0017
3VE	0.0532	0.0015	0.0881	0.0413	0.0881	0.4599	0.0730
4VE	0	0	0	0	0		0
2SE	0.0035	0.0011	0.0054	0.0017	0.0054	0.1486	0.0037
3SE	0.2823	0.2279	0.3780	0.2345	0.3099	0.4974	0.2698
4SE	0.0013	0.0000	0.0007	0.0001	0.0007	0.0729	0.0004
3DE	0.0376	0.0000	0.0678	0.0097	0.0640	0.1987	0.0408
4DE	0	0.0000	0.0007	0.0000	0.0007	0.0729	0.0003
1CW	0.0065	0.0065	0.0065	0.0065	0.0065		0.0065
2CW	0.0117	0.0101	0.0101	0.0101	0.0101		0.0101
3CW	0.3281	0.3342	0.3342	0.3342	0.3342		0.3342
4CW	0.0052	0.0052	0.0052	0.0052	0.0052		0.0052
1VW	0	0	0	0	0		0
2VW	0.0030	0.0000	0.0043	0.0006	0.0043	0.1486	0.0026
3VW	0.0389	0.0000	0.0865	0.0000	0.0467	0.4599	0.0150
4VW	0	0	0	0	0		0
2SW	0.0052	0.0040	0.0083	0.0040	0.0076	0.1486	0.0057
3SW	0.0618	0.0000	0.0866	0.0062	0.0813	0.1331	0.0459
4SW	0	0	0	0	0		0
3DW	0.0009	0.0000	0.0859	0.0000	0.0590	0.3134	0.0257
4DW	0	0	0	0	0		0

4.8 Other Approaches to the Assessment of Uncertainty

So far, uncertainty has been investigated under the likelihood principle. Other different approaches to uncertainty assessment have been developed. In this section, these approaches are briefly introduced, leaving details to the relevant literature.

4.8.1 The consistent approach

Kadane (1978) first introduced the notion of uncertainty in the statistical matching problem. In contrast to the available literature on statistical matching at that time (see references in Kadane, 1978), he claimed that some parameters in the statistical matching problem are inestimable. In particular, in the normal case, all parameters can be estimated *consistently* but $\mathbf{\Sigma_{YZ}}$. For the latter there is the need to find an interval compatible with consistent estimates of the other parameters:

$$\hat{\Theta}^{\mathrm{SM}} = \left\{ \mathbf{\Sigma_{YZ}} : \begin{vmatrix} \tilde{\mathbf{\Sigma}}_{\mathbf{XX}} & \tilde{\mathbf{\Sigma}}_{\mathbf{XY}} & \tilde{\mathbf{\Sigma}}_{\mathbf{XZ}} \\ \tilde{\mathbf{\Sigma}}_{\mathbf{YX}} & \tilde{\mathbf{\Sigma}}_{\mathbf{YY}} & \mathbf{\Sigma}_{\mathbf{YZ}} \\ \tilde{\mathbf{\Sigma}}_{\mathbf{ZX}} & \mathbf{\Sigma}_{\mathbf{ZY}} & \tilde{\mathbf{\Sigma}}_{\mathbf{ZZ}} \end{vmatrix} \geq 0 \right\}, \tag{4.61}$$

where $\tilde{\mathbf{\Sigma}}$ denotes a covariance matrix estimated consistently in $A \cup B$. Actually, this is what was studied in Section 4.4 for general distributions: the estimable parameters were estimated via ML estimators ($\tilde{\mathbf{\Sigma}} = \hat{\mathbf{\Sigma}}$). Furthermore, the notion of the uncertainty distribution and the study of the reduction of the uncertainty with the help of constraints were introduced.

In order to obtain completed synthetic data sets according to the different plausible values of $\mathbf{\Sigma_{YZ}}$, Kadane suggests using a mixed procedure such as MM5* in Section 3.6.1. This is done for various $\mathbf{\Sigma_{YZ}}$ in the set (4.61).

The results in Kadane (1978) are the basis for other papers on statistical matching. Particularly important are those by Moriarity and Scheuren (2001, 2003, 2004). They also consider only the normal case, and the estimable parameters are again estimated consistently. In particular, they use the unbiased estimators in the complete data subsets:

$$\tilde{\mathbf{\Sigma}}_{\mathbf{XX}} = S_{\mathbf{XX};A\cup B},\ \tilde{\mathbf{\Sigma}}_{\mathbf{YY}} = S_{\mathbf{YY};A},\ \tilde{\mathbf{\Sigma}}_{\mathbf{ZZ}} = S_{\mathbf{ZZ};B},$$
$$\tilde{\mathbf{\Sigma}}_{\mathbf{XY}} = S_{\mathbf{XY};A},\ \tilde{\mathbf{\Sigma}}_{\mathbf{XZ}} = S_{\mathbf{XZ};B},$$

as already shown in Remarks 2.5 and 3.3. The estimated set (4.61) of plausible covariance matrices $\mathbf{\Sigma_{YZ}}$ is still a consistent estimator of Θ^{SM}. However, the use of ML estimates computed on the overall sample $A \cup B$ seems to have an important property: these estimates are coherent with the estimated regression parameters of $\mathbf{Y}$ on $(\mathbf{X}, \mathbf{Z})$ and of $\mathbf{Z}$ on $(\mathbf{X}, \mathbf{Y})$, because they are actually computed *after* the regression parameters have been estimated under the CIA (see Section 4.4). This is not true in the results obtained by Moriarity and Scheuren: in fact, their residual covariance matrix with respect to regression, used for the random imputation of the missing items in $A \cup B$, happens to be negative definite. In those cases, Moriarity and Scheuren (2004) suggest using the RIEPS method (Rässler, 2002; see also Remark 3.2), although they point out some of its weaknesses.

4.8.2 The multiple imputation approach

Rubin (1986, 1987) is the first to propose the application of multiple imputation techniques to statistical matching. The main aim is to carry out a sensitivity analysis with respect to different assumptions on the conditional association parameters in the multinormal setting, in particular $\boldsymbol{\rho}_{\mathbf{YZ}|\mathbf{X}}$.

Later Rässler (2002, 2003, 2004) extended multiple imputation to statistical matching in order to assess uncertainty of the matching procedure. Rässler's objective was to find the upper ($\boldsymbol{\rho}^{\mathrm{U}}_{\mathbf{YZ}}$) and lower ($\boldsymbol{\rho}^{\mathrm{L}}_{\mathbf{YZ}}$) bounds for the unconditional association parameters that, apart from some simple cases where they are analytically tractable, are not easily computable. Multiple imputation consists in imputing the missing values m times, thus obtaining m complete imputed data sets (see Rubin 1987, for more details). Inferences on each complete single data set can be performed. Then, a final estimate and the corresponding estimator variance (including the part due to the imputation mechanism) can be obtained, combining the results according to the following rules. Let $\tilde{\theta}^{(t)}$ be the value of the estimator of a parameter θ on one of the completed data sets $A \cup B$. Then the multiple imputation point estimate for θ is

$$\bar{\theta} = \frac{1}{m}\sum_{t=1}^{m}\tilde{\theta}^{(t)}. \tag{4.62}$$

Let $\widetilde{\mathrm{Var}}(\tilde{\theta})$ be the estimate of the $\tilde{\theta}$ variance. It can be proved that this estimated variance has the form:

$$\widetilde{\mathrm{Var}}(\tilde{\theta}) = WV + \frac{m+1}{m}BV,$$

where BV represents the between imputation variance

$$BV = \frac{1}{m-1}\sum_{t=1}^{m}(\tilde{\theta}^{(t)} - \bar{\theta})^2,$$

while WV is the within imputation variance

$$WV = \frac{1}{m}\sum_{t=1}^{m}\widetilde{\mathrm{Var}}(\tilde{\theta}^{(t)}).$$

Imputations must be proper, as defined in Rubin (1987). It is often difficult to define if an imputation procedure is proper. As introduced in Schafer (1997), in a Bayesian context an imputation procedure based on independent draws from the posterior predictive distribution $f_{\tilde{\mathbf{U}}|\mathbf{U}}(\mathbf{u}|\mathbf{v})$ (as defined in (2.57)) satisfies most of the conditions of a proper imputation procedure.

Rässler describes a procedure, based on multiple imputation, for computing the upper ($\boldsymbol{\rho}^{\mathrm{U}}_{\mathbf{YZ}}$) and lower ($\boldsymbol{\rho}^{\mathrm{L}}_{\mathbf{YZ}}$) bounds for the unconditional association parameters. This procedure consists of the following steps.

- Assign a plausible initial $Q \times R$ matrix $\boldsymbol{\rho}_{\mathbf{YZ}|\mathbf{X}}$.
- Complete the concatenated data set $A \cup B$ m times by generating m independent values for each missing item in the $n_A + n_B$ records from the predictive distribution $f_{(\mathbf{X},\mathbf{Y},\mathbf{Z})_{\mathrm{mis}}|(\mathbf{X},\mathbf{Y},\mathbf{Z})_{\mathrm{obs}}}(\mathbf{u}|\mathbf{v})$.

- Estimate $\boldsymbol{\rho}_{\mathbf{YZ}}$ with (4.62) applied to the m complete data sets.
- Repeat the above three steps k times using k different values assigned to the initial matrix $\boldsymbol{\rho}_{\mathbf{YZ|X}}$, in order to find the bounds of the unconditional correlation $\boldsymbol{\rho}_{\mathbf{YZ}}$.

As suggested in Rässler (2003), the k values of the last step can be either arbitrarily assigned, or drawn randomly from the prior distribution of $\boldsymbol{\rho}_{\mathbf{YZ}}$ such as the uniform on the hypercube $[-1, 1]^{QR}$, where Q is the dimension of $\mathbf{Y}$ and R is the dimension of $\mathbf{Z}$. In this last case, the values drawn may be incompatible with the other estimated parameters. A first remark concerns the estimation procedures previously described. They are carried out on the Fisher z-transformation in order to satisfy the normality hypothesis of the estimator assumed by the multiple imputation procedures; for details, see Rässler (2002).

In addition to the general multiple imputation algorithms such as NORM (Schafer, 1997) based on data augmentation (see Section A.1.2) and MICE (Van Buuren and Oudshoorn, 1999), an algorithm to perform multiple imputation in the statistical matching framework is introduced in Rässler (2002). It is named NIBAS (Non-Iterative Bayesian Approach to Statistical Matching) and is essentially based on the model described in Sections 2.7 and 3.8, i.e. the m imputations are random draws from the predictive distribution as described in the last two steps of Example 2.11. This model exploits the special pattern of missing data of statistical matching problems, which allows the direct estimation of the observed-data posterior distribution $\pi(\boldsymbol{\theta}|(\mathbf{x}, \mathbf{y}, \mathbf{z})_{\text{obs}})$, thus avoiding the use of iterative algorithms such as data augmentation; see Section A.1.2 for further details.

A final remark concerns the computational burden of multiple imputation for the study of uncertainty in statistical matching. In multiple imputation, a small m is generally considered sufficient to draw inferences for the parameters of interest. However, this is a cause of an increase in the computational burden. In fact, in statistical matching, m should be considered jointly with the number of iterations needed to study uncertainty, thus leading to $m \times k$ iterations.

The general measure for assessing uncertainty proposed by Rässler in this context is

$$\frac{1}{2QR} \sum_{q=1}^{Q} \sum_{r=1}^{R} \left(\tilde{\rho}^{\text{U}}_{Y_q Z_r} - \tilde{\rho}^{\text{L}}_{Y_q Z_r} \right).$$

The study of uncertainty through multiple imputation is usually justified by the following issues.

(a) It is usually considered as a micro approach. In fact, it aims at the construction of m synthetic data sets.

(b) Since it is based on the predictive distribution (2.54), it seems that uncertainty analysis takes into account variability of the r.v. $\boldsymbol{\theta}$ through the posterior distribution $\pi(\boldsymbol{\theta}|A \cup B)$.

These points are discussed more in detail in the rest of this subsection.

Multiple imputation as a micro approach

Imputing a partially observed data set m times, computing the usual estimators of $\boldsymbol{\theta}$ on the m completed data sets and computing a final estimate (4.62) as the average of the m estimates has the objective of approximating the expectation of the usual estimator on the completed data set with respect to the predictive distribution of the missing variables given the observed ones. Hence, letting m go to infinity, the final estimates (4.62) should converge to (using the notation in Rubin 1987)

$$E\left[Q\left((\mathbf{x},\mathbf{y},\mathbf{z})_{\mathrm{mis}}\right)|(\mathbf{x},\mathbf{y},\mathbf{z})_{\mathrm{obs}}\right], \tag{4.63}$$

computed with respect to the predictive distribution of (2.54), where Q assumes different forms according to the parameter of interest: for $\boldsymbol{\mu}$ it is the sample average of each single variable, for $\boldsymbol{\Sigma}$ it is the sample variance or covariance, or similar estimators, each computed on the completed data sets and transformed so that the estimator can be considered as normally distributed. Hence, the micro approach is not the real objective of multiple imputation. It is simultaneously a device for computing the expectations (4.63) and giving information on the reliability of the estimates (4.62).

Multiple imputation, uncertainty and variability

For the sake of simplicity, let us consider the normal $(X, Y, \mathbf{Z})$ case, with $\mathbf{Z}$ bivariate, as in Example 4.6. In this case, uncertainty of ρ_{YZ_1} and ρ_{YZ_2} via the ML approach is described by all the points in an ellipse as in Figure 4.2, where the estimable parameters are fixed at their ML estimates.

In the multiple imputation approach the result is similar. Again, let m diverge. According to the multiple imputation procedures, the average of the estimates of the m completed data sets of the estimable parameters (means, variances, almost all correlations with the exception of ρ_{YZ_1} and ρ_{YZ_2}) converges to the value given by (4.63). The random generation of $\rho_{YZ_1|X}$ and $\rho_{YZ_2|X}$ from a uniform distribution in the square with edges between $[-1, 1]$ has the following aim: to find all the parameter values for the couple ρ_{YZ_1} and ρ_{YZ_2} compatible with the estimable parameters fixed at the expectations (4.63). As a result, when m goes to infinity, the result of the multiple imputation approach is again an uncertain region given by an ellipse like the one in Figure 4.2, but with the estimable parameters fixed at different values (the ML estimates and the expectations defined by (4.63) usually do not coincide).

Hence, the uncertain region has similar properties to that defined via the ML approach, although justified under the Bayesian paradigm. Note that the ellipse is composed of all those points corresponding to $\tilde{\theta}$ of (4.62), i.e. $\tilde{\rho}_{YZ_1}$ and $\tilde{\rho}_{YZ_2}$. Each of these has associated a within (BW) and between (BV) variance. It is still an open question how to combine the variances associated with each point in the ellipse in order to understand the variability associated with the multiple imputation uncertainty region, as it still is an open question under the ML approach. Rässler (2002) uses these variances to compare the estimates obtained by different multiple imputation procedures with respect to each single value postulated for $\boldsymbol{\rho}_{\mathbf{YZ|X}}$. The ranges of the parameters, which for a finite m are just a blurred image of the ranges defined by the ellipse, are computed on the finite m estimates $\tilde{\rho}_{YZ_1}$ and $\tilde{\rho}_{YZ_2}$. Further studies for both the ML and multiple imputation approaches are necessary.

4.8.3 The de Finetti coherence approach

A different approach to statistical matching can be defined by means of the concept of *de Finetti coherence*; see de Finetti (1974) and Coletti and Scozzafava (2002).

Usually classic probability theory refers to a Kolmogorovian setting. Thus it requires the specification of probabilities (say, Q) for all the events (Λ) in the σ-algebra ($\mathcal{A}$) on which the probability measure is defined. In some situations one may be forced to work with partial assignments of probabilities because of partial knowledge of the phenomenon. In such cases, the collection of all events for which probabilities are known need not have any algebraic structure (i.e. does not form an algebra or a σ-algebra). The question is whether there is a probability space (Λ, $\mathcal{A}$, Q) such that $\mathcal{A}$ contains all the events of interest and Q assigns probabilities to them equal to the partial assignments. The de Finetti coherence allows one to work in this context. Roughly, it is a syntactic approach for dealing with partial probability assessments (it also has a semantic interpretation based on a betting scheme).

More formally, the coherence can be defined, for the more general case of a conditional probability distribution, as follows (de Finetti, 1974).

Definition 4.1 *Given an arbitrary set $\mathcal{F}$ of conditional events, a real function $P(a|c)$ with $(a|c) \in \mathcal{F}$ is a coherent conditional probability assignment if there exists $\mathcal{G} \supseteq \mathcal{F}$ with $\mathcal{G} = \mathcal{B} \times \mathcal{H}$ (where $\mathcal{B}$ is an algebra, and $\mathcal{H}$ is an additive set with $\mathcal{H} \subseteq \mathcal{B}$, and the impossible event $\phi \notin \mathcal{H}$) such that there exists a conditional probability $P'(b|h)$ on $\mathcal{G}$ (i.e. $(b|h) \in \mathcal{B} \times \mathcal{H}$) extending P.*

It is worth mentioning that the conditional probability in Definition 4.1 is in the sense of de Finetti (1974) and Dubins (1975). Differently from the classic definition, it allows one to deal with zero probability events, in addition to structural zeros. Actually, as shown in Coletti and Scozzafava (2002), coherence of an assignment can be checked by solving linear optimization problems consisting of the construction of all the possible compatible distributions. In other words, it aims to check/look for probability distributions agreeing with the given assessments.

Vantaggi (2005) claims that this paradigm is particularly useful for the statistical matching problem, which is based on the partial conditional probability assessments suggested by $A \cup B$. It is particularly effective when constraints, such as structural zeros, are assumed. The following results are based on Vantaggi (2005), and assume that X, Y and Z are categorical r.v.s.

Let the partial conditional assignments $(\tilde{\boldsymbol{\theta}}_X, \tilde{\boldsymbol{\theta}}_{Y|X}, \tilde{\boldsymbol{\theta}}_{Z|X})$ be an estimate of $(\boldsymbol{\theta}_X, \boldsymbol{\theta}_{Y|X}, \boldsymbol{\theta}_{Z|X})$ by an appropriate paradigm (e.g. ML estimation as in Section 2.1.3). This estimate is always coherent in the sense of de Finetti if no structural zeros are imposed. In other words, many probability distributions are associated with the estimate $(\tilde{\boldsymbol{\theta}}_X, \tilde{\boldsymbol{\theta}}_{Y|X}, \tilde{\boldsymbol{\theta}}_{Z|X})$. Note that among all these distributions there is the one under the CIA.

When structural zeros (of the form $\theta_{ijk} = 0$ or $\theta_{.jk} = 0$) are imposed, it is no longer true that any estimate $(\tilde{\boldsymbol{\theta}}_X, \tilde{\boldsymbol{\theta}}_{Y|X}, \tilde{\boldsymbol{\theta}}_{Z|X})$ is coherent. More precisely, the pair $(\boldsymbol{\theta}_{Y|X}, \boldsymbol{\theta}_{Z|X})$ may be problematic. Incoherent combinations can be found by means of the algorithm in Capotorti and Vantaggi (2002). This algorithm finds the minimal subset of indices $\mathcal{I}$ such that the corresponding parameters are incompatible with the imposed constraints. Let

$$\mathcal{E} = \left\{\tilde{\theta}_{j|i}, \tilde{\theta}_{k|i} \text{ for } (i, j, k) \in \mathcal{I}\right\} \tag{4.64}$$

be the minimal set of incompatible parameters, given the imposed structural zeros, for the estimate $(\tilde{\boldsymbol{\theta}}_X, \tilde{\boldsymbol{\theta}}_{Y|X}, \tilde{\boldsymbol{\theta}}_{Z|X})$.

The parameters that do not belong to the set $\mathcal{E}$ can be used to determine the subsystem of equations:

$$\begin{cases} \sum_{k=1}^{K} \theta_{ijk} = \tilde{\theta}_{j|i}\tilde{\theta}_{i..}, & \tilde{\theta}_{j|i} \notin \mathcal{E}, \\ \sum_{j=1}^{J} \theta_{ijk} = \tilde{\theta}_{k|i}\tilde{\theta}_{i..}, & \tilde{\theta}_{k|i} \notin \mathcal{E}, \\ \sum_{j=1}^{J}\sum_{k=1}^{K} \theta_{ijk} = \tilde{\theta}_{i..}, & i = 1, \ldots, I, \\ \boldsymbol{\theta} \in \Omega. \end{cases} \tag{4.65}$$

This system allows one to find all the values of $\boldsymbol{\theta}$ compatible with the imposed structural zeros and the values of the parameters of $(\tilde{\boldsymbol{\theta}}_X, \tilde{\boldsymbol{\theta}}_{Y|X}, \tilde{\boldsymbol{\theta}}_{Z|X})$ which are not in $\mathcal{E}$.

Let $(\tilde{\boldsymbol{\theta}}_X, \tilde{\boldsymbol{\theta}}_{Y|X}, \tilde{\boldsymbol{\theta}}_{Z|X}) = (\hat{\boldsymbol{\theta}}_X, \hat{\boldsymbol{\theta}}_{Y|X}, \hat{\boldsymbol{\theta}}_{Z|X})$, obtained via the ML estimates (2.33)–(2.35). The system (4.65) is a subsystem of (4.60) where the relevant equations suggested by $\mathcal{E}$ are eliminated.

The set of all the solutions of the system (4.65) is already a description of uncertainty. It may be further refined by searching for new values for $(\tilde{\theta}_{j|i}, \tilde{\theta}_{k|i}) \in \mathcal{E}$ that are compatible with the others. Different strategies can be adopted. In the following, two procedures are described.

(a) For those parameters in $\mathcal{E}$, find the nearest values to the old incompatible unrestricted ML estimates with respect to the L_1 distance, i.e. minimize the function

$$\min_{\boldsymbol{\theta} \in \Theta} \sum_{i \in \mathcal{I}} \left(\sum_{j \in \mathcal{I}} \left| \frac{\sum_k \theta_{ijk}}{\hat{\theta}_{i..}} - \hat{\theta}_{j|i} \right| + \sum_{k \in \mathcal{I}} \left| \frac{\sum_j \theta_{ijk}}{\hat{\theta}_{i..}} - \hat{\theta}_{k|i} \right| \right) \tag{4.66}$$

under the system (4.65) (with $\hat{\theta}_{i..} > 0$).

(b) For those parameters in $\mathcal{E}$, find those that maximize the likelihood function under the system (4.65).

Both approaches show that some of the unconstrained estimates remain unchanged, while the others (i.e. the ones in $\mathcal{E}$) are modified in order to minimize the L_1 distance or maximize the likelihood function. In both approaches, the solutions may not be unique; in other words, some solutions θ_{ijk} will not be a single value but will lie in an interval $[\theta^{\mathrm{L}}_{ijk}, \theta^{\mathrm{U}}_{ijk}]$. The ridge of solutions is composed of all those parameters θ_{ijk} which satisfy the constraints. The different objective functions in (a) and (b) generate different values for the parameters $(\tilde{\theta}_{j|i}, \tilde{\theta}_{k|i}) \in \mathcal{E}$ to be changed, and consequently different bounds for the relevant θ_{ijk}.

Note that the second approach is a nonlinear optimization problem which is more difficult to solve than the first, which is linear.

Remark 4.14 Approach (b) is similar to that described in Section 4.5 for the estimation of the likelihood ridge. There are some slight differences. Procedure (b) modifies only the parameters in $\mathcal{E}$, while that in Section 4.5 may modify all the given distributions involved with $\boldsymbol{\theta}_{YZ|X=i}$ for $i \in \mathcal{I}$. Hence, the latter approach will give a solution whose likelihood is not inferior to that of the former.

Example 4.13 Vantaggi (2005) applies the de Finetti coherence method to the example in D'Orazio *et al.* (2005b), also discussed in Section 4.7. The minimal set $\mathcal{E}$ of parameters incoherent with the imposed structural zeros (S1 in Section 4.7) is composed of those referring to the conditional events: $(PRO = M \mid AGE = 4)$, $(PRO = E \mid \mathrm{AGE} = 4)$ and $(EDU = C \mid AGE = 4)$:

$$\mathcal{E} = \left\{\theta_{\mathrm{M}|4}, \theta_{\mathrm{E}|4}, \theta_{\mathrm{C}|4}\right\}.$$

A comparison between the intervals $[\hat{\theta}^{\mathrm{L}}_{ijk}, \hat{\theta}^{\mathrm{U}}_{ijk}]$ obtained through ML as in Section 4.5.1 and the interval $[\tilde{\theta}^{\mathrm{L}}_{ijk}, \tilde{\theta}^{\mathrm{U}}_{ijk}]$ determined via the solution of (4.66) subject to the system (4.65) starting with the unrestricted ML estimates under the CIA, i.e.

$$\hat{\theta}_{j|i} = \frac{n^A_{ij.}}{n^A_{i..}}, \qquad i = 1, \ldots, I, \ j = 1, \ldots, J$$

$$\hat{\theta}_{k|i} = \frac{n^B_{i.k}}{n^B_{i..}}, \qquad i = 1, \ldots, I, \ k = 1, \ldots, K$$

$$\hat{\theta}_{i..} = \frac{n^A_{i..} + n^B_{i..}}{n_A + n_B}, \qquad i = 1, \ldots, I,$$

is given in Table 4.12. Note that both approaches satisfy the imposed constraints, consequently only parameters other than structural zeros are illustrated. Note that the two methods find different intervals for $\theta_{4\mathrm{SM}}$, $\theta_{4\mathrm{DM}}$, $\theta_{4\mathrm{SE}}$, $\theta_{4\mathrm{DE}}$, and $\theta_{4\mathrm{CW}}$. In fact, they are concerned with the incoherent parameters in $\mathcal{E}$.

Table 4.12 Intervals of equally plausible values of the parameters θ_{ijk} under the maximum likelihood approach and under the de Finetti coherence approach with the norm of L_1

ijk	$\hat{\theta}^{L}_{ijk}$	$\hat{\theta}^{U}_{ijk}$	$\tilde{\theta}^{L}_{ijk}$	$\tilde{\theta}^{U}_{ijk}$
3SM	0.0186	0.1550	0.0186	0.1550
4SM	0.0024	0.0031	0.0009	0.0011
3DM	0.0000	0.1363	0.0000	0.1363
4DM	0.0013	0.0021	0.0002	0.0014
2VE	0.0000	0.0043	0.0000	0.0043
3VE	0.0014	0.0881	0.0014	0.0881
4VE	0	0	0	0
2SE	0.0011	0.0054	0.0011	0.0054
3SE	0.1591	0.3776	0.1591	0.3821
4SE	0.0000	0.0007	0.0000	0.0011
3DE	0.0000	0.1364	0.0000	0.1364
4DE	0.0000	0.0007	0.0000	0.0011
1CW	0.0065	0.0065	0.0065	0.0065
2CW	0.0101	0.0101	0.0101	0.0101
3CW	0.3342	0.3342	0.3342	0.3342
4CW	0.0052	0.0052	0.0069	0.0069
1VW	0	0	0	0
2VW	0.0000	0.0043	0.0000	0.0043
3VW	0.0000	0.0866	0.0000	0.0866
4VW	0	0	0	0
2SW	0.0040	0.0083	0.0040	0.0083
3SW	0.0000	0.0866	0.0000	0.0866
4SW	0	0	0	0
3DW	0.0000	0.0855	0.0000	0.0866
4DW	0	0	0	0

The other parameters must have the same interval bounds. For the sake of completeness, it is possible to note that $\theta_{3\text{SE}}$ and $\theta_{3\text{DW}}$, though not involved in the incoherent set $\mathcal{E}$, have different bounds. However, since the difference is very low and it is just in one of the two bounds, it is caused by a numerical approximation problem of the EMH algorithm.

It is also worth noting how the difference in the two methods may lead to very different intervals. For instance, the intersection of the intervals obtained by the two methods for $\theta_{4\text{SM}}$ and $\theta_{4\text{DM}}$ is almost empty.

5

Statistical Matching and Finite Populations

Some statistical matching procedures described under the CIA (Chapter 2) and with the help of auxiliary information (Chapter 3) may be extended to the case where samples are drawn from a finite population $\mathcal{P}$ of size N according to complex survey designs. Despite the great importance of this topic, only a few methodological results are available: specifically, Rodgers (1984), Rubin (1986) and Renssen (1998). Rubin (1986) and Renssen (1998) reflect two different ways of approaching the problem of integration. While Renssen's techniques mainly aim to fuse estimates from samples A and B coherently, Rubin focuses on the construction of a unique sample $A \cup B$ (also called a concatenated file).

In this chapter, Renssen's results are shown in detail. First (Section 5.1), the statistical framework for samples drawn from finite populations is defined. That is to say, statistical matching of two complete archives (registers) relative to the same population is considered: this framework allows a particular definition of CIA suitable for the finite population framework. Hence, some statistical matching procedures for sample surveys from a finite population are illustrated (Sections 5.3 and 5.4). These procedures are the finite population counterparts of parametric procedures described respectively in Sections 2.1.1 and 3.2. These parametric procedures may be used for both the micro and the macro approaches.

Rubin's results mainly concern the appropriateness of sample weights when jointly analysing the two sample surveys. These ideas are outlined in Section 5.5.

Nonparametric procedures are especially used for the micro approach. The methods already illustrated in Section 2.4 and 3.5 can be still used in this framework, respectively under the CIA and with the help of auxiliary information. The main difference is in the treatment of sample weights induced by the sampling designs of respectively A and B. Some results are shown in Section 5.6.

Statistical Matching: Theory and Practice M. D'Orazio, M. Di Zio and M. Scanu

5.1 Matching Two Archives

Let A and B be two different registers containing respectively the variables (X, Y) and (X, Z) on all the population units, $s = 1, \ldots, N$. From now on, X, Y and Z are assumed to be univariate categorical variables, with respectively I, J and K categories. The continuous case would consider the same procedures, under constraints that will be highlighted in appropriate remarks. The reference population of the two registers is assumed to be the same. However, we also suppose there is no unit identifier in the two archives. Furthermore, the common information, i.e. the variable X collected in both the registers, is not able to identify each single unit.

The problem of matching two complete registers has a long history. An exhaustive exposition of this problem is given in DeGroot (1987); see also Goel and Ramalingam (1989) and references therein. The matching of two registers is beyond the scope of this book, and will not be further explained. The situation of availability of the registers A and B is considered here because it corresponds to complete knowledge of the bivariate variables (X, Y) and (X, Z). While this knowledge allows the direct computation of the (X, Y) and (X, Z) tables, the (Y, Z) table is uncertain. In other words, this framework shows that the (X, Y) and (X, Z) tables are estimable when A and B are just two samples, while the (Y, Z) table needs particular assumptions or external auxiliary information.

In order to easily compute contingency tables, it is useful to represent the variable categories assumed by each unit by appropriate vectors of indicator functions:

$$\gamma_s^X = \begin{pmatrix} I_s^X(1) \\ \vdots \\ I_s^X(i) \\ \vdots \\ I_s^X(I) \end{pmatrix}, \quad \gamma_s^Y = \begin{pmatrix} I_s^Y(1) \\ \vdots \\ I_s^Y(j) \\ \vdots \\ I_s^Y(J) \end{pmatrix}, \quad \gamma_s^Z = \begin{pmatrix} I_s^Z(1) \\ \vdots \\ I_s^Z(k) \\ \vdots \\ I_s^Z(K) \end{pmatrix},$$

$s = 1, \ldots, N$, where

$$I_s^X(i) = \begin{cases} 1 & x_s = i, \\ 0 & \text{otherwise}, \end{cases} \qquad i = 1, \ldots, I, \ s = 1, \ldots, N,$$

$$I_s^Y(j) = \begin{cases} 1 & y_s = j, \\ 0 & \text{otherwise}, \end{cases} \qquad j = 1, \ldots, J, \ s = 1, \ldots, N,$$

$$I_s^Z(k) = \begin{cases} 1 & z_s = k \\ 0 & \text{otherwise}, \end{cases} \qquad k = 1, \ldots, K, \ s = 1, \ldots, N.$$

Note that each vector has only one '1'. The one-way contingency tables (i.e. vectors of appropriate dimension) and two-way contingency tables (i.e. matrices of appropriate dimension) observed on the two archives are easily computed from the two registers.

- The one-way contingency tables are given by

$$\mathbf{N}_X = \sum_{s=1}^{N} \boldsymbol{\gamma}_s^X, \ \mathbf{N}_Y = \sum_{s=1}^{N} \boldsymbol{\gamma}_s^Y, \ \mathbf{N}_Z = \sum_{s=1}^{N} \boldsymbol{\gamma}_s^Z$$

Note that $\mathbf{N}_X$ can be computed equivalently on A or B, $\mathbf{N}_Y$ only on A and $\mathbf{N}_Z$ only on B.

- The two-way contingency tables are given by

$$\mathbf{N}_{XY} = \sum_{s=1}^{N} \boldsymbol{\gamma}_s^X \boldsymbol{\gamma}_s^{Y'}, \ \mathbf{N}_{XZ} = \sum_{s=1}^{N} \boldsymbol{\gamma}_s^X \boldsymbol{\gamma}_s^{Z'},$$

where $\mathbf{N}_{XY}$ can be computed only on A and $\mathbf{N}_{XZ}$ can be computed only on B.

Tables where Y and Z are jointly analysed cannot be computed from the two registers. Renssen (1998) considers the following macro and micro approaches.

(i) Macro approach. Find the (Y, Z) contingency table

$$\mathbf{N}_{YZ} = \sum_{s=1}^{N} \boldsymbol{\gamma}_s^Y \boldsymbol{\gamma}_s^{Z'}.$$

(ii) Micro approach. Find plausible values for the vectors $\boldsymbol{\gamma}_s^Z$, $s = 1, \ldots, N$, to associate to each record s in A (or, equivalently, find plausible $\boldsymbol{\gamma}_s^Y$, $s = 1, \ldots, N$, to associate to each record s in B).

These two objectives are pursued in the following sections. Before going into the statistical matching problem of two samples drawn from a finite population, let us describe a particular assumption that allows the direct computation of joint information of Y and Z from A and B (when it holds).

5.1.1 Definition of the CIA

Renssen (1998) considers a particular form of statistical relationship between the variables of interest: a *linear* dependence among the indicator vectors $\boldsymbol{\gamma}^X$, $\boldsymbol{\gamma}^Y$ and $\boldsymbol{\gamma}^Z$. Due to the lack of joint information on X, Y and Z, one will inevitably resort to the computation of the linear relationship of Y on X in register A,

$$\hat{\boldsymbol{\gamma}}_s^Y = \boldsymbol{\beta}_{YX} \boldsymbol{\gamma}_s^X, \qquad s = 1, \ldots, N, \tag{5.1}$$

and the linear relationship of Z on X in register B,

$$\hat{\boldsymbol{\gamma}}_s^Z = \boldsymbol{\beta}_{ZX} \boldsymbol{\gamma}_s^X, \qquad s = 1, \ldots, N. \tag{5.2}$$

The (population) regression parameters $\boldsymbol{\beta}_{YX}$ and $\boldsymbol{\beta}_{ZX}$ are respectively $J \times I$ and $K \times I$ matrices computed by the least squares method by means of the following normal equations:

$$\boldsymbol{\beta}_{YX}\left(\sum_{s=1}^{N}\boldsymbol{\gamma}_s^X\boldsymbol{\gamma}_s^{X'}\right)=\sum_{s=1}^{N}\boldsymbol{\gamma}_s^Y\boldsymbol{\gamma}_s^{X'}, \tag{5.3}$$

$$\boldsymbol{\beta}_{ZX}\left(\sum_{s=1}^{N}\boldsymbol{\gamma}_s^X\boldsymbol{\gamma}_s^{X'}\right)=\sum_{s=1}^{N}\boldsymbol{\gamma}_s^Z\boldsymbol{\gamma}_s^{X'}. \tag{5.4}$$

The linear model, although not suited for binary variables, has important properties. The most important is that population totals on the archives are preserved. Let $\mathbf{1}_h$ be a vector of 1s of length h. By definition,

$$\mathbf{1}_I'\boldsymbol{\gamma}_s^X=1,\ \mathbf{1}_J'\boldsymbol{\gamma}_s^Y=1,\ \mathbf{1}_K'\boldsymbol{\gamma}_s^Z=1, \tag{5.5}$$

for every $s = 1, \ldots, N$. Then, from the normal equations (5.3) and (5.4),

$$\sum_{s=1}^{N}\hat{\boldsymbol{\gamma}}_s^Y=\boldsymbol{\beta}_{YX}\left(\sum_{s=1}^{N}\boldsymbol{\gamma}_s^X\boldsymbol{\gamma}_s^{X'}\right)\mathbf{1}_I=\sum_{s=1}^{N}\boldsymbol{\gamma}_s^Y\boldsymbol{\gamma}_s^{X'}\mathbf{1}_I=\sum_{s=1}^{N}\boldsymbol{\gamma}_s^Y=\mathbf{N}_Y,$$

$$\sum_{s=1}^{N}\hat{\boldsymbol{\gamma}}_s^Z=\boldsymbol{\beta}_{ZX}\left(\sum_{s=1}^{N}\boldsymbol{\gamma}_s^X\boldsymbol{\gamma}_s^{X'}\right)\mathbf{1}_I=\sum_{s=1}^{N}\boldsymbol{\gamma}_s^Z\boldsymbol{\gamma}_s^{X'}\mathbf{1}_I=\sum_{s=1}^{N}\boldsymbol{\gamma}_s^Z=\mathbf{N}_Z.$$

The regression models in (5.1) and (5.2) may be used for an approximate computation of the marginal contingency table for (Y, Z), using the linear relationship of Y on X and of Z on X respectively:

$$\sum_{s=1}^{N}\left(\boldsymbol{\beta}_{YX}\boldsymbol{\gamma}_s^X\right)\left(\boldsymbol{\beta}_{ZX}\boldsymbol{\gamma}_s^X\right)'=\boldsymbol{\beta}_{YX}\left(\sum_{s=1}^{N}\boldsymbol{\gamma}_s^X\boldsymbol{\gamma}_s^{X'}\right)\boldsymbol{\beta}_{ZX}'. \tag{5.6}$$

This approximation is induced by the assumption of *linear independence* between Y and Z given X, given that the regression parameters of Y on Z given X (and vice versa) in (5.1) and (5.2) are assumed null. The relationship between the true, and unknown, contingency table $\mathbf{N}_{YZ}$ and that in (5.6) is

$$\sum_{s=1}^{N}\boldsymbol{\gamma}_s^Y\boldsymbol{\gamma}_s^{Z'}=\sum_{s=1}^{N}\left(\boldsymbol{\beta}_{YX}\boldsymbol{\gamma}_s^X\right)\left(\boldsymbol{\beta}_{ZX}\boldsymbol{\gamma}_s^X\right)'$$
$$+\sum_{s=1}^{N}\left(\boldsymbol{\gamma}_s^Y-\boldsymbol{\beta}_{YX}\boldsymbol{\gamma}_s^X\right)\left(\boldsymbol{\gamma}_s^Z-\boldsymbol{\beta}_{ZX}\boldsymbol{\gamma}_s^X\right)'. \tag{5.7}$$

If at least one of the regression models of Y on X and of Z on X is without error for all $s = 1, \ldots, N$, i.e. $\hat{\boldsymbol{\gamma}}_s^Y=\boldsymbol{\gamma}_s^Y$ or $\hat{\boldsymbol{\gamma}}_s^Z=\boldsymbol{\gamma}_s^Z$ for all $s = 1, \ldots, N$, than contingency table (5.6) is the true one. This can be considered as the definition of CIA. Note that the contingency table (5.6) has the property of preserving the marginal Y and Z distributions, as observed respectively in A and B. In fact,

$$\mathbf{1}'_J \hat{\boldsymbol{\gamma}}_s^Y = \mathbf{1}_J \left(\sum_{s=1}^N \boldsymbol{\gamma}_s^Y \boldsymbol{\gamma}_s^{X'} \right) \left(\sum_{s=1}^N \boldsymbol{\gamma}_s^X \boldsymbol{\gamma}_s^{X'} \right)^{-1} \boldsymbol{\gamma}_s^X$$

$$= \mathbf{1}'_I \left(\sum_{s=1}^N \boldsymbol{\gamma}_s^X \boldsymbol{\gamma}_s^{X'} \right) \left(\sum_{s=1}^N \boldsymbol{\gamma}_s^X \boldsymbol{\gamma}_s^{X'} \right)^{-1} \boldsymbol{\gamma}_s^X = \mathbf{1}'_I \boldsymbol{\gamma}_s^X = 1;$$

hence

$$\mathbf{1}'_J \sum_{s=1}^N \left(\boldsymbol{\beta}_{YX} \boldsymbol{\gamma}_s^X\right) \left(\boldsymbol{\beta}_{ZX} \boldsymbol{\gamma}_s^X\right)' = \sum_{s=1}^N \left(\boldsymbol{\beta}_{ZX} \boldsymbol{\gamma}_s^X\right)' = \mathbf{N}'_Z.$$

Similar results hold for the marginal Y distribution.

Remark 5.1 Renssen (1998) extends the previous arguments to the case of continuous variables when the following generalization of (5.5) holds: there should exist three vectors of constants, $\mathbf{v}_1$, $\mathbf{v}_2$ and $\mathbf{v}_3$ such that

$$\mathbf{v}'_1 \boldsymbol{\gamma}_s^X = \mathbf{v}'_2 \boldsymbol{\gamma}_s^Y = \mathbf{v}'_3 \boldsymbol{\gamma}_s^Z = 1,$$

for all $s = 1, \ldots, N$. Note that, if at least one of X, Y or Z is multivariate, and at least one component is categorical, the existence of the corresponding vector $\mathbf{v}$ is straightforward. Without loss of generality, let $\mathbf{X}$ be a multivariate variable whose first r variables are continuous and whose last one is categorical with I categories (this last categorical variable may be defined as the combination of a set of categorical variables). Then $\mathbf{v}_1$ can be defined as a vector of dimension $r + I$ whose first r elements are set to zero, and to one elsewhere.

Remark 5.2 Contingency table (5.6) may be considered as a macro result when matching two registers A and B. The regression equations (5.1) and (5.2) may be considered as predictive functions for the micro approach to statistical matching, as in Section 2.2. The predicted files are consistent with the macro approach.

5.2 Statistical Matching and Sampling from a Finite Population

Let A and B be two samples drawn from the finite population $\mathcal{P}$ according to sampling designs with positive first- and second-order inclusion probabilities:

$$P(A \ni s) = \pi_s^A, \qquad P(A \ni \{s_1, s_2\}) = \pi_{s_1 s_2}^A,$$

$$P(B \ni s) = \pi_s^B, \qquad P(B \ni \{s_1, s_2\}) = \pi_{s_1 s_2}^B$$

(see Appendix D for more details on finite survey sampling). We assume that only a subset of the variables observed in samples A and B is in common. As usual, these variables are denoted by the symbol **X**. However, in the finite population context **X** may be characterized by two different types of common variables:

(i) **V**, common variables for which population totals are known;

(ii) **U**, common variables whose population totals are unknown.

Usually, **V** are variables collected on archives, and they are mostly used for the definition of the sampling designs for A and B. It might happen that sampling designs are not defined according to some of the archive variables **V**. In other words, the sample weights

$$\omega_s^A = \frac{1}{\pi_s^A}, \qquad \omega_s^B = \frac{1}{\pi_s^B}$$

are such that the corresponding Horvitz–Thompson estimator of the marginal population distribution of **V**, $\mathbf{N}_{\mathbf{V}}$, is unable to reproduce the archive population totals. In this case, calibration of the sample weights may be considered (see Appendix D for more details).

From now on, both **V** and **U** will be considered as categorical univariate variables, respectively V and U. Finally, let X be the univariate categorical variable whose categories are given by the Cartesian product of the categories of U and V.

5.3 Parametric Methods under the CIA

First, it is necessary to make the sample weights in A and B consistent. The linear relationship among the indicator vectors γ^X, γ^Y and γ^Z is particularly suitable for adapting the sample weights according to the calibration estimator. Actually, more than one calibration of sample weights is necessary in the statistical matching problem, as described in Renssen (1998).

(i) The population distribution $\mathbf{N}_V$ of V is already known.

(ii) Sample weights ω_a^A, $a = 1, \ldots, n_A$, and ω_b^B, $b = 1, \ldots, n_B$, are calibrated a first time to the new weights ω_a^{A1} and ω_b^{B1} respectively, subject to the constraint

$$\sum_{a=1}^{n_A} \omega_a^{A1} \gamma_a^V = \sum_{b=1}^{n_B} \omega_b^{B1} \gamma_b^V = \mathbf{N}_V.$$

These weights are used to compute *preliminary* estimates of the population distribution of U, $\mathbf{N}_U$, from the two samples A and B:

$$\hat{\mathbf{N}}_U^{(A)} = \sum_{a=1}^{n_A} \omega_a^{A1} \gamma_a^U, \tag{5.8}$$

$$\hat{\mathbf{N}}_U^{(B)} = \sum_{b=1}^{n_B} \omega_b^{B1} \gamma_b^U. \tag{5.9}$$

(iii) These preliminary estimates may be different due to sampling variability. Consequently, a final estimate for $\mathbf{N}_U$ is obtained by pooling together the estimates (5.8) and (5.9):

$$\hat{\mathbf{N}}_U = \lambda \sum_{a=1}^{n_A} \omega_a^{A1} \boldsymbol{\gamma}_a^U + (1-\lambda) \sum_{b=1}^{n_B} \omega_b^{B1} \boldsymbol{\gamma}_b^U, \tag{5.10}$$

for some appropriate λ, $0 \leq \lambda \leq 1$.

(iv) The calibrated sample weights ω_a^{A1} and ω_b^{B1} are calibrated a second time in order to reproduce the known $\mathbf{N}_V$ and the estimated $\hat{\mathbf{N}}_U$. Let ω_a^{A2} and ω_b^{B2} be the final recalibrated sample weights.

The final recalibrated sample weights may be used for the macro and micro approaches under the CIA.

5.3.1 The macro approach when the CIA holds

Let the estimation of the contingency table $\mathbf{N}_{YZ}$ for (Y, Z) be the objective. When only samples A and B are known, only the contingency table (5.6) can be estimated, i.e. only the contingency table for (Y, Z) under the CIA described in Section 5.1.1.

(i) Estimate $\sum_{s=1}^{N} \boldsymbol{\gamma}_s^X \boldsymbol{\gamma}_s^{X'}$ in (5.6) by combining the estimates that can be computed with samples A and B with the final sample weights:

$$\delta \sum_{a=1}^{n_A} \omega_a^{A2} \boldsymbol{\gamma}_a^X \boldsymbol{\gamma}_a^{X'} + (1-\delta) \sum_{b=1}^{n_B} \omega_b^{B2} \boldsymbol{\gamma}_b^X \boldsymbol{\gamma}_b^{X'}, \tag{5.11}$$

for some appropriate δ, $0 \leq \delta \leq 1$.

(ii) The regression coefficients $\boldsymbol{\beta}_{YX}$ and $\boldsymbol{\beta}_{ZX}$ are estimated from the two different samples A and B according to the final recalibrated sample weights:

$$\hat{\boldsymbol{\beta}}_{YX} = \left(\sum_{a=1}^{n_A} \omega_a^{A2} \boldsymbol{\gamma}_a^Y \boldsymbol{\gamma}_a^{X'} \right) \left(\sum_{a=1}^{n_A} \omega_a^{A2} \boldsymbol{\gamma}_a^X \boldsymbol{\gamma}_a^{X'} \right)^{-1}, \tag{5.12}$$

$$\hat{\boldsymbol{\beta}}_{ZX} = \left(\sum_{b=1}^{n_B} \omega_b^{B2} \boldsymbol{\gamma}_b^Z \boldsymbol{\gamma}_b^{X'} \right) \left(\sum_{b=1}^{n_B} \omega_b^{B2} \boldsymbol{\gamma}_b^X \boldsymbol{\gamma}_b^{X'} \right)^{-1}. \tag{5.13}$$

(iii) The regression parameter estimates (5.12) and (5.13) and the estimated matrix in (5.11) allow the estimation of the contingency table (5.6), henceforth $\mathbf{N}_{YZ}^{\text{CIA}}$:

$$\hat{\mathbf{N}}_{YZ}^{\text{CIA}} = \hat{\boldsymbol{\beta}}_{YX} \left[\delta \sum_{a=1}^{n_A} \omega_a^{A2} \boldsymbol{\gamma}_a^X \boldsymbol{\gamma}_a^{X'} + (1-\delta) \sum_{b=1}^{n_B} \omega_b^{B2} \boldsymbol{\gamma}_b^X \boldsymbol{\gamma}_b^{X'} \right] \hat{\boldsymbol{\beta}}_{ZX}'. \tag{5.14}$$

5.3.2 The predictive approach

For the sake of simplicity, let A be the recipient file. The variable to be imputed is Z. As in Section 2.5, imputations can be performed through a two-step procedure.

(a) A preliminary imputed value can be obtained by means of the regression model (5.2), where $\boldsymbol{\beta}_{YX}$ is estimated in (5.13):

$$\tilde{\boldsymbol{\gamma}}_a^Z = \hat{\boldsymbol{\beta}}_{ZX}\boldsymbol{\gamma}_a^X, \qquad a = 1, \ldots, n_A.$$

(b) Most of the time, the previous predictions $\tilde{\boldsymbol{\gamma}}_a^Z$, $a = 1, \ldots, n_A$, are unrealistic. A live value can be obtained from file B by an appropriate distance hot deck method (Section 2.4.3).

If B is to be completed, a similar method may be used in order to predict the missing Y values. When both samples are completed, the second step may consider a distance hot deck method with distances between (X, Y, Z) values or (Y, Z) values.

5.4 Parametric Methods when Auxiliary Information is Available

Renssen (1998) follows the framework offered by the Dutch Household Survey on Living Conditions; see Winkels and Everaers (1998) and Example 6.1 below. Actually, that survey is suited to the approaches described in Section 5.3, which can be heavily biased if the CIA does not hold. The bias is well described by (5.7), and is given by the term

$$\sum_{s=1}^{N} \left(\boldsymbol{\gamma}_s^Y - \boldsymbol{\beta}_{YX}\boldsymbol{\gamma}_s^X\right)\left(\boldsymbol{\gamma}_s^Z - \boldsymbol{\beta}_{ZX}\boldsymbol{\gamma}_s^X\right)'.$$

Renssen (1998) considers the situation where a third reliable and complete sample C is available. According to the notation in Section 5.3, C is of size n_C and all the variables under study, i.e. U, V, Y and Z, are observed. Again, let $\boldsymbol{\gamma}_c^U$, $\boldsymbol{\gamma}_c^V$, $\boldsymbol{\gamma}_c^Y$, and $\boldsymbol{\gamma}_c^Z$ be the observed indicator vectors, $c = 1, \ldots, n_C$, and ω_c^C the corresponding sample weights.

5.4.1 The macro approach

Renssen (1998) shows that it is possible to define two alternative approaches for the estimation of $\mathbf{N}_{YZ}$. He calls them *incomplete two-way stratification* and *synthetic two-way stratification*. Both consider the final weights ω_a^{A2}, $a = 1, \ldots, n_A$, and ω_b^{B2}, $b = 1, \ldots, n_B$, determined in Section 5.3.

Incomplete two-way stratification

This strategy computes an estimate of $\mathbf{N}_{YZ}$ from the only available sample with joint information on the pair (Y, Z), i.e. C. The other two samples are used only in order to appropriately calibrate sample weights of the units in C. The marginal population distributions that the new sample weights should reproduce are the marginal estimated distributions of Y and Z:

$$\hat{\mathbf{N}}_Y = \sum_{a=1}^{n_A} \omega_a^{A2} \boldsymbol{\gamma}_a^Y, \qquad \hat{\mathbf{N}}_Z = \sum_{b=1}^{n_B} \omega_b^{B2} \boldsymbol{\gamma}_b^Z.$$

(i) Calibrate the initial weights ω_c^C, $c = 1, \ldots, n_C$, to those (ω_c^{C1}) that satisfy the constraints:

$$\sum_{c=1}^{n_C} \omega_c^{C1} \boldsymbol{\gamma}_c^Y = \hat{\mathbf{N}}_Y, \qquad \sum_{c=1}^{n_C} \omega_c^{C1} \boldsymbol{\gamma}_c^Z = \hat{\mathbf{N}}_Z.$$

(ii) The estimate of $\mathbf{N}_{YZ}$ obtained via incomplete two-way stratification is:

$$\hat{\mathbf{N}}_{YZ}^{\mathrm{inc}} = \sum_{c=1}^{n_C} \omega_c^{C1} \boldsymbol{\gamma}_c^Y \boldsymbol{\gamma}_c^{Z'}.$$

Synthetic two-way stratification

This procedure is based on the decomposition (5.7). The first term on the right-hand side has already been estimated in (5.14): $\hat{\mathbf{N}}_{YZ}^{\mathrm{CIA}}$. In order to estimate the second term, the weights ω_c^C are calibrated to the new weights ω_c^{C2} according to the following constraint:

$$\sum_{c=1}^{n_C} \omega_c^{C2} \left[\boldsymbol{\gamma}_c^Y \boldsymbol{\gamma}_c^{Z'} - \left(\boldsymbol{\gamma}_c^Y - \hat{\boldsymbol{\beta}}_{YX} \boldsymbol{\gamma}_c^X\right) \left(\boldsymbol{\gamma}_c^Z - \hat{\boldsymbol{\beta}}_{ZX} \boldsymbol{\gamma}_c^X\right)'\right] = \hat{\mathbf{N}}_{YZ}^{\mathrm{CIA}}.$$

Hence, the final estimate of $\mathbf{N}_{YZ}$ is again obtained by weighting units in C with the new weights ω_c^{C2}, but assumes a new interpretation:

$$\begin{aligned} \hat{\mathbf{N}}_{YZ}^{\mathrm{syn}} &= \sum_{c=1}^{n_C} \omega_c^{C2} \left(\boldsymbol{\gamma}_c^Y \boldsymbol{\gamma}_c^{Z'}\right) \\ &= \hat{\mathbf{N}}_{YZ}^{\mathrm{CIA}} + \sum_{c=1}^{n_C} \omega_c^{C2} \left(\boldsymbol{\gamma}_c^Y - \hat{\boldsymbol{\beta}}_{YX} \boldsymbol{\gamma}_c^X\right) \left(\boldsymbol{\gamma}_c^Z - \hat{\boldsymbol{\beta}}_{ZX} \boldsymbol{\gamma}_c^X\right)'. \end{aligned}$$

Renssen (1998) emphasizes that this estimator is equal to that under the CIA plus an adjustment term that makes $\hat{\mathbf{N}}_{YZ}^{\mathrm{syn}}$ approximately unbiased.

5.4.2 The predictive approach

Again, let A be the recipient file. The complete regression model to consider in order to predict the missing Z values would be:

$$\tilde{\gamma}_a^Z = \boldsymbol{\beta}_{ZX.Y}\boldsymbol{\gamma}_a^X + \boldsymbol{\beta}_{ZY.X}\boldsymbol{\gamma}_a^Y, \qquad a = 1, \ldots, n_A.$$

Equations (3.13) and (3.14) lead to the following regression parameter estimates when C is available:

$$\hat{\boldsymbol{\beta}}_{ZY.X} = \left[\sum_{c=1}^{n_C} \omega_c^{C2}\left(\boldsymbol{\gamma}_c^Z - \hat{\boldsymbol{\beta}}_{ZX}\boldsymbol{\gamma}_c^X\right)\left(\boldsymbol{\gamma}_c^Y - \hat{\boldsymbol{\beta}}_{YX}\boldsymbol{\gamma}_c^X\right)'\right]$$
$$\times \left[\sum_{a=1}^{n_A} \omega_a^{A2}\left(\boldsymbol{\gamma}_a^Y - \hat{\boldsymbol{\beta}}_{YX}\boldsymbol{\gamma}_a^X\right)\left(\boldsymbol{\gamma}_a^Y - \hat{\boldsymbol{\beta}}_{YX}\boldsymbol{\gamma}_a^X\right)'\right]^{-1}$$
$$\hat{\boldsymbol{\beta}}_{ZX.Y} = \hat{\boldsymbol{\beta}}_{ZX} - \hat{\boldsymbol{\beta}}_{ZY.X}\hat{\boldsymbol{\beta}}_{YX},$$

where $\hat{\boldsymbol{\beta}}_{YX}$ and $\hat{\boldsymbol{\beta}}_{ZX}$ are the regression parameter estimates (5.12) and (5.13). Hence, again by the mixed approach, imputations can be performed by the following two steps.

(a) Impute a preliminary value to the missing Z in A with:

$$\tilde{\gamma}_a^Z = \hat{\boldsymbol{\beta}}_{ZX.Y}\boldsymbol{\gamma}_a^X + \hat{\boldsymbol{\beta}}_{ZY.X}\boldsymbol{\gamma}_a^Y$$
$$= \hat{\boldsymbol{\beta}}_{ZX}\gamma_a^X + \hat{\boldsymbol{\beta}}_{ZY.X}\left(\boldsymbol{\gamma}_a^Y - \hat{\boldsymbol{\beta}}_{YX}\boldsymbol{\gamma}_a^X\right), \qquad a = 1, \ldots, n_A.$$

(b) Change the preliminary value to a live one with an appropriate hot deck method (Section 2.4).

5.5 File Concatenation

A different approach is explained in Rubin (1986). This approach is used for the so-called *file concatenation*, i.e. the aim is to consider the overall sample $A \cup B$ as the basis for the statistical analysis. The creation of a unique sample $A \cup B$ from two different samples A and B is not an easy task in the finite population context. In fact, it requires the definition of a synthetic sampling scheme for $A \cup B$ derived from the sampling scheme used for drawing A and B separately. In other words, a new coherent system of weights ω_a^{AB}, for $a = 1, \ldots, n_A$, and ω_b^{AB} for $b = 1, \ldots, n_B$, must be computed.

Following Section 5.2, the probability that an observation s is included in $A \cup B$ is

$$\pi_s^{A\cup B} = P(A \cup B \ni s) = P(A \ni s) + P(B \ni s) - P(A \cap B \ni s).$$

Rubin assumes that the probability that a unit is observed in both samples is negligible, and thus

$$\pi_s^{A\cup B} = \pi_s^A + \pi_s^B.$$

Then the new weight to assign to each unit in A is

$$\omega_a^{AB} = \frac{1}{\pi_a^{A\cup B}} = \frac{1}{1/\omega_a^A + 1/\omega_a^B}, \qquad a = 1, \ldots, n_A,$$

while the new weight to assign to each unit in B is

$$\omega_b^{AB} = \frac{1}{\pi_b^{A\cup B}} = \frac{1}{1/\omega_b^A + 1/\omega_b^B}, \qquad b = 1, \ldots, n_B.$$

This approach requires the following information.

(i) It is possible to compute which sample weight would have been assigned to each unit a, $a = 1, \ldots, n_A$, under the sampling scheme of sample B. Let ω_a^B, $a = 1, \ldots, n_A$, be this new sample weight.

(ii) It is possible to compute which sample weight would have been assigned to each unit b, $b = 1, \ldots, n_B$, under the sampling scheme of sample A. Let ω_b^A, $b = 1, \ldots, n_B$, be this new sample weight.

Once the new weights have been computed, the concatenated file $A \cup B$ (i.e. the $n_A + n_B$ weighted records of Table 1.1) can be used for the estimation of any (X, Y, Z) parameter under the CIA. Rubin (1986) proves that the new weights lead to unbiased estimators of the mean of X, Y or Z or their squares, products, etc., where these variables are assumed continuous. Let us extend this result for the distribution of a categorical variable, say X. As a matter of fact, it is necessary to resort to the indicator vectors $\boldsymbol{\gamma}^X$. Let N be the size of the reference population. The concatenated sample would give the following estimator:

$$\sum_{s=1}^{N} \omega_s^{AB}(I_A(s) + I_B(s))\boldsymbol{\gamma}_s^X, \tag{5.15}$$

where $I_A(s) = 1$ if s is in sample A and $I_A(s) = 0$ otherwise, and a similar definition holds for $I_B(s)$, $s = 1, \ldots, N$. The expectation of (5.15) with respect to both sampling schemes is:

$$\sum_{s=1}^{N} \omega_s^{AB} \left(\frac{1}{\omega_s^A} + \frac{1}{\omega_s^B} \right) \boldsymbol{\gamma}_s^X = \sum_{s=1}^{N} \boldsymbol{\gamma}_s^X = \mathbf{N}_X.$$

Rubin (1986) also notes that the sum of the concatenated sample weights may not exactly be equal to N. In that case, appropriate ratio estimators can be considered.

For models other than the CIA, auxiliary information is necessary. When auxiliary information is not available, Rubin (1986) adopts the multiple imputation framework of Section 4.8.2. This is the only result that has the aim of assessing uncertainty when samples are drawn according to complex survey sampling designs.

Remark 5.3 Note that it is quite difficult to have all the information necessary for computing the weights ω^{AB}. For instance, assume that the sampling design in A is a simple stratified design with respect to a variable Φ. The weights ω_b^A would be difficult to find unless B is also drawn according to a simple stratified sample design where Φ is one of the stratification variables.

5.6 Nonparametric Methods

Hot deck procedures (Sections 2.4) are among the most commonly used in practice. Random, conditional random, rank and distance hot deck procedures can still be applied as described in Sections 2.4.1–2.4.3. One issue remains unresolved: what sample weights should be assigned to the records of the synthetic data file?

When A is a recipient file, a first approach would assign the pair (a, b) (a recipient and b donor records) the weight ω_a^A, $a = 1, \ldots, n_A$. This approach would leave unchanged the sample marginal and joint distributions for $(\mathbf{X}, \mathbf{Y})$ as observed in A. However, it may bias the sample distribution of $\mathbf{Z}$ in B. As shown in Section 2.4.3, the problem of the preservation of the $\mathbf{Z}$ distribution is solved by means of a constrained approach. When samples are drawn according to complex survey designs, the constrained statistical matching procedure assigns new sample weights ω_{ab} to each pair of donors and recipients under the following constraints:

$$\sum_{b=1}^{n_B} \omega_{ab} = \omega_a^A, \qquad a = 1, \ldots, n_A, \tag{5.16}$$

$$\sum_{a=1}^{n_A} \omega_{ab} = \omega_b^B, \qquad b = 1, \ldots, n_B. \tag{5.17}$$

When distance hot deck is used, the pairs of recipient and donor records are found by minimizing the objective function

$$\sum_{a=1}^{n_A} \sum_{b=1}^{n_B} d_{ab}\omega_{ab}$$

under the constraints (5.16) and (5.17). This is actually a generalization of the problem of minimizing (2.48) under the constraints (2.49) and (2.50).

Rank hot deck also needs to be modified in order to preserve information observed in A and B. This approach is due to a set of constraints adopted by Statistics Canada for creating the Social Policy Simulation Data Base: use all records from the two files, maintaining the observed distributions in A and B in a parsimonious way; see Liu and Kovacevic (1994) for the details. The following approach, also called the *weight-split algorithm*, is based on Liu and Kovacevic (1994). Let X be a continuous variable (similar results hold for categorical ordered variables). For the sake of simplicity, assume that the units in the two samples have already been put in ascending order according to X:

$$x_a^A \leq x_{a+1}^A, \qquad a = 1, \ldots, n_A;$$

$$x_b^B \leq x_{b+1}^B, \qquad b = 1, \ldots, n_B.$$

Finally, assume that the sample weights have been normalized in the two samples, so that

$$\sum_{a=1}^{n_A} \omega_a^A = 1, \qquad \sum_{b=1}^{n_B} \omega_b^B = 1.$$

The empirical cumulative distribution function of X in A is given by

$$\hat{F}_X^A(x) = \sum_{a:x_a \leq x} \omega_a, \qquad x \in \mathbb{R}.$$

The corresponding estimate in B is

$$\hat{F}_X^B(x) = \sum_{b:x_b \leq x} \omega_b, \qquad x \in \mathbb{R}.$$

Liu and Kovacevic (1994) define the following rank hot deck imputation procedure for Z in A.

(i) Impute z_b to all the records a in A such that

$$\hat{F}_X^B(x_{b-1}) < \hat{F}_X^A(x_a) \leq \hat{F}_X^B(x_b).$$

(ii) For those b records in B whose quantiles do not coincide with any quantile of the records in A, i.e. there does not exist an a such that

$$\hat{F}_X^A(x_a) = \hat{F}_X^B(x_b), \tag{5.18}$$

impute z_b to the first a record in A such that

$$\hat{F}_X^A(x_a) > \hat{F}_X^B(x_b).$$

It is easy to see that all the matched records with the rank hot deck procedure of Section 2.4.2 are included. However, while the procedure of Section 2.4.2 resulted in a completed A file, with n_A records, this approach produces a data set of $n_A + n_B - T$ records, where T is the number of pairs (a, b) for which (5.18) holds. Order the $i = 1, \ldots, n_A + n_B - T$ completed records by the cumulative weight

$$\hat{F}_i = \min\left\{\hat{F}_X^A(x_a); \hat{F}_X^B(x_b)\right\},$$

where a and b are the recipient and donor records for the ith completed record. Note that the completed sample is still ordered according to X, and according to the cumulative weight $\hat{F}_i$. Then, the relative weight for the ith record is

$$\omega_i = \hat{F}_i - \hat{F}_{i-1}, \qquad i = 1, \ldots, n_A + n_B - T,$$

where $\hat{F}_0 = 0$. This procedure has the following properties.

(a) The marginal and joint distributions of (X, Y) are still those observed in A.

(b) The marginal distribution of Z is still that observed in B.

This procedure may produce a very large data set. Consequently, Liu and Kovacevic (1994) show how to reduce its size without losing too much information.

6

Issues in Preparing for Statistical Matching

In the previous chapters, A and B have always been assumed to be integrable sources, meaning that they are homogeneous in terms of their concepts and definitions. Actually, this is a highly fortuitous event which rarely happens. In fact, matching A and B implies a great preliminary effort in terms of time and resources for their homogenization. Some aspects of these issues are illustrated in Section 6.1. Once the homogenization step has been performed, another preliminary step should be considered. The two sources A and B may be characterized by many common variables. The choice of which common variables should be used as matching variables is a very delicate point, which is covered in some detail in Section 6.2.

6.1 Reconciliation of Concepts and Definitions of Two Sources

Nowadays, large numbers of data sets are available in many different forms. The integration of these sources must necessarily involve reconciling their concepts and definitions. Although this problem is usually dealt with in a very general framework (the sources can either be statistical sources, e.g. sample surveys and census results, or administrative sources, e.g. institutional archives; see Ruggles, 1999), we will consider only the case of two sample surveys, as usual in the statistical matching problem.

Even when two independent sample surveys are conducted by the same organization, they may be affected by many incompatibilities. Generally speaking, the possibility of making these surveys compatible should be considered in advance. This is the case of the Dutch Household Survey on Living Conditions.

Statistical Matching: Theory and Practice M. D'Orazio, M. Di Zio and M. Scanu

Example 6.1 Renssen (1998) cites the Dutch Household Survey on Living Condition (POLS) as an ideal situation for statistical matching; see Bakker and Winkels (1998) and Winkels and Everaers (1998). As a matter of fact, this is an example of integrated survey design. In other words, this survey is composed of many different subsurveys or modules, each focused on a particular aspect of household living conditions. These modules have the important characteristic of already being integrated as far as definitions and methods of observation are concerned. The modules consist of (Winkels and Everaers, 1998):

- a joint questionnaire on demographic (age, sex, marital status, nationality, place of birth) and socio-economic (education, socio-economic status, net household income, and main source of income) issues;
- a joint questionnaire with a few screening questions on every relevant aspect of living conditions;
- a more exhaustive questionnaire on living conditions.

The first two parts are to be answered by all the units in the sample. The last one is split into subquestionnaires, so that each unit receives one subquestionnaire. This kind of questionnaire reduces the response burden. In order to give a complete sketch on the household living conditions, the overall sample is split into as many subsamples as there are subquestionnaires. Each subsample is associated with one subquestionnaire. Hence, the first two parts of the questionnaire consist of the common variables $\mathbf{X}$, while the variables of the subquestionnaires of the third part play the role of the $\mathbf{Y}$ and $\mathbf{Z}$ variables in statistical matching.

When the two sources A and B have not been planned to be homogeneous, different actions should be carried out. van der Laan (2000) describes in detail which actions should be performed for the harmonization of two distinct sources, among those needed for their integration:

(a) harmonization of the definition of units;

(b) harmonization of reference periods;

(c) completion of populations;

(d) harmonization of variables;

(e) harmonization of classifications;

(f) adjustment for measurement errors (accuracy);

(g) adjustment for missing data;

(h) derivation of variables.

Actions (a)–(e) are typical of the harmonization phase, actions (f)–(g) are needed when A and B are affected by nonsampling errors, so that the matching step is performed on accurate data sets, and action (h) is carried out in order to provide some new variables from those observed in A and B. While actions (f) and (g) can refer to a wide statistical literature, actions (a)–(e) are quite *ad hoc*. In the following, we will refer to two distinct problems and to some *ad hoc* solutions for them (see also Section 7.2.1 for an application):

(i) the two sources are *biased* – refers to actions (a), (b), and (c);

(ii) the two sources are *inconsistent* – refers to actions (d) and (e).

6.1.1 Reconciliation of biased sources

Two sources can be considered *biased* if their generating model is different. This can happen when:

- A and B are samples characterized by different reference periods;
- A and B, even if sharing the same reference period, are drawn from different populations.

The first case is inevitable unless the ideal situation of Example 6.1 obtains. The second case is more linked to the definition of population in the two surveys. In particular,

- two surveys may refer to distinct subpopulations, partially overlapping, of a more general population;
- population units are defined differently.

Example 6.2 Unfortunately, although at first sight two sample surveys refer to the same unit, they can actually be defined differently. For instance, this is the case for the unit 'household'. It will be seen in Section 7.2.1 that different institutions (in that case the Bank of Italy and the Italian National Statistical Institute) conduct distinct surveys on households using different definitions of household.

Reconciliation of two biased sources due to differences in time can be done according to the following assumption.

Assumption 6.1 *The generating model does not change from the first survey to the second.*

When two sources refer to different populations, two alternative assumptions can be considered.

Assumption 6.2 *The populations from which A and B are drawn can be considered as subpopulations of a more general set of units. The distribution of* $(\mathbf{X}, \mathbf{Y}, \mathbf{Z})$ *is the same in the two subpopulations.*

Assumption 6.3 *The two populations are almost completely overlapping. In other words, the distinct units of the two populations, due to differences in unit definitions, are a rare event. Hence, the two samples A and B can be considered as a sample from the intersection of the two populations.*

Note that Assumptions 6.1 and 6.2 are untestable hypotheses for $A \cup B$. On the other hand, Assumption 6.3 can sometimes be verified (see Section 7.2.1 for an example of this situation).

Remark 6.1 A more formal treatment would not distinguish between Assumptions 6.1, 6.2 and 6.3. In this case, the situation outlined in Remark 1.3 should be considered. More precisely, this is a case where the missing data generating model can be suitably defined through pattern mixture models, and the objective is a mixture of the data generating distributions in A and B.

Let the selection of samples A and B be characterized by an additional pair of events, respectively E_A and E_B, with probability $P(E_A)$ and $P(E_B)$ such that

$$P(E_A) \geq 0, \quad P(E_B) \geq 0, \quad P(E_A) + P(E_B) = 1. \tag{6.1}$$

These events represent the presence of bias between the reference populations for A and B (e.g. the target populations of the two surveys are subpopulations of a more general population), but also differences in data collection methods. Assumptions 6.1, 6.2 and 6.3 are equivalent to the following assumption.

Assumption 6.4 $(\mathbf{X}, \mathbf{Y}, \mathbf{Z})$ *is independent of* E_A *and* E_B.

This is generally not true. Denoting by $f_A(\mathbf{x}, \mathbf{y}, \mathbf{z})$ and $f_B(\mathbf{x}, \mathbf{y}, \mathbf{z})$ the distribution of $(\mathbf{X}, \mathbf{Y}, \mathbf{Z})$ given respectively E_A and E_B, the target distribution would be

$$f(\mathbf{x}, \mathbf{y}, \mathbf{z}) = P(E_A) f_A(\mathbf{x}, \mathbf{y}, \mathbf{z}) + P(E_B) f_B(\mathbf{x}, \mathbf{y}, \mathbf{z}). \tag{6.2}$$

When sampling from finite populations, this situation resembles that illustrated by the file concatenation approach (see Section 5.5). The events E_A and E_B are the different sampling schemes. When the samples have enough information, these probabilities can be derived and the file concatenated.

In general, distribution (6.2) cannot be estimated from A and B. In fact, these samples give information on the *conditional* distributions of $(\mathbf{X}, \mathbf{Y}, \mathbf{Z})$ given E_A and given E_B. On the other hand, the two surveys are unable to estimate $P(A)$ and $P(B)$ (the surveys are conducted conditional on these events), and nothing is known about them apart from constraints (6.1). Considerations such as those in Manski (1995) or Coletti and Scozzafava (2002) for an assessment of all the plausible values of the distribution $(\mathbf{X}, \mathbf{Y}, \mathbf{Z})$ may be helpful. Further research on this topic is necessary.

6.1.2 Reconciliation of inconsistent definitions

Even if the two sources are not biased, they can be inconsistent because of differences in the definition of variables and categorizations. In contrast to the previous situation, these inconsistencies cannot be resolved through particular assumptions. If they are not resolved in advance, particular care and a lot of work should be devoted to the reconciliation of inconsistent definitions. Sometimes, no amount of effort can produce an interesting result.

For the statistical matching problem, this aspect of the harmonization phase involves all the common variables in A and B. Generally speaking, different strategies can be considered for these variables.

(i) Some variables cannot be harmonized. These variables will not be considered for the matching phase.

(ii) Some variables are substituted by new variables, by transformation of the available information.

(iii) Some variables are recoded and recategorized.

An example of these actions will be discussed in Section 7.2.1.

6.2 How to Choose the Matching Variables

In an applied setting, two independent sample surveys A and B may observe many r.v.s in common. For instance, although social surveys may have different objectives (economic, social, health, leisure, cultural), they almost always observe variables such as gender, age, place of residence, occupational status, educational status, and so on. In some cases (as in Example 6.1) a whole set of variables constitutes the core in common between many social surveys, and this core can be used for matching the samples. Should all the common variables be used in the matching process, or just some of them? In other words, is it necessary/useful to divide the common variables into two sets of variables, the *matching variables*, $\mathbf{X}$, and the discarded variables, say Ψ, and if so, how?

As a matter of fact, the fewer the matching variables in $\mathbf{X}$, the easier (from the computational point of view) are the matching procedures. Furthermore, as usual in a multivariate framework, inferential procedures are at the same time less efficient and interpretable when the number of variables increases.

A first approach in determining $\mathbf{X}$, described by Singh *et al.* (1988) and Cohen (1991), quite naturally consists in disregarding all those variables which are not statistically connected with $\mathbf{Y}$ or $\mathbf{Z}$. More formally:

(i) let Ψ_A consist of all the common variables such that Ψ_A is independent of $\mathbf{Y}$ given the other common variables, in A;

(ii) let Ψ_B consist of all the common variables such that Ψ_B is independent of $\mathbf{Z}$ given the other common variables, in B;

(iii) let $\Psi = \Psi_A \bigcap \Psi_B$; then the other common variables define $\mathbf{X}$.

Hence, variables in Ψ are useless: they do not contain any further statistical information on $\mathbf{Y}$ or $\mathbf{Z}$ in addition to that contained in $\mathbf{X}$ when $A \cup B$ is available.

Remark 6.2 More correctly, the selection of $\mathbf{X}$ should have been done so that $(\mathbf{Y}, \mathbf{Z})$ were independent of Ψ given $\mathbf{X}$; see also Remark 2.15. However, this search is not allowed in $A \cup B$.

In order to find those variables which are directly linked with $\mathbf{Y}$ or $\mathbf{Z}$, different procedures can be applied. The easiest, albeit the least precise, consist in computing all the bivariate measures of association among each common variable and each variable in $\mathbf{Y}$, in A, and in $\mathbf{Z}$, in B. Table 6.1 lists some nonparametric association measures that can be computed according to the measurement scale of the variables under study. Note that all of them are test statistics which, under the null hypothesis of independence between the pairs of r.v.s, are (at least approximately) distributed according to known distributions. Some of them are listed in the following. For more details, see Agresti (1990).

Table 6.1 Some nonparametric measures of association, by variable measurement scale

X measurement scale	Y Measurement scale	Association measure
Nominal	Nominal	χ^2
		Φ
		Contingency coefficient
		V (Cramér)
		Uncertainty coefficient
		Concentration coefficient (Goodman and Kruskal's τ)
		Λ (Goodman and Kruskal)
Ordinal	Ordinal	Γ
		d (Somer)
		τ_b (Kendall)
		τ_c (Stuart)
Ordinal	Interval	η (Pearson)
		Point biserial (if X is dichotomous)
Interval	Interval	Pearson's correlation coefficient
		Spearman rank correlation coefficient

- When two variables, say X and Y, are both categorical, various measures of association are based on the classical Pearson chi-squared statistic:

$$\chi^2 = \sum_{i=1}^{I} \sum_{j=1}^{J} \frac{\left(n_{ij} - m_{ij}\right)^2}{m_{ij}},$$

  where n_{ij} and m_{ij} represent respectively the observed and the expected cell frequencies, $i = 1, \ldots, I$, $j = 1, \ldots, J$. One of the most useful is the Cramér's V:

$$V = \sqrt{\frac{\chi^2/n}{\min\{I - 1, J - 1\}}}.$$

  In fact, this measure has the important characteristic of assuming values between 0 (absence of association between X and Y) and 1 (perfect association between X and Y).

- For ordinal variables, Somer's d is based on the comparison between *concordant* (C) and *discordant* (D) pairs of units:

$$d = \frac{C - D}{\frac{1}{2}n(n-1) - U},$$

  where n is the sample size and U is the number of untied pairs for one of the variables.

- When one variable, say X, is categorical, and the other, say Y, is an interval scale variable, one of the most commonly used coefficients is the correlation ratio, i.e. Pearson's η^2:

$$\eta^2 = \frac{1}{\sigma_Y^2} \frac{\sum_{i=1}^{I} (\bar{y}_i - \bar{y})^2 n_i}{n}, \qquad (6.3)$$

  where n is the sample size, n_i is the number of units in the sample with $X = i$, $i = 1, \ldots, I$, $\bar{y}$ is the overall sample average for Y, and $\bar{y}_i$ is the conditional sample average given $X = i$.

- Finally, for continuous variables X and Y, linear independence can be tested with the usual correlation coefficient ρ. A more robust measure is the Spearman rank correlation coefficient, i.e. the correlation coefficient among the X and Y ranks. Note that this measure is based on ranks, and consequently can also be applied when the variables are ordinal, but its robustness makes this measure the usual nonparametric alternative to ρ.

More sophisticated analyses would consider the dependence relationship between multivariate r.v.s. A strategy to reduce the number of matching variables, without losing too much useful information, can choose between the following alternatives.

- When $\mathbf{Y}$ and $\mathbf{Z}$ are continuous, they can be regressed against common variables by means of a stepwise method. $\mathbf{X}$ will be determined by the subset of best predictors.

- Although some common variables may have high explanatory power with respect to $\mathbf{Y}$ or $\mathbf{Z}$, they may be affected by multicollinearity. In this case, hierarchical cluster analysis can be performed by using measures of association such as similarity measures. Dendrogram analysis can be a very effective tool in choosing matching variables.

- If a nonlinear relationship is believed to exist between a univariate Y and $\mathbf{X}$, classification and regression trees (CART) may represent a powerful tool Breiman *et al.* (1984). This methodology is based on binary recursive partitioning of units in order to form groups of units with homogeneous values for the response variable. The final output is in the form of a tree. Variables that appear in the higher part of the tree are usually those with greater explanatory power. CART can deal with both categorical and continuous response variables. At the same time, mixed mode predictors can be included. Moreover, it is completely nonparametric in the sense that no assumptions on the underlying distribution of predictors are required.

- Canonical correlation analysis among the chosen common variables and $\mathbf{Y}$ and $\mathbf{Z}$ (separately) can be performed. In this case, it is possible to restrict $\mathbf{X}$ to those variables with the highest weight in the canonical correlation.

A more precise way of determining Ψ_A and Ψ_B may be based on the concept of the *Markov boundary* of a set of variables. For the sake of simplicity, let us consider the determination of Ψ_A. In the joint multivariate distribution of the common variables and $\mathbf{Y}$, a *Markov blanket* of $\mathbf{Y}$, $M(\mathbf{Y})$, is any set of variables not in $\mathbf{Y}$ such that $\mathbf{Y}$ is independent of the other variables given $M(\mathbf{Y})$. A Markov boundary of $\mathbf{Y}$, $B(\mathbf{Y})$, is any $M(\mathbf{Y})$ such that none of its subsets of variables is a Markov blanket of $\mathbf{Y}$. Hence, all the common variables not in $B(\mathbf{Y})$ form Ψ_A. A similar definition holds for Ψ_B. Both the Markov blanket and Markov boundary of a set of variables are concepts strictly related to the graphical representation of a joint multivariate distribution by means of a Bayesian network. See Neapolitan (2004) and references therein.

For discrete (or discretized) variables, under the CIA, a particular approach has been described in Singh *et al.* (1988). This approach is defined when one file, say A, is the recipient (see Remark 2.10). In this case, $\mathbf{Z}$ should be imputed in A. Hence, the objective is to choose the set of variables $\mathbf{X}$ which are the best common variables for this imputation process. The association between $\mathbf{Z}$ and $\mathbf{X}$ is measured through the computation of the relative entropy distance between the observed distribution $\tilde{f}_{\mathbf{XZ}}$ of $(\mathbf{Z}, \mathbf{X})$ in B and the distribution we would obtain if the variables were independent:

$$I(\mathbf{Z};\mathbf{X}) = \frac{1}{4}\sum_{\mathbf{x},\mathbf{z}} \tilde{f}_{\mathbf{XZ}}(\mathbf{x},\mathbf{z}) \ln \frac{\tilde{f}_{\mathbf{XZ}}(\mathbf{x},\mathbf{z})}{\tilde{f}_{\mathbf{X}}(\mathbf{x})\tilde{f}_{\mathbf{Z}}(\mathbf{z})},$$

where $0\ln(0) = 0$. This index is asymptotically equivalent to the Hellinger distance.

Disregarding the constant, it is also usually named the *mutual information*, and it is used to measure the quantity of information that one r.v. explains about another.

Variables in $\mathbf{X}$ can be chosen in the following way.

(i) Fix a threshold value $\delta > 0$; a suggested threshold is $\delta = 0.05$.

(ii) Determine all those r.v.s with $I(\mathbf{Z};\mathbf{X}) \geq \delta$. Let $\mathbf{W}$ denote all those variables, say $\mathbf{W} = (X_1, \ldots, X_T)$.

(iii) Determine the contribution of each of the variables in $\mathbf{W}$ through the relative entropy property.

For instance, let $T = 2$, i.e. $\mathbf{W} = (X_1, X_2)$. Then, in order to decide if it is convenient to take both the r.v.s or discard one, the conditional contribution of each variable X_t, $t = 1, 2$, is computed. It is possible to compute the conditional contribution $I(Z; X_t|X_v)$ by means of the relative entropy property

$$I(Z; X_1, \ldots, X_t) = \sum_{v=1}^{t} I(X_v; Z|X_{v-1}, \ldots, X_1).$$

Hence, first the marginal contributions $I(Z; X_t)$ for $t = 1, 2$ are computed, and then the conditional contribution

$$I(Z; X_t|X_v) = I(Z; X_1, X_2) - I(Z; X_v)$$

for $t, v = 1, 2$, with $t \neq v$, can be evaluated and compared to a given threshold δ.

7

Applications

7.1 Introduction

There is a rich variety of statistical matching applications, mostly in the fields of microsimulation, marketing research and official statistics.

Microsimulation

Microsimulation is generally used at the decision making level, where it is useful to determine the impact of policy changes on individual units of interest. This is done by applying models to simulate what happens to such units. Microsimulation is based on complex models that require data sets particularly rich in information, in terms of both the number of variables and micro level details. Such information is usually not available in a unique data set, but is collected piecewise from different sources. For this reason, a statistical matching procedure (at micro level) is needed for fusing different sources so as to give a unique database for the microsimulation model application (Martini and Trivellato, 1997). A number of microsimulation models are described in Anagnoson (2000), while microsimulation applications based on statistical matching are dealt with in Cohen (1991).

Microsimulation models are frequently used in the area of tax–benefit analysis. Related to this issue, early applications using statistical matching can be found in Budd (1971) and Okner (1972, 1974).

The experience of Statistics Canada is of considerable interest. A synthetic file called the Social Policy Simulation Database (SPSD) was constructed by combining a number of sources: the Survey of Consumer Finances (contributing income distribution data), a sample of tax return data, a sample of unemployment insurance claim histories, and the Family Expenditure Survey. This database is used by the Social Policy Simulation Model (SPSM) to analyse tax and transfer policy changes. It is possible, for example, to perform macro analyses examining

Statistical Matching: Theory and Practice M. D'Orazio, M. Di Zio and M. Scanu

total program costs and gross redistributive flows, and micro analyses of effects on individuals with given characteristics. A detailed description of the techniques used for the creation of SPSD and for the SPSM is given by Wolfson *et al.* (1987, 1989) and Bordt *et al.* (1990). Similar experiences are described in Ettlinger *et al.* (1996) and Szivós *et al.* (1998). In the latter application, the Social Research Informatics Center of Budapest (TÁRKI) developed a tax–benefit microsimulation model based on a database constructed by the fusion of income and demographic variables from the Hungarian Household Panel Survey, consumption variables from the Household Budget Survey of the Hungarian Central Statistical Office and tax variables from administrative tax records.

The analysis of taxes and benefits using the POLIMOD microsimulation model (Redmond *et al.*, 1998), by matching the UK Family Expenditure Survey and the Family Resources Survey (Sutherland *et al.*, 2001), is the aim of a project carried out by the Microsimulation Unit of the Department of Applied Economics at the University of Cambridge. Moore (2004) describes the creation of a synthetic pension data set called PenSync for use with a microsimulation model (MINT, Modelling Income in the Near Term) developed by the Social Security Administration, in order to evaluate a defined benefit pension plan in meeting the income needs of pensioners.

Abello and Phillips (2004) analyse the statistical matching of the 1998–99 Household Expenditure Survey and the 2001 National Health Survey carried out by the Australian Bureau of Statistics in order to explore the effects of the 'Pharmaceutical Benefits Scheme' through microsimulation modelling.

Further applications deal with the assessment of tax–benefit plans; see, for example, Citoni *et al.* (1991) and Decoster and Van Camp (1998).

Marketing research

Another context where statistical matching has been used is that of marketing research; see, for instance, Baker *et al.*, (1989) O'Brien (1991) and Jephcott and Bock (1991). Advertisers and media planners study customer behaviours by fusing information describing people's characteristics, product and brand usage, and media exposure. Such information enables media planners and marketing researchers to pursue such objectives as increasing sales by formulating the right campaign and selecting the most appropriate media for it. An example is the fusion of the Broadcasters Audience Research Board (BARB) and the Target Group Index (TGI) in the United Kingdom. BARB is a database of information on television exposure. Information is provided by a panel equipped with TV set meters. The TGI database concerns purchasing and media usage. Information is collected by questionnaire and relates to users of more than 400 types of products; see Adamek (1994) and Baker (1990). Further studies and applications related to this topic can be found in Roberts (1994), Buck (1989), Wiegand (1986), Antoine (1987) and Antoine and Santini (1987).

Official statistics

The large number of surveys conducted by national statistical institutes is an appealing area of application for statistical matching procedures. For instance, Radner *et al.* (1980) describe a set of experiences in the United States. A more recent instance is DIECOFIS (see Denk and Hackl, 2003, and references therein), an EU-funded international research project that aims to create a system of indicators of competitiveness and fiscal impact on enterprise performance. To this purpose, one of the aims is the construction of an integrated multisource enterprise database, obtained by the fusion of economic, tax and social insurance sources. Yoshizoe and Araki (1999) describe a Japanese application, in which statistical matching involves the fusion of the Family Income and Expenditure Survey and the Family Savings Survey.

Another application in official statistics is described in detail in the following section. It addresses the problem of the construction of a social accounting matrix.

7.2 Case Study: The Social Accounting Matrix

The new system of national accounts (also known as the European System of Accounts or ESA95) is a source of very detailed information on the economic behaviour of all economic agents, such as households and enterprises. A very important role in ESA95 is played by the social accounting matrix (SAM).

The SAM is a system of statistical information containing economic and social variables in a matrix formatted data framework. The matrix includes economic indicators such as per capita income and economic growth, reporting incomes and expenditures of economic agents grouped by appropriate characteristics (for a more detailed definition see United Nations, 1993). Despite the similarity, an SAM is broader than an input–output table and typical national accounts: it shows more detail about all kinds of transactions within an economy. In fact, an SAM has two main objectives: first, to organize information about the economic and social structure of a country over a period of time (usually a year), and second, to provide a statistical basis for the creation of a plausible economic model capable of presenting a static image of the economy as well as simulating the effects of policy interventions in the economy.

The richness of information that an SAM can provide has led many countries and statistical institutes to try to estimate their own. With some remarkable exceptions (such as the Netherlands), the estimation of an SAM is not straightforward, given that the necessary information has to be collected from many different and independent sources. In the following, we will refer just to one module of the SAM, that related to households. This problem has been studied in Coli and Tartamella (2000a,b) and Coli *et al.* (2005).

The SAM module for households is a matrix containing, for each type of household, the expenditures (broken down into a large number of categories) and incomes (employee income, self-employed income, interest, dividends, rents, social security transfers). Households may be categorized in different ways, for example the area (region) of residence of the household or the primary income source of the household. In Italy, the main sources used for estimating the income and expenditures of households are the Banca d'Italia Survey on Household Income and Wealth (SHIW), and the Italian National Statistical Institute (Istat) Household Budget Survey (HBS).

The two surveys are independent and carried out by different institutes. Integration of information from the two surveys is needed in order to gather information on household expenditures from the HBS and information on household incomes from the SHIW through the information on the socio-economic characteristics common to the two samples. Hence, statistical matching techniques seem appropriate for dealing with this problem. According to all the previous chapters, the matching process should consist of the following steps:

(i) checking the consistency of the two surveys and, if necessary, harmonization of the two surveys (as in Section 6.1);

(ii) definition of the statistical framework for the two samples–this aspect is mainly concerned with the data generating model for the records of the two samples (parametric or nonparametric), and the amount of auxiliary information at hand;

(iii) application of an appropriate statistical matching method (either micro or macro).

In the following sections, some of the work carried out by a group of Istat and Banca d'Italia researchers, on the construction of the SAM in Italy, will be outlined. Full results are given in Coli *et al.* (2005). The following analysis is limited to the year 2000. Data sets are those provided within the workgroup. In particular, the Banca d'Italia data sets are downloadable from the Banca d'Italia website (`http://www.bancaditalia.it/`). The steps are described in a simplified fashion. Only total household income and expenditures will be considered, together with a few socio-economic variables. It will be shown (Section 7.2.4) that the application of statistical matching procedures under the CIA leads to results inconsistent with economic theory. The partial use of auxiliary information in terms of proxy variables is able to recover some consistency (Section 7.2.5). Finally, assessment of uncertainty without the use of any untestable assumption is shown in Section 7.2.6.

7.2.1 Harmonization step

The two main surveys used for estimating the SAM (the SHIW and HBS) are affected by many inconsistencies. These inconsistencies must be resolved in order to make the surveys comparable and suitable for the matching process. As explained in Sections 6.1.1 and 6.1.2, harmonization of two sample surveys generally consists of two major phases: harmonization of (i) population and unit definitions, and (ii) definitions of variables. Both phases should be carried out with care. Note that there is no optimal procedure for the harmonization phase. Indeed, most of the time the harmonization phase implies some form of simplification of the surveys' key characteristics, as will be outlined in the following. This simplification may induce transformations in the original meaning of some variables, changes in the population of interest, and consequently changes in the initial informative power of the sources. Statistical matching results will be heavily affected by these preparatory operations. A good rule of thumb is to change as little as possible during the harmonization step.

Population and unit harmonization

The target population of the two surveys consists of the set of all Italian households resident at a particular day of the year. Unfortunately, the two populations may differ due to inconsistent definitions of the units of the population: the households. The HBS assumes that a household is a set of cohabitants, linked by marriage, familiarity, affinity, adoption, tutelage and affection. The SHIW, on the other hand, assumes that a household consists of a set of people who, regardless of familiarity, completely or partially combine their income. This inconsistency is difficult to resolve, due to the lack of information in the two surveys to enable the transformation, whenever necessary, of a household in one survey into an actual household in the other. Nevertheless, this has been considered to be a minor problem. More precisely, the two populations overlap almost always, i.e. the fraction of the SHIW population of households inconsistent with the HBS, and vice versa, is very low. A check on the data sets has reveated that inconsistent households have not been sampled. As a result, the two samples are considered as samples from a unique population given by the intersection of the two previous population definitions.

Variable harmonization

The two surveys investigate two different aspects of the economic situation of households. Nevertheless, they contain many common variables. Broadly speaking, these variables may be divided into three groups: socio-demographic variables, household expenditure variables and household income variables. These variables are usually inconsistent either in their definition or in their categorization. The harmonization phase adopts different strategies.

(a) Some variables cannot be harmonized. These variables are not considered for the matching phase. However, they can be used after matching has been done.

(b) Some variables are substituted by new variables, by transformation of the available information.

(c) Some variables are recoded.

One variable in the first group is the 'head of household'. This variable is extremely important because a number of characteristics of the head of household (age, gender, education and occupational status) are taken into account in the definition of the socio-economic household groups in the SAM. The use of such information in the definition of socio-economic household groups is justified by the fact that it makes sense that information on the household characteristics can be correlated with both the household expenditures and income. The problem is that the two surveys define the head of household differently. The HBS takes the head of the household to be whoever is already registered in the public archives. The SHIW assumes that the head of the household is the person responsible for the household finances. These two definitions cannot be harmonized through the information available in the two surveys. Consequently, the head of household and his/her characteristics were disregarded during the harmonization phase. Once the matching was completed, the characteristics of the head of household as per the HBS were retained for use in the analyses. Note that this procedure implies an assumption of conditional independence between the head of household characteristics and the SHIW variables not in common with the HBS, given the matching variables.

As far as the second group of transformations is concerned, some variables were created in order to harmonize the characteristics of households. Given that the head of household characteristics could not be considered as information in common for the two surveys, the transformed variables describe the socio-economic characteristics of the different household members. These characteristics were considered at a common unit level, i.e. at the household level. As a consequence, additional variables were computed on the two surveys: among the others, number of household members aged over 64 (categories: 0,1,2 2+), number of employed members (0,1,2,3,3+), number of members who are university graduates (0,1,2,2+), number of females (0,1,2,3,4,4+), and so on. Most of these variables were considered as matching variables (**X**) during the matching process; see Section 7.2.3.

As far as the last group of variables is concerned, many variables in the two surveys are categorized differently. As a consequence, it is necessary to redefine the categories as the largest categories that are in common. An example is shown in Table 7.1 for the variable 'main occupational activity'. In some cases (e.g. for industry and commerce) the HBS has a more detailed classification. This inconsistency can be resolved by amalgamating the corresponding categories. In another case, the HBS category 'other public and private services' is partly consistent with two distinct SHIW categories, i.e. 'public administration' and 'private services'. Hence, a unique category for both public administration and private services is needed.

Remark 7.1 The SHIW and HBS are characterized by two different sample designs. In the next subsections, sample weights will not be considered. In fact, some of the approaches used in finding matching variables were not affected by sample weights. Furthermore, some of the design variables (used for the application of stratified sample designs) will be used as matching variables in random hot deck with donation classes. In this case, sample weights can be disregarded.

Table 7.1 Codification of the variable 'main occupational activity' for the matching of HBS and SHIW. The numbers in parentheses correspond to the old categories of the two surveys

	Category	SHIW	HBS
1	Agriculture	Agriculture (1)	Agriculture (1)
2	Industry	Industry (2)	Energy (2) + Industry (3)
3	Constructions	Constructions (3)	Constructions (4)
4	Commerce	Commerce (4)	Workshops (5) + Commerce (6)
5	Transportation	Transportation (5)	Transportation (7)
6	Banks and insurance	Banks and insurance (6)	Banks and insurance (8)
7	Public administration and private services	Public administration (7) + Private services (8)	Public administration (0) + Other public or private services (9)

7.2.2 Modelling the social accounting matrix

Roughly speaking, an SAM looks like Table 7.2. Columns are of two different kinds:

(i) $\mathbf{C} = (C_1, \ldots, C_u, \ldots, C_U)$ represents different expenditure categories, e.g. food expenditures, durable goods expenditures, and so on;

(ii) $\mathbf{M} = (M_1, \ldots, M_v, \ldots, M_V)$ denotes different income categories, e.g. salaries, dividends and interests, and so on.

Rows correspond to the different household categories of interest, T_w, $w = 1, \ldots, W$. These categories are a function of socio-demographic variables not necessarily in common for the two sources, e.g. head of household educational level,

Table 7.2 A social accounting matrix schema. Values of c_{wu} and m_{wv} should be estimated from the available data sets

	C_1	...	C_u	...	C_U	M_1	...	M_v	...	M_V
T_1	c_{11}	...	c_{1u}	...	c_{1U}	m_{11}	...	m_{1v}	...	m_{1V}
...			...					...		
T_w	c_{w1}	...	c_{wu}	...	c_{wU}	m_{w1}	...	m_{wv}	...	m_{wV}
...			...					...		
T_W	c_{W1}	...	c_{Wu}	...	c_{WU}	m_{W1}	...	m_{Wv}	...	m_{WV}

number of members, primary income source. The table reports, for each household category T_w, $w = 1, \ldots, W$, the amount of expenditure c_{wu} for each expenditure category C_u, as well as the amount of income m_{wv}, for each income category M_v, $v = 1, \ldots, W$.

The HBS consists of socio-demographic variables $\mathbf{X}$ in common with the SHIW, a much more detailed vector of expenditure variables (e.g. if C_1 represents 'food consumption', it can be considered as a combination of such HBS variables as consumption of meat, eggs, fish, vegetables and so on), and finally an incomes variable, TM$^\bullet$. TM$^\bullet$ is the monthly total of household incomes, represented as a categorical variable with 14 categories. Note that some of the household categories T_w (such as primary income source) cannot be derived in the HBS. Hence, some, but not all, of the terms c_{wu} can be directly estimated from the HBS from the available information on the derivable T_w and $\mathbf{C}$.

The SHIW consists of the same socio-demographic variables $\mathbf{X}$, a detailed list of income variables from which the variables M_v, $v = 1, \ldots, V$, can be reconstructed, and a few generic questions on spending (e.g. the amount of expenditure on durable goods or on food). Hence, this survey collects enough information for the estimation of all the terms m_{wv}.

At first sight, part of the SAM can be directly estimated: i.e. those rows where T_w is available in the HBS. However, for these rows there is also a problem. The two independent surveys sometimes produce inconsistent results. In other words, even after reconciliation of definitions and concepts of the two surveys, sample variability produces estimates of the table entries which are incompatible with current economic theory.

Example 7.1 One of the most important indicators when studying incomes and expenditures is the *propensity to consume*. This is defined as consumption as a fraction of the overall income of a household or a group of households. One key characteristic of the propensity to consume is that it is a decreasing function of income. When the components of Table 7.2 are estimated directly from the SHIW and HBS, it turns out that the propensity to consume does not satisfy the previous rule (e.g. when $T_1, \ldots, T_W$ represent the geographical disaggregation in the 20 Italian regions). This and other unpleasant results are investigated in Coli *et al.* (2005).

A natural solution is the joint analysis of $\mathbf{C}$, $\mathbf{M}$, and $\mathbf{X}$ from the two available samples, i.e. statistical matching of HBS and SHIW. Usually, this matching aims at the reconstruction of joint micro information on $\mathbf{C}$, $\mathbf{M}$, and $\mathbf{X}$ (Coli and Tartamella, 2000a).

Remark 7.2 The micro approach is particularly attractive because it furnishes a synthetic sample of important socio-economic variables on the households that are not yet collected in any survey. This has not been and will not be the aim of this application. Our goal is simply the estimation of Table 7.2. The micro approach, used in the next sections, is suitable for this purpose.

As already explained in the previous chapters, even when a micro approach is considered, statistical matching should first estimate a plausible (parametric or nonparametric) model on the variables of interest, and then impute the missing variables in the recipient file. The model is the joint distribution of $\mathbf{X}$, $\mathbf{M}$ and $\mathbf{C}$, which can be decomposed via the following factorization:

$$P(\mathbf{X}, \mathbf{M}, \mathbf{C}|T_w) = P(\mathbf{C}|\mathbf{X}, \mathbf{M}, T_w)P(\mathbf{X}, \mathbf{M}|T_w), \quad w = 1, \ldots, W. \tag{7.1}$$

The joint distribution of $\mathbf{X}$ and $\mathbf{M}$ can easily be estimated from the SHIW. More problematic is the estimate of $P(\mathbf{C}|\mathbf{X}, \mathbf{M}, T_w)$.

- This cannot be estimated from the SHIW, although this survey collects some of the $\mathbf{C}$ variables. In fact, these expenditure variables are not as reliable and detailed as those in the HBS. The main problem is that they may be affected by memory problems: respondents are supposed to remember the amount of expenditure on very large groups of products in a week in the past, and may confuse the week, aggregate expenditures for more than one week, or move a very important purchase (e.g. purchase of a car) in the reference week. This problem does not affect the HBS, which is based on the compilation of a very detailed diary during the reference week.
- Neither can it be estimated from the HBS. In fact, this survey observes $\mathrm{TM}^\bullet$, and not the detailed vector $\mathbf{M}$. Furthermore, even the total amount of income $\mathrm{TM}^\bullet$ is not as reliable as in the SHIW. On the one hand, the SHIW can give the total amount of entries TM instead of just a category. On the other hand, the SHIW computes TM by aggregating the amount of the components of $\mathbf{M}$. Asking directly for the total amount of income, as in the HBS, usually leads to it being underreported. Finally, some household categories are not observed in the HBS.

Hence, the possibilities are the following.

(a) Use only information on $\mathbf{X}$ and $\mathbf{C}$ in the HBS and $\mathbf{X}$ and $\mathbf{M}$ in the SHIW. This option corresponds to performing statistical matching under the CIA.

(b) Make partial use of the unreliable information available in the samples. Actually, although unreliable, the collected information can still be precious for matching purposes. In fact, it can suggest some auxiliary information on the parameters of interest. These suggestions are still untestable hypotheses.

(c) Disregard any sort of untestable assumption and verify how uncertain the parameters of interest are.

For the sake of simplicity, the problem of the estimation of equation (7.1) will be restricted to the following case: instead of $\mathbf{M}$, only the total incomes TM will be considered; instead of $\mathbf{C}$ only the total consumption TC will be considered. Hence, the statistical matching problem is characterized by the following elements.

(i) The SHIW plays the role of A, where $\mathbf{X}$ and TM are observed (TM $= Y$).

(ii) The HBS plays the role of B, where $\mathbf{X}$ and TC are observed (TC $= Z$). Note that this survey also collects data on an additional variable $\text{TM}^{\bullet}$, which is a categorical variable that may play the role of a proxy variable of TM under suitable hypotheses.

Finally, the problem is restricted to the case where T_w is a function of the common variables $\mathbf{X}$. Hence, T_w can be disregarded in (7.1), which becomes:

$$P(\mathbf{X}, \text{TM}, \text{TC}) = P(\text{TC}|\mathbf{X}, \text{TM})P(\mathbf{X}, \text{TM}) = P(\text{TC}, \text{TM}|\mathbf{X})P(\mathbf{X}). \qquad (7.2)$$

The more detailed case where the complete vector variables $\mathbf{C}$ and $\mathbf{M}$ are analysed is described in Coli *et al.* (2005).

Remark 7.3 The two factorizations in (7.2) correspond to two different approaches.

The first factorization is appropriate for the micro approach, when a complete synthetic file is to be obtained. The usual practice considers $A =$ SHIW as the recipient file, and $B =$ HBS as the donor file. This practice is based on the fact that the HBS is a much larger file than the SHIW ($n_A = 8001$ and $n_B = 23\,728$). This factorization will be the reference one in Sections 7.2.4 and 7.2.5.

The second factorization is particularly useful for the macro approach. More precisely, it will be the reference factorization for assessing uncertainty. This approach is based on the fact that $\mathbf{X}$ is reliable in both surveys, and disregarding one of them reduces the amount of available information.

Remark 7.4 It has already been remarked that some of the household categories T_w, such as the 'primary income source', are observed in the SHIW. However, they cannot be derived in the HBS. This problem may be only solved under the following assumption: the variables 'primary income source' and 'consumption' are independent given the common information $\mathbf{X}$. This form of CIA does not generally hold, as emphasized in Coli *et al.* (2005). Further work on this topic is needed.

7.2.3 Choosing the matching variables

The common variables in both surveys are demographic variables and wealth indicators. The demographic variables refer to the number of household members, age, sex, educational level, occupational status, and so on. Wealth indicators are those focused on domestic characteristics such as house ownership, house surface area, ownership of a second home, and so on. Some of these common variables obtained after the harmonization phase are listed in Table 7.3. In order to avoid the case of a functional dependence among common variables, e.g.

$$\text{NCOMP} = \text{MASCHI} + \text{FEMMINE}$$

Table 7.3 Some of the common variables in the SHIW and HBS

Description	Name	Categories
Geographic area of residence	AREA5	1='North-West' 2='North-East' 3='Central' 4='South' 5='Islands'
No. of household members	NCOMP	1,2,3,...
No. of males	MASCHI	0,1,2,...
No. of females	FEMMINE	0,1,2,...
No. of members aged < 18	AGE017	0,1,2,...
No. of members aged 18–34	AGE1834	0,1,2,...
No. of members aged 35–49	AGE3549	0,1,2,...
No. of members aged 50–64	AGE5064	0,1,2,...
No. of members aged ≥ 65	AGE65	0,1,2,...
No. of children	FIGLI	0,1,2,...
Presence of children aged < 15	BAM	1='Presence' 0='Absence'
Presence of members with age > 75	ANZI	1='Presence' 0='Absence'
No. of members with no school qualifications	NOTIT	0,1,2,...
No. of members with up to 8 years' schooling	OBBLIG0	0,1,2,...
No. of members with 9–13 years' schooling	DIPLOMA	0,1,2,...
No. of members with a university degree	LAUREA	0,1,2,...
No. of members with a job	OCC	0,1,2,...
No. of members employed	DIP	0,1,2,...
No. of members self-employed	INDIP	0,1,2,...
No. of members retired	PENS	0,1,2,...
No. of members of working age who are economically inactive	OTHER	0,1,2,...
House renting	AFFCASA	1='Yes' 2='other'
Total surface area of house (m^2)	SUPERFT	

or

$$\text{NCOMP} = \text{DIP} + \text{INDIP} + \text{PENS} + \text{OTHER},$$

the following variables have been eliminated: MASCHI, AGE65, NOTIT, OBBLIG0, INDIP, OTHER.

As far as the SHIW is concerned, the average monthly 'household net income' (in euro) in 2000 is considered as the total amount of income (TM.SHIW).

Moreover, despite their reliability, we decided to analyse some consumption variables: the average monthly 'household food consumption' (TF.SHIW) and the average monthly 'total household consumption' (TC.SHIW), both in euro. 'Household food consumption' is considered because at first sight a comparison of this variable in the two surveys shows few discrepancies. However, a more refined analysis shows that food consumption and other consumption variables observed in the SHIW are affected by rounding, see Battistin *et al.* (2003).

The HBS observes many consumption categories. To simplify, we restricted our attention to average monthly food consumption (TF.HBS) and average monthly overall consumption (TC.HBS). As far as income is considered, the HBS observes only categories of average monthly income (TM•.HBS). There are 14 categories: '1'= 'Up to € 310', ..., '14'='More than € 6 198'.

The main results on these two surveys can be found in Istat (2002) and Banca d'Italia (2002).

Various techniques can be used to identify the subset of those common variables that best explain income and consumption (see Section 6.2). TM and TC being continuous variables, a first analysis can take into consideration the study of a linear relation between them on **X**. In this case, it is a common practice to apply a logarithmic transform to both TM and TC. A subset of the common variables is then derived by analysing the table of the estimated regression coefficients or by letting variables selected by means of a stepwise regression.

As far as the SHIW is concerned, we have regressed log(TM.SHIW) on a subset of socio-demographic variables. In this application, counting variables have been considered as categorical variables, after some suitable truncation (e.g. NCOMP5 refers to NCOMP truncated so that '5' stands for 5 and more household members) and therefore a generalized linear model has been fitted by means of the R function `glm`.

```
> glmf.tm.shiw<-glm(appo.shiw$y[ ,"tm"] ~ .,data=appo.
      shiw$x)
> summary(glmf.tm.shiw, correlation=F)

Call:
glm(formula = appo.shiw$y[, "tm"] ~ ., data = appo.
      shiw$x)

Deviance Residuals:
     Min        1Q     Median        3Q       Max
-6.14314  -0.22718   0.02089   0.28074   3.10271

Coefficients:
                Estimate Std. Error t value Pr(>|t|)
(Intercept)   6.5086494  0.0375582 173.295  < 2e-16 ***
area52       -0.0287229  0.0202620  -1.418 0.156353
area53       -0.1124298  0.0201655  -5.575 2.55e-08 ***
```

```
area54      -0.3550997  0.0201098 -17.658  < 2e-16 ***
area55      -0.3696007  0.0243135 -15.201  < 2e-16 ***
superft      0.0022236  0.0001343  16.560  < 2e-16 ***
affcasa     -0.3871204  0.0171657 -22.552  < 2e-16 ***
ncomp52      0.2505843  0.0276929   9.049  < 2e-16 ***
ncomp53      0.2413520  0.0478758   5.041 4.73e-07 ***
ncomp54      0.3235971  0.0687737   4.705 2.58e-06 ***
ncomp55      0.3043979  0.0898564   3.388 0.000708 ***
femmine41   -0.1063617  0.0298981  -3.557 0.000377 ***
femmine42   -0.1404394  0.0350878  -4.003 6.32e-05 ***
femmine43   -0.1864614  0.0421792  -4.421 9.97e-06 ***
femmine44   -0.0936023  0.0702709  -1.332 0.182892
age0171      0.0050051  0.0301753   0.166 0.868266
age0172     -0.1074214  0.0496704  -2.163 0.030595 *
age0173      0.0140428  0.0843767   0.166 0.867823
age0174     -0.4472977  0.1875576  -2.385 0.017109 *
age0175     -0.2660207  0.3224264  -0.825 0.409363
age0176      0.4932686  0.6076527   0.812 0.416953
age18341    -0.1542386  0.0254107  -6.070 1.34e-09 ***
age18342    -0.3514850  0.0423862  -8.292  < 2e-16 ***
age18343    -0.4138880  0.0846925  -4.887 1.04e-06 ***
age18344    -0.2501429  0.1714880  -1.459 0.144698
age18345    -0.6483267  0.3618822  -1.792 0.073245 .
age18346    -0.9567347  0.6064768  -1.578 0.114714
age35491    -0.1265838  0.0254135  -4.981 6.46e-07 ***
age35492    -0.1540725  0.0395309  -3.898 9.80e-05 ***
age35493    -0.3639750  0.2489511  -1.462 0.143772
age35494    -0.3499535  0.6037024  -0.580 0.562148
age50641    -0.0090927  0.0215514  -0.422 0.673105
age50642    -0.0491191  0.0309297  -1.588 0.112306
age50643     0.0068277  0.2481726   0.028 0.978052
figli41     -0.0712858  0.0300311  -2.374 0.017633 *
figli42     -0.0257224  0.0462960  -0.556 0.578495
figli43     -0.0713597  0.0664255  -1.074 0.282729
figli44     -0.0464591  0.1044761  -0.445 0.656559
bam          0.0117564  0.0288529   0.407 0.683680
anzi        -0.0184400  0.0226151  -0.815 0.414876
diploma41    0.1942655  0.0168678  11.517  < 2e-16 ***
diploma42    0.3365010  0.0223213  15.075  < 2e-16 ***
diploma43    0.4846376  0.0387511  12.506  < 2e-16 ***
diploma44    0.5228207  0.0672939   7.769 8.89e-15 ***
laurea31     0.3256796  0.0219685  14.825  < 2e-16 ***
laurea32     0.5722453  0.0378279  15.128  < 2e-16 ***
laurea33     0.6793931  0.1105918   6.143 8.48e-10 ***
```

```
occ41          0.8020079  0.0287927  27.855  < 2e-16 ***
occ42          1.1893779  0.0380773  31.236  < 2e-16 ***
occ43          1.4205565  0.0633873  22.411  < 2e-16 ***
occ44          1.5836739  0.1134310  13.962  < 2e-16 ***
dip41         -0.0027170  0.0233363  -0.116 0.907316
dip42         -0.0222083  0.0319790  -0.694 0.487411
dip43          0.0090986  0.0711376   0.128 0.898230
dip44          0.1505702  0.1460156   1.031 0.302482
pens31         0.4341889  0.0223661  19.413  < 2e-16 ***
pens32         0.6379675  0.0330352  19.312  < 2e-16 ***
pens33         0.9559579  0.0969353   9.862  < 2e-16 ***
---
Signif. codes:  0 '***' 0.001 '**' 0.01 '*' 0.05 '.'
     0.1 ' ' 1

(Dispersion parameter for gaussian family taken
to be 0.3510427)

    Null deviance: 5938.1  on 7981  degrees of freedom
Residual deviance: 2781.7  on 7924  degrees of freedom
AIC: 14356

Number of Fisher Scoring iterations: 2
```

A further stepwise regression was carried out to automatically choose the subset of independent variables that best explain the average monthly income. The R function `step` based on the AIC criterion was used. The following shows the list of the regressors excluded by the model.

```
> fin.glm.tm.shiw <- step(glmf.tm.shiw, trace=F)
> fin.glm.tm.shiw$anova

       Step Df   Deviance Resid. Df Resid. Dev      AIC
1             NA         NA      7924   2781.662 14355.79
2     - dip4  4 0.66559754      7928   2782.328 14349.70
3 - age5064  3 1.04252611      7931   2783.370 14346.69
4     - anzi  1 0.05245522      7932   2783.423 14344.84
5      - bam  1 0.07189275      7933   2783.494 14343.04
6   - figli4  4 2.50568854      7937   2786.000 14342.22
```

To identify which regressor best explains the variability of TM.SHIW, an ANOVA should be carried out. We decided to compute η^2 (see equation (6.3)) and partial η_p^2, where the denominator of η^2 is substituted by the sum of the deviance explained by the regressor and the residual deviance.

The coefficient η^2 can be interpreted as the portion of deviance of the dependent variable explained by the chosen common variable. It has the advantage of being able to capture the nonlinear association among the variables. To compute η^2 the *ad hoc* function `eta.fcn` was developed (code is given in Section E.5):

```
> eta.fcn(fin.glm.tm.shiw)

$dev.tot
[1] 5938.141

$dev.res
[1] 2786

$anova
          Df  Deviance         eta        eta.p
superft    1 645.30967 0.108672000 0.188065115
ncomp5     4 514.29593 0.086608909 0.155833270
area5      4 488.32820 0.082235870 0.149138435
occ4       4 480.73781 0.080957626 0.147161425
affcasa    1 299.24839 0.050394287 0.096993286
laurea3    3 259.37305 0.043679165 0.085169545
pens3      3 189.76991 0.031957797 0.063771698
diploma4   4 145.55040 0.024511106 0.049649632
age017     6  74.24528 0.012503119 0.025957662
femmine4   4  26.18363 0.004409399 0.009310783
age3549    4  15.05432 0.002535191 0.005374519
age1834    6  14.04440 0.002365117 0.005015776
```

The `eta` column shows the proportion of TM.SHIW deviance explained by each regressor. The regressors are ordered by their importance in explaining the variability of TM.SHIW. SUPERFT tops the list, explaining 10.9% of total deviance. The number of household members (NCOMP5) the geographic area of residence of the households (AREA5) and the number of members with a job (OCC4) are able to explain at least 8% of the total deviance each.

To explore the nonlinear relation among income and the matching variables another possibility is that of using a regression tree (CART). In R this can be done by means of the function `tree` contained in the R library `tree`.

```
> tree.tm.shiw<-tree(appo.shiw$y[,"tm"]~.,data=appo.
      shiw$x)
> summary(tree.tm.shiw)

Regression tree:
tree(formula = appo.shiw$y[, "tm"] ~ ., data = appo.
      shiw$x)
Variables actually used in tree construction:
[1] 'occ4' 'pens3' 'affcasa' 'area5' 'superft' 'laurea3'
```

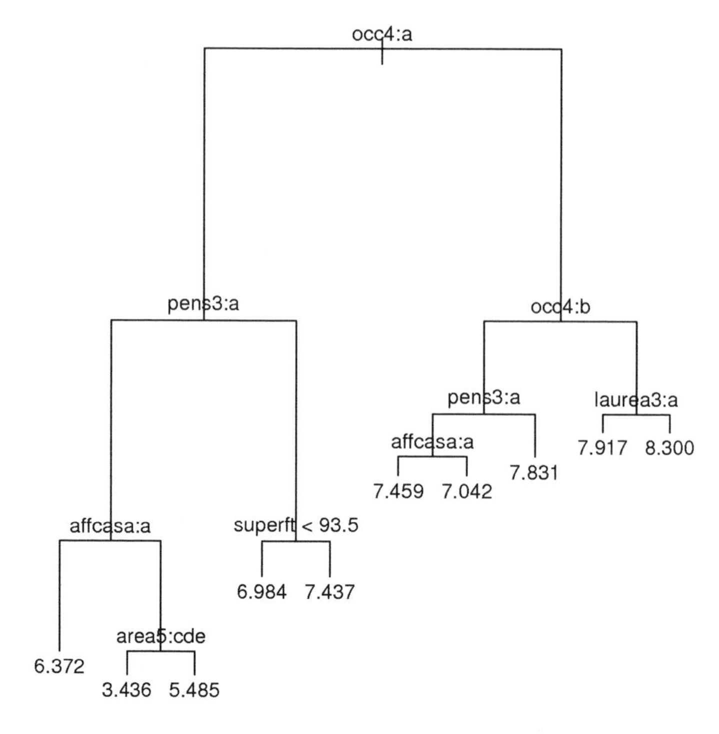

Figure 7.1 Estimated regression tree

```
Number of terminal nodes:  10
Residual mean deviance:  0.3626 = 2891 / 7972
Distribution of residuals:
      Min.     1st Qu.      Median        Mean   3rd Qu.
-5.000e+00 -2.861e-01 -3.665e-03 1.102e-17 2.980e-01
                                                    Max.
                                               3.534e+00
```

It can easily be observed that the variables used for the construction of the regression tree are the first seven variables with the highest value of η^2 with the exception of the number of components. Figure 7.1 shows the estimated regression tree.

The same analysis was carried out on the consumption variables observed in the SHIW, to get a general idea of their relation with the socio-demographic matching variables. Clearly, results have to be treated with caution given that the consumption variables observed in the SHIW are not reliable (the consumption variables were transformed logarithmically before the regression was carried out). In the following we report only the values of η^2.

As far as the average monthly total consumption TC.SHIW is considered the following results are obtained:

```
> eta.fcn(fin.glm.tc.shiw)

$dev.tot
[1] 2484.704

$dev.res
[1] 1137.103

$anova
          Df    Deviance          eta        eta.p
superft    1 396.1150570 1.594214e-01 0.2583553614
ncomp5     4 383.0814804 1.541759e-01 0.2519967362
area5      4 254.7987876 1.025470e-01 0.1830580463
laurea3    3 116.7504869 4.698769e-02 0.0931133546
diploma4   4  69.5799284 2.800331e-02 0.0576621559
occ4       4  35.2816880 1.419956e-02 0.0300939563
affcasa    1  29.3686290 1.181977e-02 0.0251773237
age5064    3  16.1767488 6.510534e-03 0.0140267367
anzi       1  11.9974101 4.828507e-03 0.0104406995
femmine4   4  10.3243011 4.155144e-03 0.0089977838
age3549    4   9.0194114 3.629975e-03 0.0078695022
pens3      3   7.7456194 3.117321e-03 0.0067656287
age017     6   4.6218383 1.860117e-03 0.0040481200
dip4       4   2.0944223 8.429264e-04 0.0018385072
figli4     4   0.4437108 1.785770e-04 0.0003900595
bam        1   0.2013572 8.103873e-05 0.0001770478
```

Notice that the most important regressors are more or less the same as for TM.SHIW. For average monthly food consumption (TF.SHIW), we have the following results.

```
> eta.fcn(fin.glm.tf.shiw)

$dev.tot
[1] 2405.812

$dev.res
[1] 1412.944

$anova
          Df   Deviance          eta        eta.p
ncomp5     4 686.490175 0.2853466163 0.3269881605
superft    1 127.535640 0.0530114846 0.0827895563
area5      4  58.396832 0.0242732365 0.0396895293
laurea3    3  38.802310 0.0161285745 0.0267280208
```

```
diploma4  4  21.795419 0.0090594873 0.0151912019
age5064   3  21.028461 0.0087406934 0.0146644786
occ4      4  10.394308 0.0043204998 0.0073027662
affcasa   1   6.382165 0.0026528115 0.0044966153
femmine4  4   5.899023 0.0024519890 0.0041576288
pens3     3   5.514741 0.0022922583 0.0038878401
dip4      4   3.027080 0.0012582366 0.0021378119
figli4    4   1.961494 0.0008153147 0.0013863069
bam       1   1.913703 0.0007954502 0.0013525763
age1834   6   1.354948 0.0005631981 0.0009580353
age3549   4   1.255609 0.0005219066 0.0008878582
anzi      1   1.115435 0.0004636418 0.0007888173
```

It is interesting to observe that NCOMP5 alone explains 28.5% of the variability of food consumption, while the second variable in the list, SUPERFT, accounts for only 5.3% of overall variability.

The previous analyses of the dependence relationship between TM and the common variables in the SHIW, confirmed also by the analysis on the dependence relationship between consumption variables and the common variables, led us to choose as suggested matching variables those in Table 7.4.

Obviously, it is necessary to see if the previous analyses are also confirmed in the HBS. In this case, it is of primary importance to study the dependence relationship between the consumption and the common variables. As far as average monthly overall consumption is concerned, the following result is obtained when a stepwise procedure is used to choose which common variables to consider as regressors:

```
> glmf.tc.hbs<-glm(appo.hbs$ytn[,"tc"]~.,data=appo.
       hbs$x)
> fin.glmf.tc.hbs<-step(glmf.tc.hbs, trace=F)
> summary(fin.glmf.tc.hbs)

Call:
glm(formula = appo.hbs$ytn[,"tc"] ~ area5 + superft
    + affcasa + ncomp5 + femmine4 + age017 + age1834
```

Table 7.4 Suggested matching variables chosen with the analysis of dependence relationship between TM and common variables in the SHIW

Dependent variable	Suggested matching variables
TM	AREA5, NCOMP5, AFFCASA, SUPERFT, OCC4, PENS3, LAUREA3

```
    + age3549 + age5064 + figli4 + bam + anzi + diploma4
    + laurea3 + occ4 + dip4 + pens3, data = appo.hbs$x)

Deviance Residuals:
     Min         1Q     Median         3Q        Max
-2.11164  -0.33238  -0.03029   0.29583   2.49383

Coefficients:
              Estimate Std. Error t value Pr(>|t|)
(Intercept)  6.8452909  0.0185647 368.727  < 2e-16 ***
area52      -0.0345951  0.0100392  -3.446  0.00057 ***
area53      -0.1605737  0.0102123 -15.724  < 2e-16 ***
area54      -0.3177787  0.0096196 -33.035  < 2e-16 ***
area55      -0.3407705  0.0125771 -27.095  < 2e-16 ***
superft      0.0025123  0.0000807  31.132  < 2e-16 ***
affcasa     -0.1467882  0.0087581 -16.760  < 2e-16 ***
ncomp52      0.2238655  0.0138543  16.159  < 2e-16 ***
ncomp53      0.2580898  0.0245608  10.508  < 2e-16 ***
ncomp54      0.2608409  0.0343406   7.596 3.17e-14 ***
ncomp55      0.2571744  0.0452540   5.683 1.34e-08 ***
femmine41   -0.0607178  0.0147003  -4.130 3.63e-05 ***
femmine42   -0.0503898  0.0173126  -2.911  0.00361 **
femmine43   -0.0390848  0.0206862  -1.889  0.05885 .
femmine44   -0.0718506  0.0330204  -2.176  0.02957 *
age0171      0.1145290  0.0141489   8.095 6.02e-16 ***
age0172      0.2097433  0.0229549   9.137  < 2e-16 ***
age0173      0.3136309  0.0402564   7.791 6.93e-15 ***
age0174      0.3683167  0.0826348   4.457 8.34e-06 ***
age0175      0.7043700  0.1550908   4.542 5.61e-06 ***
age0176      0.6184123  0.5053056   1.224  0.22103
age18341     0.0943412  0.0121870   7.741 1.02e-14 ***
age18342     0.1636143  0.0201316   8.127 4.61e-16 ***
age18343     0.1536178  0.0393141   3.907 9.35e-05 ***
age18344     0.0586779  0.0817994   0.717  0.47317
age18345     0.3496619  0.1987072   1.760  0.07847 .
age18346     0.3449234  0.2578000   1.338  0.18093
age18348     0.3312053  0.5085980   0.651  0.51492
age35491     0.1038359  0.0123591   8.402  < 2e-16 ***
age35492     0.1998795  0.0194373  10.283  < 2e-16 ***
age35493     0.2333317  0.1107848   2.106  0.03520 *
age35494     0.2194125  0.2567522   0.855  0.39280
age50641     0.0754488  0.0108269   6.969 3.28e-12 ***
age50642     0.1496586  0.0154981   9.657  < 2e-16 ***
age50643     0.1431271  0.1206525   1.186  0.23553
```

```
age50644      0.2786358  0.3580266    0.778  0.43643
figli41      -0.0917404  0.0163223   -5.621 1.92e-08 ***
figli42      -0.1484182  0.0239513   -6.197 5.86e-10 ***
figli43      -0.2027771  0.0341995   -5.929 3.09e-09 ***
figli44      -0.2111099  0.0533700   -3.956 7.66e-05 ***
bam          -0.0422143  0.0135285   -3.120  0.00181 **
anzi         -0.1387619  0.0114096  -12.162  < 2e-16 ***
diploma41     0.1622769  0.0083586   19.414  < 2e-16 ***
diploma42     0.2150501  0.0106606   20.173  < 2e-16 ***
diploma43     0.2857706  0.0192512   14.844  < 2e-16 ***
diploma44     0.3599232  0.0334207   10.769  < 2e-16 ***
laurea31      0.2258502  0.0111784   20.204  < 2e-16 ***
laurea32      0.3715385  0.0187909   19.772  < 2e-16 ***
laurea33      0.4184028  0.0450026    9.297  < 2e-16 ***
occ41         0.2548662  0.0141206   18.049  < 2e-16 ***
occ42         0.3400445  0.0177759   19.129  < 2e-16 ***
occ43         0.4195959  0.0307959   13.625  < 2e-16 ***
occ44         0.4492974  0.0555396    8.090 6.27e-16 ***
dip41        -0.0628915  0.0110996   -5.666 1.48e-08 ***
dip42        -0.0639507  0.0150244   -4.256 2.08e-05 ***
dip43        -0.0714607  0.0344919   -2.072  0.03829 *
dip44         0.0100156  0.0756351    0.132  0.89465
pens31        0.0449552  0.0105546    4.259 2.06e-05 ***
pens32        0.1187042  0.0158951    7.468 8.43e-14 ***
pens33        0.3780902  0.0542352    6.971 3.22e-12 ***
---
Signif. codes: 0 '***' 0.001 '**' 0.01 '*' 0.05 '.'
          0.1 ' ' 1

(Dispersion parameter for gaussian family taken
to be 0.2516786)

    Null deviance: 10376.0  on 23684  degrees of freedom
Residual deviance:  5945.9  on 23625  degrees of freedom
AIC: 34601

Number of Fisher Scoring iterations: 2
```

Also in this context, it is useful to consider the estimates of the η^2 coefficient:

```
> eta.fcn(fin.glmf.tc.hbs)

$dev.tot
[1] 10376.04
```

```
$dev.res
[1] 5945.907

$anova
          Df    Deviance          eta        eta.p
ncomp5     4 1703.010939 1.641291e-01 0.2226472945
superft    1 1224.894747 1.180503e-01 0.1708169809
area5      4  356.544050 3.436224e-02 0.0565722818
laurea3    3  249.761618 2.407099e-02 0.0403122929
age5064    4  179.802918 1.732866e-02 0.0293521753
age3549    4  156.844604 1.511603e-02 0.0257006360
diploma4   4  155.341224 1.497114e-02 0.0254605635
occ4       4  113.565287 1.094495e-02 0.0187417777
affcasa    1   82.660909 7.966516e-03 0.0137115327
age1834    7   53.880402 5.192770e-03 0.0089803848
femmine4   4   48.515089 4.675683e-03 0.0080933718
anzi       1   47.971841 4.623328e-03 0.0080034717
pens3      3   21.198824 2.043055e-03 0.0035526139
figli4     4   17.427417 1.679582e-03 0.0029224281
age017     6    9.767385 9.413401e-04 0.0016400132
dip4       4    8.094017 7.800679e-04 0.0013594248
bam        1    0.853699 8.227598e-05 0.0001435570
```

The first two variables, NCOMP5 and SUPERFT, explain respectively 16.4% and 11.8% of the total deviance of the average monthly overall consumption. Note that, this output is similar, if we limit attention to the first four variables, to the results obtained for total consumption in the SHIW. In this case, however, AREA5 and LAUREA3 have a smaller explanatory power than in the SHIW.

The results change slightly if a regression tree is fitted to the data. In fact, now AREA5 and LAUREA3 are replaced by OCC4 and DIPLOMA4.

```
> tree.tc.hbs <- tree(appo.hbs$ytn[,"tc"]~ .,
                      data=appo.hbs$x)
> summary(tree.tc.hbs)

Regression tree:
tree(formula = appo.hbs$ytn[, "tc"] ~ ., data = appo.
     hbs$x)
Variables actually used in tree construction:
[1] "occ4"     "ncomp5"   "diploma4" "superft"
Number of terminal nodes:  6
Residual mean deviance:  0.3063 = 7253 / 23680
```

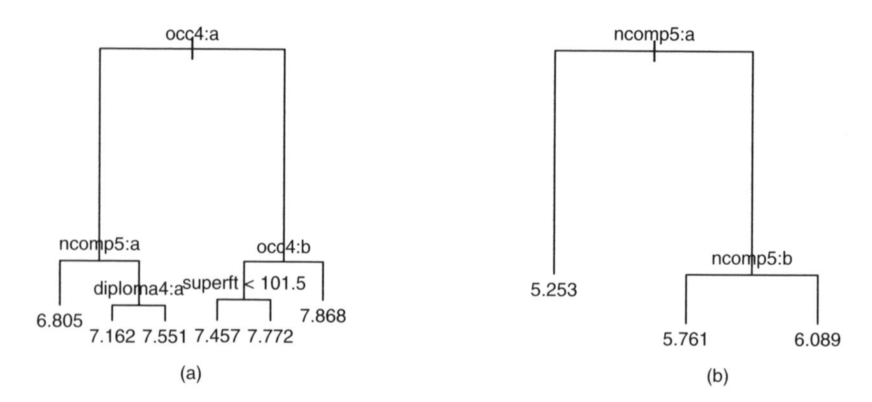

Figure 7.2 Estimated regression trees: (a) TC; (b) TF

```
Distribution of residuals:
      Min.     1st Qu.      Median        Mean   3rd Qu.
-2.238e+00 -3.662e-01 -2.082e-02 -4.459e-17 3.357e-01
                                                    Max.
                                               2.657e+00
```

The estimated regression tree is shown in Figure 7.2(a).

For HBS food consumption data, the values of η^2 after discarding less important variables are the following.

```
> eta.fcn(fin.glmf.tf.hbs)

$dev.tot
[1] 11208.46

$dev.res
[1] 8403.657

$anova
         Df    Deviance          eta         eta.p
ncomp5    4 2172.949136 0.1938669113 0.2054486213
superft   1  419.748181 0.0374492353 0.0475721307
area5     4   96.544352 0.0086135266 0.0113578901
age5064   4   46.175245 0.0041196786 0.0054646345
occ4      4   11.574689 0.0010326745 0.0013754451
pens3     3   11.447742 0.0010213485 0.0013603802
anzi      1   10.346886 0.0009231320 0.0012297221
femmine4  4   10.125400 0.0009033713 0.0012034302
affcasa   1    8.268614 0.0007377120 0.0009829633
bam       1    7.347518 0.0006555333 0.0008735601
```

```
age1834    7     6.813282 0.0006078697 0.0008100952
diploma4   4     3.460034 0.0003086985 0.0004115602
```

NCOMP5 explains 19.4% of the overall deviance, but SUPERFT, the second variable in terms of deviance explanation, accounts for a modest 3.7%. It appears that the number of household members is the only important variable in predicting average monthly food consumption. The same findings are determined when a regression tree is fitted (Figure 7.2(b)):

```
> tree.tf.hbs<-tree(appo.hbs$ytn[,"tf"]~.,data=appo.
       hbs$x)
> summary(tree.tf.hbs)

Regression tree:
tree(formula = appo.hbs$ytn[, "tf"] ~ ., data = appo.
       hbs$x)
Variables actually used in tree construction:
[1] "ncomp5"
Number of terminal nodes:  3
Residual mean deviance:  0.3692 = 8742 / 23680
Distribution of residuals:
      Min.    1st Qu.    Median       Mean   3rd Qu.
-6.089e+00 -3.102e-01 3.332e-02 1.310e-16 3.646e-01
                                                 Max.
                                            2.453e+00
```

The results of the analyses carried out on income data in the HBS are not really significant. Given that the observed income in the HBS, TM• (denoted as `tmcat` in the R computations), is a categorical variable with 14 response categories, we tried to fit a classification tree:

```
> tree.tmcat.hbs <- tree(appo.hbs$yfac[,"tmcat"]~.,
+                        data=appo.hbs$x)
> summary(tree.tmcat.hbs)

Classification tree:
tree(formula = appo.hbs$yfac[, "tmcat"] ~ ., data =
       appo.hbs$x)
Variables actually used in tree construction:
[1] "occ4"    "ncomp5" "pens3"
Number of terminal nodes:  5
Residual mean deviance:  3.787 = 89680 / 23680
Misclassification error rate: 0.7042 = 16678 / 23685
```

The procedure uses only three variables, OCC4, NCOMP5 and PENS3, to classify units in terms of monthly income categories, but the high misclassification rate indicates that the model does not fit the data.

Table 7.5 Suggested matching variables chosen with the analysis of dependence relationship between TC and common variables in the HBS

Dependent variable	Suggested matching variables
TC	AREA5, NCOMP5, SUPERFT, OCC4, LAUREA3, DIPLOMA4

Table 7.5 reproduces the suggested matching variables from the analyses on the HBS. These results substantially confirm what was already found in Table 7.4 (note that only AFFCASA and PENS3 are not considered as important variables in the matching phase). From the results of Section 6.2, we considered as matching variables those listed in Table 7.5.

7.2.4 The SAM under the CIA

For the present problem, the CIA takes the following form.

Assumption 7.1 TC *and* TM *are independent given* **X**.

Under this assumption, equation (7.2) becomes:

$$P(\mathbf{X}, \mathrm{TM}, \mathrm{TC}) = P(\mathrm{TC}|\mathbf{X})P(\mathbf{X}, \mathrm{TM}). \tag{7.3}$$

The two factors that can be estimated from the two surveys are, in obvious notation:

- $\tilde{P}^{(\mathrm{SHIW})}(\mathbf{X}, \mathrm{TM})$;
- $\tilde{P}^{(\mathrm{HBS})}(\mathrm{TC}|\mathbf{X})$.

If the CIA holds, a synthetic complete data set can be obtained by using a hot deck matching technique (Section 2.4). We decided to carry out random and nearest neighbour hot deck by using the SHIW as recipient (see Remark 7.3). Random hot deck was carried out with two different possible household stratifications: by (i) geographic area, AREA5, and (ii) both geographic area, AREA5, and number of household members, NCOMP5. In case (i), for each household in a given geographic area we randomly chose an HBS household in the same area without regard to the number of household members (NCOMP5). In case (ii), the further constraint of equal number of members is introduced. The R functions used for the analysis are those reported in Section E.2.

Table 7.6 reports the ratios of some descriptive statistics for the two variables TC.HBS and TF.HBS computed in the filled data set and in the original data set (HBS) in case (i) and (ii), respectively.

Random hot deck preserves the summary characteristics of consumption variables quite well. However, it is possible to observe that the estimated correlations

Table 7.6 Comparison of summary statistics for food and total consumption computed on the final synthetic data set obtained by means of random hot deck in case (i) and (ii) with respect to the corresponding estimates in the donor file HBS (estimates in HBS=100; unweighted data)

	(i)	(ii)
$\hat{\mu}_{TC}$	99.97	100.08
$\hat{\sigma}_{TC}$	99.91	99.75
$\hat{\mu}_{TF}$	99.76	99.83
$\hat{\sigma}_{TF}$	101.88	101.87
$\hat{\rho}_{TC,TF}$	98.11	96.80

Table 7.7 Estimated correlations between some income and consumption variables (observed in the SHIW and imputed) in the final synthetic file after random hot deck matching (i) within the same geographic area, and (ii) within the same geographic area and for equal numbers of household members

	(i)	(ii)
$\hat{\rho}_{TM.SHIW,\widetilde{TC}}$	0.0511	0.2154
$\hat{\rho}_{TC.SHIW,\widetilde{TC}}$	0.0563	0.2750
$\hat{\rho}_{TF.SHIW,\widetilde{TC}}$	0.0206	0.2942
$\hat{\rho}_{TM.SHIW,\widetilde{TF}}$	−0.0250	0.1407
$\hat{\rho}_{TC.SHIW,\widetilde{TF}}$	−0.0334	0.1832
$\hat{\rho}_{TF.SHIW,\widetilde{TF}}$	−0.0295	0.2526

between the imputed variables ($\widetilde{TC}$ and $\widetilde{TF}$) and the observed ones are negligible (Table 7.7). This is a clear effect of the CIA.

Correlation estimates increase slightly when donation classes are formed by two variables, AREA5 and NCOMP5, as shown in the second column of Table 7.7. However, these estimates show still negligible correlations among the variables.

In the application of nearest neighbour distance matching, different subsets of the chosen matching variables were considered. In particular, AREA5 and NCOMP5 were always used to define the donation classes while, for units within the same donation classes, distance was computed with different combinations of the matching variables (see Table 7.8). In this case, we have used Gower distance (see Appendix C). The corresponding R code is reported at the end of Section E.2.

Table 7.8 Different subsets of the matching variables used to perform nearest neighbour matching within the same geographic area and for equal numbers of household members

Case	Matching Variables
(d1)	SUPERFT
(d2)	SUPERFT, OCC4
(d3)	SUPERFT, OCC4, LAUREA3
(d4)	SUPERFT, OCC4, LAUREA3, DIPLOMA4
(d5)	SUPERFT, OCC4, LAUREA3, DIPLOMA4, PENS3
(d6)	SUPERFT, OCC4, LAUREA3, DIPLOMA4, PENS3, AFFCASA

Table 7.9 Comparison of summary statistics for food and total consumption computed on the final synthetic data set obtained by means of nearest neighbour distance with various subsets of the matching variables with respect to the corresponding estimates in the donor file HBS (estimates in HBS=100; unweighted data)

	(d1)	(d2)	(d3)	(d4)	(d5)	(d6)
$\hat{\mu}_{\mathrm{TC}}$	100.00	99.70	99.49	99.17	99.08	97.93
$\hat{\sigma}_{\mathrm{TC}}$	100.01	99.15	99.88	99.69	101.72	120.75
$\hat{\mu}_{\mathrm{TF}}$	100.05	99.65	99.68	99.35	99.08	98.31
$\hat{\sigma}_{\mathrm{TF}}$	101.42	100.11	99.93	100.77	99.06	105.79
$\hat{\rho}_{\mathrm{TC,TF}}$	100.60	99.90	103.49	102.46	105.90	109.23

Table 7.9 reports the ratios of some summary statistics computed for total and food consumption respectively in the HBS and SHIW data sets obtained at the end of the matching procedure for each of the subsets of the matching variables considered.

Table 7.9 shows that with the first two hypotheses, (d1) and (d2), the final synthetic data set also preserves the average and the standard deviation of TC and TF and their correlation quite well. A tendency to overestimate the correlation between TC and TF can be observed in (d3), (d4), (d5) and (d6). In this last case (where AFFCASA is added) a significant worsening is observed in the estimates computed on the final synthetic data set.

Table 7.10 reports some summary results concerning correlation among income observed in the SHIW and consumption variables donated from the HBS.

It can be observed that the highest correlation among income and total consumption is achieved with subset (d2) when only two matching variables are used to compute the distances between SHIW records and those from the HBS belonging

Table 7.10 Estimated correlations among income and consumption in the fused SHIW data set after nearest neighbour matching within the same geographic area and for equal numbers of household members

	$\hat{\rho}_{\mathrm{TM},\widetilde{\mathrm{TC}}}$	$\hat{\rho}_{\mathrm{TM},\widetilde{\mathrm{TF}}}$
(d1)	0.3037	0.1764
(d2)	0.3345	0.1791
(d3)	0.3084	0.1617
(d4)	0.3009	0.1432
(d5)	0.2779	0.1372
(d6)	0.2531	0.1269

to the same donation classes (identified by AREA5 and NCOMP5). The estimated correlation decreases by adding further variables into the distance computation. In general, the estimated correlation between income and total consumption does not take values greater than 0.33.

7.2.5 The SAM and auxiliary information

It has been already claimed that the HBS includes a categorical variable $\mathrm{TM}^\bullet$ on the total monthly household entries. This variable usually underreports TM, and this is why it is usually disregarded. Nevertheless, $\mathrm{TM}^\bullet$ may contain precious information for matching SHIW and HBS. In the following, the use of this variable as a source of auxiliary information is illustrated. Note that this approach resembles those of Chapter 3, where auxiliary information is in parametric form and suggested by proxy variables, although it does not come from auxiliary sources but from one of the samples to be matched. This idea is the basis for the actual estimation of the SAM, as outlined in Coli *et al.* (2005).

Given that the main problem of $\mathrm{TM}^\bullet$ is that the true total household entry is usually underreported, we assume that:

Assumption 7.2 *The rank of each household according to* $\mathrm{TM}^\bullet$ *is a reliable indication of the rank of the household according to* TM.

Note that $\mathrm{TM}^\bullet$ is a categorical variable, with 14 categories. Furthermore, it is absent from the SHIW. According to Assumption 7.2, $\mathrm{TM}^\bullet$ can be derived in the SHIW by the following procedure.

(i) Order the categories of $TM^\bullet$ from the category of the poorest households ($TM^\bullet = 1$) to the category of the richest households ($TM^\bullet = 14$). Let:

$$\alpha_j = \tilde{P}^{(HBS)}(TM^\bullet = j),$$

be the estimated fraction of households in category j, $j = 1, \ldots, 14$.

(ii) Derive $TM^\bullet$ in the SHIW, by ranking the households in the SHIW according to TM (which is a numeric variable), and assigning the value $TM^\bullet = 1$ to the first $100\alpha_1\%$ ranked households, $TM^\bullet = 2$ to the remaining $100\alpha_2\%$ households with the lowest TM, and so on.

The distribution of $TM^\bullet$ in the SHIW obviously satisfies the following equation:

$$\tilde{P}^{(SHIW)}(TM^\bullet = j) = \tilde{P}^{(HBS)}(TM^\bullet = j) = \alpha_j, \qquad j = 1, \ldots, 14.$$

We again emphasize that $TM^\bullet$ was used only as an indication of the relative position of the households with respect to the total entries, from 1 to 14. If $TM^\bullet = 1$ corresponds to the total entries class '0–310' in the HBS, the corresponding class in the SHIW gathers households with a totally different range of total entries. $TM^\bullet$ was used only as a source of information on the rank, and not on the amount of total entry.

The use of the additional variable $TM^\bullet$ changes equation (7.2) in the following way:

$$P(\mathbf{X}, TM, TM^\bullet, TC) = P(TC|\mathbf{X}, TM, TM^\bullet)P(\mathbf{X}, TM, TM^\bullet). \qquad (7.4)$$

It is still not possible to estimate the first factor in (7.4). The following assumption, although untestable and based on the notion of conditional independence, is particularly useful.

Assumption 7.3 TC *and* TM *are independent given* ($\mathbf{X}$, $TM^\bullet$).

Note that this assumption is much milder than Assumption 7.1: conditional independence is established given important information on the entries in $TM^\bullet$.

Remark 7.5 Assumption 7.2 is suggested by some analyses on the SHIW; see Coli *et al.* (2005). More precisely, this data set can be used in order to understand which kind of relationship relates TM.SHIW (Y) and TC.SHIW (Z). If for example, SUPERFT (X) is used as common variable, it is possible to compute $\rho_{YZ|X}$. Considering five different situations, characterized respectively by the relative stratification,

(i) AREA;

(ii) NCOMP5;

(iii) AREA5 × NCOMP5;

(iv) income categorization induced by $TM^\bullet$;

(v) AREA5 and income categorization induced by $TM^\bullet$;

we have observed that only cases (iv) and (v) have partial correlation coefficient close to the CIA (within strata).

Assumption 7.3 implies a simplification of the first factor:

$$P(\mathrm{TC}|\mathbf{X}, \mathrm{TM}, \mathrm{TM}^{\bullet}) = P(\mathrm{TC}|\mathbf{X}, \mathrm{TM}^{\bullet}). \quad (7.5)$$

The right-hand side of (7.5) can now be estimated from the HBS:

$$\tilde{P}(\mathrm{TC}|\mathbf{X}, \mathrm{TM}^{\bullet}). \quad (7.6)$$

Many different methods can be used. In our application, we preferred not to assume any parametric model. More precisely, random hot deck of TC in the SHIW was performed in the donation classes ($\mathbf{X}$, $\mathrm{TM}^{\bullet}$). This procedure, following Remark 2.11, corresponds to estimating (7.6) via its empirical cumulative distribution function in the HBS, and then drawing a value at random from the estimated distribution for each record in the SHIW by donation class.

Given the large number of matching variables (Section 7.2.3) and the presence of a continuous variable (SUPERTF), the number of HBS households in some donation classes is too low. Hence, donation classes were constructed with the following strategy.

- 'Geographical area' and $\mathrm{TM}^{\bullet}$ are considered as 'stratification variables'. In other words, the donor and the recipient records must coincide for these two variables.

- As far as the other matching variables are considered, given the results obtained by nearest neighbour distance hot deck under the CIA, we limit our attention to two subsets of matching variables: NCOMP5 plus variables in subset (d1) in Table 7.8 and NCOMP5 plus variables in subset (d2) in Table 7.8. The values of the matching variables should not strictly coincide in the donor and recipient records, but were allowed to have a distance not greater than a fixed threshold. For these variables, the Gower distance was used.

The final synthetic sample, i.e. the completed SHIW sample, produces results which are more consistent with economic theory, because it recovers part of the correlation between consumption and income.

Table 7.11 shows the summary characteristics of consumption variables in the HBS and in the SHIW final fused data set. It can be seen that there do not seem to be great differences in the average and standard deviation of food and total consumption (here expressed in logarithmic terms).

As far as correlations are considered, the estimated correlation among TM.SHIW and TC.HBS reaches values around 0.5 (Table 7.12), greater than the largest value, 0.33, obtained under the CIA.

Table 7.11 Summary statistics for food and total consumption in the donor HBS (unweighted data) and the final synthetic data set by means of nearest neighbour matching within the same geographic area and TM class

	NCOMP5 + (d1)	NCOMP5 + (d2)
$\hat{\mu}_{\text{TC}}$	99.96	101.01
$\hat{\sigma}_{\text{TC}}$	101.90	101.39
$\hat{\mu}_{\text{TF}}$	100.00	100.02
$\hat{\sigma}_{\text{TF}}$	101.37	99.14
$\hat{\rho}_{\text{TC,TF}}$	99.56	103.33

Table 7.12 Estimated correlations among income and consumption in the fused SHIW data set after nearest neighbour matching within the same geographic area and TM class

	NCOMP5 + (d1)	NCOMP5 + (d2)
$\hat{\rho}_{\text{TM.SHIW},\widetilde{\text{TC}}}$	0.5262	0.5328
$\hat{\rho}_{\text{TC.SHIW},\widetilde{\text{TC}}}$	0.5464	0.5503
$\hat{\rho}_{\text{TF.SHIW},\widetilde{\text{TC}}}$	0.4211	0.4221
$\hat{\rho}_{\text{TM.SHIW},\widetilde{\text{TF}}}$	0.2367	0.2618
$\hat{\rho}_{\text{TC.SHIW},\widetilde{\text{TF}}}$	0.2637	0.2923
$\hat{\rho}_{\text{TF.SHIW},\widetilde{\text{TF}}}$	0.2917	0.3168

7.2.6 Assessment of uncertainty for the SAM

Sections 7.2.4 and 7.2.5 describe the results of the construction of the SAM respectively under Assumptions 7.1 and 7.2. In this subsection, the problem is analysed without assumptions. Thus, as explained in Chapter 4, the objective will be the description of uncertainty. In other words, all the models compatible with the sample will be considered as acceptable. The uncertain parameters are those related to the variables observed separately in the SHIW and HBS: TC and TM. The analysis is focused on the parameter $\rho_{\text{TC,TM}}$. The variable SUPERFT (ST) is chosen as common variable X. In this section, uncertainty is analysed by means of the maximum likelihood approach as described in Section 4.4.1, under the hypothesis that, after a logarithmic transformation, the variables (TC, TM, ST) are normally distributed.

In Section 4.3.1, it is shown that for assessing uncertainty, when the variables (X, Y, Z) are all univariate, it is sufficient to consider only the width of the interval

Table 7.13 Computation of uncertainty interval $[\hat{\rho}^{\mathrm{L}}_{\mathrm{TC,TM}}, \hat{\rho}^{\mathrm{U}}_{\mathrm{TC,TM}}]$ for the inestimable parameter $\rho_{\mathrm{TC,TM}}$, and maximum likelihood estimates $\hat{\rho}_{\mathrm{ST,TC}}$, $\hat{\rho}_{\mathrm{ST,TM}}$, and the corresponding value under the CIA, $\hat{\rho}^{\mathrm{CIA}}_{\mathrm{TC,TM}}$, on the overall sample $A \cup B$

$\hat{\rho}_{\mathrm{ST,TC}}$	$\hat{\rho}_{\mathrm{ST,TM}}$	$\hat{\rho}^{\mathrm{CIA}}_{\mathrm{TC,TM}}$	$\hat{\rho}^{\mathrm{L}}_{\mathrm{TC,TM}}$	$\hat{\rho}^{\mathrm{U}}_{\mathrm{TC,TM}}$
0.3808	0.3956	0.1506	−0.6986	0.9999

$[\hat{\rho}^{\mathrm{L}}_{\mathrm{TC,TM}}, \hat{\rho}^{\mathrm{U}}_{\mathrm{TC,TM}}]$, i.e. the likelihood ridge. The estimated boundaries are given by:

$$\hat{\rho}^{\mathrm{L}}_{\mathrm{TC,TM}} = \hat{\rho}_{\mathrm{ST,TC}}\hat{\rho}_{\mathrm{ST,TM}} - \sqrt{\left(1 - \hat{\rho}^2_{\mathrm{ST,TC}}\right)\left(1 - \hat{\rho}^2_{\mathrm{ST,TM}}\right)}, \tag{7.7}$$

$$\hat{\rho}^{\mathrm{U}}_{\mathrm{TC,TM}} = \hat{\rho}_{\mathrm{ST,TC}}\hat{\rho}_{\mathrm{ST,TM}} + \sqrt{\left(1 - \hat{\rho}^2_{\mathrm{ST,TC}}\right)\left(1 - \hat{\rho}^2_{\mathrm{ST,TC}}\right)}. \tag{7.8}$$

Table 7.13 shows the estimated boundaries (7.7) and (7.8).

Uncertainty is high, since the estimated range of $\rho_{\mathrm{TC,TM}}$ is wide. This means that data are compatible with a very large class of models. This is typically the case when a point estimate under a particular model (e.g. the CIA) would not be justifiable. Only further knowledge assumed reliable (as in Section 7.2.5, justified in Remark 7.5) would allow a point estimate.

Appendix A

Statistical Methods for Partially Observed Data

A.1 Maximum Likelihood Estimation with Missing Data

Let $\mathbf{X} = (X_1, \ldots, X_P)$ be a P-dimensional vector of r.v.s. with density $f(\mathbf{x}; \boldsymbol{\theta})$, for $\mathbf{x} \in \mathcal{X}$ and parameter vector $\boldsymbol{\theta} \in \boldsymbol{\Theta}$. Let $\boldsymbol{x}$ denote the sample consisting of n i.i.d. observations generated from $f(\mathbf{x}; \boldsymbol{\theta})$, and let some items of $\boldsymbol{x}$ be missing. The observed components of $\boldsymbol{x}$ are denoted by $\boldsymbol{x}_{\text{obs}}$, while the missing components are $\boldsymbol{x}_{\text{mis}}$, thus $\boldsymbol{x} = (\boldsymbol{x}_{\text{obs}}, \boldsymbol{x}_{\text{mis}})$.

Maximum likelihood estimation of $\boldsymbol{\theta}$ when the sample is only partially observed has been discussed in many papers, in particular Rubin (1974, 1976); for an extensive exposition of this problem and of the possible solutions, see Little and Rubin (2002). In general, it is necessary to take into account two elements:

(i) an appropriate model that jointly describes the variables of interest and the missing data mechanism;

(ii) an appropriate method for obtaining ML estimates.

A.1.1 Missing data mechanisms

When missing items are present, it is necessary to take into account the completely observed $n \times p$ matrix $\boldsymbol{r}$ whose components r_{ij} are equal to 1 when X_j is observed in the ith unit, and 0 otherwise.

Both the matrices $\boldsymbol{x}$ and $\boldsymbol{r}$ are the result of r.v.s, respectively $\boldsymbol{X}$ and $\boldsymbol{R}$. As a matter of fact, the possibility of learning something about $\boldsymbol{\theta}$ with the use of

Statistical Matching: Theory and Practice M. D'Orazio, M. Di Zio and M. Scanu

$\boldsymbol{x}$ without resorting to further external information depends on the probabilistic relationship between both the r.v.s $(\boldsymbol{X}, \boldsymbol{R})$. Usually, this is done via the conditional distribution of the missing data mechanism given the variables of interest, $h(\boldsymbol{r}|\boldsymbol{x};\xi)$, where $\xi \in \Xi$ is the parameter vector related to the missing data mechanism.

Rubin (1976) defines three different missing data models, which are generally assumed by the analyst: missing completely at random (MCAR), missing at random (MAR) and missing not at random (MNAR).

A missing data mechanism is MCAR when $\boldsymbol{R}$ is independent of either the observed and the unobserved components of $\boldsymbol{X}$, i.e.

$$h(\boldsymbol{r}|\boldsymbol{x};\boldsymbol{\xi}) = h(\boldsymbol{r};\boldsymbol{\xi}), \quad \forall\, \boldsymbol{\xi} \in \Xi.$$

The MAR assumption states that the missing mechanism depends only on the observed part of $\boldsymbol{X}$, i.e. $\boldsymbol{X}_{\text{obs}}$,

$$h(\boldsymbol{r}|\boldsymbol{x};\boldsymbol{\xi}) = h(\boldsymbol{r}|\boldsymbol{x}_{\text{obs}}), \qquad \forall\, \boldsymbol{x}_{\text{mis}}, \ \forall\, \boldsymbol{\xi} \in \Xi.$$

Finally, the MNAR mechanism assumes that $\boldsymbol{R}$ depends also on the unobserved part of $\boldsymbol{X}$, i.e. $\boldsymbol{X}_{\text{mis}}$.

It is evident that appropriate inferences on $\boldsymbol{\theta}$ can be based on just the observed part of the data set $\boldsymbol{x}_{\text{obs}}$ only when MCAR or MAR hold. These hypotheses should, however, be complemented by an additional assumption: distinctness of the parameters $\boldsymbol{\theta}$ of the data model and the parameter $\boldsymbol{\xi}$ of the missingness mechanism. This means that the joint parameter space of $(\boldsymbol{\theta}, \boldsymbol{\xi})$ is the Cartesian cross product of the relevant spaces Θ and Ξ. When both MAR and distinctness may be assumed, the missing data mechanism is termed *ignorable*.

A.1.2 Maximum likelihood and ignorable nonresponse

Under the ignorability assumption, inferences on $\boldsymbol{\theta}$ can be performed by ignoring the missing data mechanism $\boldsymbol{R}$. Following Little and Rubin (2002, p. 119), the distribution of the observed data, i.e. of the r.v. $(\boldsymbol{X}_{\text{obs}}, \boldsymbol{R})$, can be obtained by integrating out $\boldsymbol{X}_{\text{mis}}$ from the joint distribution of $(\boldsymbol{R}, \boldsymbol{X})$:

$$f(\boldsymbol{x}_{\text{obs}}, \boldsymbol{r}; \boldsymbol{\theta}, \boldsymbol{\xi}) = \int f(\boldsymbol{x}_{\text{obs}}, \boldsymbol{x}_{\text{mis}}; \boldsymbol{\theta}) h(\boldsymbol{r}|\boldsymbol{x}_{\text{obs}}, \boldsymbol{x}_{\text{mis}}; \boldsymbol{\xi}) \mathrm{d}\boldsymbol{x}_{\text{mis}}.$$

The MAR assumption allows us to simplify this distribution:

$$f(\boldsymbol{x}_{\text{obs}}, \boldsymbol{r}; \boldsymbol{\theta}; \boldsymbol{\xi}) = h(\boldsymbol{r}|\boldsymbol{x}_{\text{obs}}; \boldsymbol{\xi}) f(\boldsymbol{x}_{\text{obs}}; \boldsymbol{\theta}). \tag{A.1}$$

Equation (A.1) determines the *full likelihood*, given the observed data $\boldsymbol{x}_{\text{obs}}$ and $\boldsymbol{r}$:

$$L_{\text{full}}(\boldsymbol{\theta}, \boldsymbol{\xi}; \boldsymbol{x}_{\text{obs}}, \boldsymbol{r}) = f(\boldsymbol{x}_{\text{obs}}, \boldsymbol{r}; \boldsymbol{\theta}, \boldsymbol{\xi})$$

In Equation (A.1) the only term depending on $\boldsymbol{\theta}$ is $f(\boldsymbol{x}_{\text{obs}}; \boldsymbol{\theta})$. Hence, when interest is confined to $\boldsymbol{\theta}$, the ignorability assumption allows the use of only the *observed likelihood function*

$$L(\boldsymbol{\theta}; \boldsymbol{x}_{\text{obs}}) = f(\boldsymbol{x}_{\text{obs}}; \boldsymbol{\theta}) = \prod_{i=1}^{n} \int f(\mathbf{x}_{i;\text{obs}}, \mathbf{x}_{i;\text{mis}}; \boldsymbol{\theta}) d\mathbf{x}_{i;\text{mis}},$$

where $\mathbf{x}_{i;\text{obs}}$ and $\mathbf{x}_{i;\text{mis}}$ are respectively the observed and missing part of the ith record in $\boldsymbol{x}$.

Maximization of the observed likelihood function with respect to $\boldsymbol{\theta}$ is usually not easy. In some special cases it is possible to resort to the usual maximization procedures for complete data sets.

Special patterns of missing data

In some situations it is possible to define an alternative and equivalent parameterization $\boldsymbol{\phi}(\boldsymbol{\theta}) = (\phi_1, \ldots, \phi_J)$ such that $(\phi_1, \ldots, \phi_J)$ are distinct parameters (the joint parameter space is the Cartesian product of each parameter space) and the observed likelihood function satisfies the following requirements.

- The observed likelihood function can be factorized so that each factor depends only on a parameter ϕ_j:

$$L(\boldsymbol{\theta}; \boldsymbol{x}_{\text{obs}}) = \prod_{j=1}^{J} L_j(\phi_j; \boldsymbol{x}_{\text{obs}}).$$

- Each component $L_j(\phi_j; \boldsymbol{x}_{\text{obs}})$ is a likelihood for a complete data set (possibly a data subset of $\boldsymbol{x}$).

This is possible for instance, when the pattern of missing data is monotone, i.e. when the variables $X_1, \ldots, X_p$ are ordered so that X_i is always observed when X_j is observed, for any $i < j$. In this case, ML estimation of $\boldsymbol{\theta}$ reduces to J distinct maximization problems, one for each ϕ_j. Each of these maximization problems can be solved with standard ML estimators on complete data sets.

General pattern of missing data

In general, the observed likelihood function can still be factorized according to the factorization lemma in Rubin (1974), showing which parameters are estimable on a complete data set, which ones need appropriate techniques for partially observed data sets, and which ones are inestimable. We show this by the following simplified arguments from Rubin (1974) (see also Little and Rubin, 2002, p. 156).

Assume that there are three r.v.s X_1, X_2 and X_3 such that:

(i) X_3 is more observed that X_1, in the sense that for any unit in which X_1 is observed, X_3 is completely observed;

(ii) X_1 and X_2 are never jointly observed;

Table A.1 Matrix $\boldsymbol{r}$ of the pattern of missing data for X_1, X_2 and X_3 according to the requirements (i), (ii) and (iii). The symbol r is arbitrarily zero or one

R_1	R_2	R_3
r	0	1
...	...	...
r	0	1
0	r	r
...	...	...
0	r	r

(iii) the rows of $\boldsymbol{R}$ given X_3 are conditionally independent and identically distributed (the observed missing data pattern $\boldsymbol{r}$ is shown in Table A.1).

In this case, the observed likelihood function can be factorized as follows:

$$L(\boldsymbol{\theta}; \boldsymbol{x}_{\text{obs}}) = f_1(\boldsymbol{x}_{1,\text{obs}}|\boldsymbol{x}_{3,\text{obs}}; \boldsymbol{\theta}_{1|3}) f_{2,3}(\boldsymbol{x}_{2,\text{obs}}, \boldsymbol{x}_{3,\text{obs}}; \boldsymbol{\theta}_{2,3}).$$

As noted in Rubin (1974), the two parameters $\boldsymbol{\theta}_{1|3}$ and $\boldsymbol{\theta}_{2,3}$ are distinct when the variables are normal or categorical. There are additional parameters that are actually distinct from $\boldsymbol{\theta}_{1|3}$ and $\boldsymbol{\theta}_{2,3}$ and that are not shown in the observed likelihood – for instance, the parameter of association of X_1 and X_2 given X_3, $\boldsymbol{\theta}_{1,2|3}$. This example reflects three different situations.

(a) The parameter $\boldsymbol{\theta}_{1|3}$ can be estimated via standard ML methods on the data subsets where both X_1 and X_3 are observed.

(b) The parameter $\boldsymbol{\theta}_{2,3}$ needs to be estimated appropriately on the data set with missing items.

(c) The parameter $\boldsymbol{\theta}_{1,2|3}$ is inestimable from the observed data set.

Rubin (1974) characterizes the parameter and the possibility of estimation on a complete or an incomplete data set via a generalization of the previous arguments to the case of multivariate X_1, X_2 and X_3.

The EM algorithm

Although the factorization given by the factorization lemma is particularly helpful, in particular for the statistical matching problem, it tends to be too informative. Most of the time, the role played by the different distinct parameters of $\boldsymbol{\theta}$ (whether they are estimable on complete or incomplete data sets) is not necessary.

A technique which is able to maximize the observed likelihood function is the expectation–maximization (EM) algorithm; see Dempster *et al.* (1977). This algorithm is an iterative algorithm that successively performs the following two steps. Let $\boldsymbol{\theta}^{(v)}$ be the latest parameter estimate of this iterative procedure.

E step. This step computes the expected complete data loglikelihood, where expectation is with respect to the distribution of $\boldsymbol{X}_{\text{mis}}$ with $\boldsymbol{\theta} = \boldsymbol{\theta}^{(v)}$:

$$Q(\boldsymbol{\theta}|\boldsymbol{\theta}^{(v)}) = \int \log\left[L(\boldsymbol{\theta}; \boldsymbol{x}_{\text{obs}}, \boldsymbol{x}_{\text{mis}})\right] f(\boldsymbol{x}_{\text{mis}}|\boldsymbol{x}_{\text{obs}}; \boldsymbol{\theta}^{(v)}) \mathrm{d}\boldsymbol{x}_{\text{mis}}.$$

M step. This step finds the new $\boldsymbol{\theta}^{(v+1)}$ which maximizes $Q(\boldsymbol{\theta}|\boldsymbol{\theta}^{(v)})$.

This algorithm simplifies when the distribution function of $\boldsymbol{X}$ belongs to the exponential family (see Little and Rubin, 2002, p. 175). For instance, this is the case for the normal and multinomial distributions.

A.2 Bayesian Inference with Missing Data

In the Bayesian approach the vectors $\boldsymbol{\theta}$ and $\boldsymbol{\xi}$ introduced in Section A.1.1 are r.v.s, and thus the definition of distinctness is no longer in terms of the Cartesian product of the relevant spaces $\boldsymbol{\Theta}$ and $\boldsymbol{\Xi}$ but in terms of independence of the r.v.s $\boldsymbol{\theta}$ and $\boldsymbol{\xi}$, i.e. $\pi(\boldsymbol{\theta}, \boldsymbol{\xi}) = \pi_{\boldsymbol{\Theta}}(\boldsymbol{\theta})\pi_{\boldsymbol{\Xi}}(\boldsymbol{\xi})$. For the sake of simplicity, the subscripts will be omitted from the distributions $\pi(\cdot)$. In this setting, parameter inference must be based on the observed posterior distribution (Schafer, 1997)

$$\pi(\boldsymbol{\theta}, \boldsymbol{\xi}|\boldsymbol{x}_{\text{obs}}, \boldsymbol{r}) = k^{-1} f(\boldsymbol{x}_{\text{obs}}, \boldsymbol{r}|\boldsymbol{\theta}, \boldsymbol{\xi})\pi(\boldsymbol{\theta}, \boldsymbol{\xi}),$$

where $\pi(\boldsymbol{\theta}, \boldsymbol{\xi})$ is a prior distribution for $(\boldsymbol{\theta}, \boldsymbol{\xi})$ and k is the normalizing constant

$$k = \iint f(\boldsymbol{x}_{\text{obs}}, \boldsymbol{r}|\boldsymbol{\theta}, \boldsymbol{\xi})\pi(\boldsymbol{\theta}, \boldsymbol{\xi}) \mathrm{d}\boldsymbol{\theta} \mathrm{d}\boldsymbol{\xi}.$$

Similarly to the likelihood approach, under the ignorability assumption inferences on $\boldsymbol{\theta}$ can be drawn by ignoring the missing data mechanism $\boldsymbol{R}$, in fact

$$\pi(\boldsymbol{\theta}|\boldsymbol{x}_{\text{obs}}, \boldsymbol{r}) \propto L(\boldsymbol{\theta}|\boldsymbol{x}_{\text{obs}})\pi(\boldsymbol{\theta}),$$

where the proportionality is up to a term that does not involve $\boldsymbol{\theta}$. Thus, all information about $\boldsymbol{\theta}$ is in a posterior distribution that ignores the missing mechanism, i.e. the observed-data posterior distribution

$$\pi(\boldsymbol{\theta}|\boldsymbol{x}_{\text{obs}}) = L(\boldsymbol{\theta}|\boldsymbol{x}_{\text{obs}})\pi(\boldsymbol{\theta}).$$

The data augmentation algorithm

Analogously to the likelihood approach, there are situations where inferences with missing data can be reduced to a series of subproblems where inferences can be performed on complete data sets (Rubin, 1974). These situations are induced by the missing data patterns described in Section A.1.2, with the further specification that the reparameterization $\boldsymbol{\phi}$ is a priori independent. For a general pattern of missing data, the posterior distribution $\pi(\boldsymbol{\theta}|\boldsymbol{x}_{\text{obs}})$ is intractable. Hence, an iterative algorithm is needed in order to handle it – the data augmentation algorithm, introduced by Tanner and Wong (1987). It is similar in spirit to the EM, in the sense that instead of solving problems with missing data, it fills in the data set and iteratively solves complete data problems. In detail: let $\boldsymbol{\theta}^{(0)}$ be an initial value for $\boldsymbol{\theta}$. The following steps are performed iteratively, for $t \geq 0$.

I step. Draw $\boldsymbol{x}_{\text{mis}}^{(t+1)}$ from $f(\boldsymbol{x}_{\text{mis}}|\boldsymbol{x}_{\text{obs}}, \boldsymbol{\theta}^{(t)})$.

P step. Draw $\boldsymbol{\theta}^{(t+1)}$ from $\pi(\boldsymbol{\theta}|\boldsymbol{x}_{\text{obs}}, \boldsymbol{x}_{\text{mis}}^{(t+1)})$.

The iteration of the I and P steps yields a stochastic sequence $(\boldsymbol{\theta}^{(t)}, \boldsymbol{x}_{\text{mis}}^{(t)})$, $t = 1, 2, \ldots$, whose stationary distribution is $f(\boldsymbol{\theta}, \boldsymbol{x}_{\text{mis}}|\boldsymbol{x}_{\text{obs}})$. Furthermore, the stationary distributions for the subsequences $(\boldsymbol{\theta}^{(t)})$ and $(\boldsymbol{x}_{\text{mis}}^{(t)})$ are respectively $\pi(\boldsymbol{\theta}|\boldsymbol{x}_{\text{obs}})$ and $f(\boldsymbol{x}_{\text{mis}}|\boldsymbol{x}_{\text{obs}})$. Hence, for large t, the values $(\boldsymbol{\theta}^{(t)})$ and $(\boldsymbol{x}_{\text{mis}}^{(t)})$ can be considered as an approximate draw from those distributions; for more details, see Schafer (1997) and Little and Rubin (2002).

Tanner and Wong (1987) referred to the I step as the *imputation step* and the P step as the *posterior step*. The advantage of using this algorithm is that the computation of the observed data posterior $\pi(\boldsymbol{\theta}|\boldsymbol{x}_{\text{obs}})$ is no longer required. Instead, the full posterior distribution $\pi(\boldsymbol{\theta}|\boldsymbol{x}_{\text{obs}}, \boldsymbol{x}_{\text{mis}})$, which can be more easily achieved by standard methods of Bayesian inference, should be computed.

Appendix B

Loglinear Models

Loglinear models are very popular statistical models for multidimensional contingency tables; see, for instance, Agresti (1990) and references therein. Their objective is the representation of the *interdependence* of the variables in the contingency table, or in other words their association structure. In the following we will refer to the case of three categorical variables (X_1, X_2, X_3), with categories (i, j, k), $i = 1, \ldots, I$, $j = 1, \ldots, J$, $k = 1, \ldots, K$, such that

$$\theta_{ijk} = P(X_1 = i, X_2 = j, X_3 = k), \qquad \forall\, i, j, k.$$

A loglinear model is a new parameterization from the probabilities θ_{ijk} to the *interaction terms*. The link between these two objects is the following. Let n be the sample size, and let η_{ijk} be the expected value of the sample counts in cell (i, j, k):

$$\eta_{ijk} = n\theta_{ijk}, \qquad \forall\, i, j, k.$$

Then, a loglinear model is defined by:

$$\log(\eta_{ijk}) = \lambda + \lambda_i^1 + \lambda_j^2 + \lambda_k^3 + \lambda_{ij}^{12} + \lambda_{ik}^{13} + \lambda_{jk}^{23} + \lambda_{ijk}^{123}, \quad \forall\, i, j, k,$$

under the constraints

$$\sum_i \lambda_i^1 = 0, \quad \sum_j \lambda_j^2 = 0, \quad \sum_k \lambda_k^3 = 0,$$

$$\sum_{ij} \lambda_{ij}^{12} = 0, \quad \sum_{ik} \lambda_{ik}^{13} = 0, \quad \sum_{jk} \lambda_{jk}^{23} = 0,$$

$$\sum_{ijk} \lambda_{ijk}^{123} = 0.$$

Statistical Matching: Theory and Practice M. D'Orazio, M. Di Zio and M. Scanu

When some of the interaction terms (i.e. the λs) are set to zero, different kinds of dependence relationships are defined. Among these are the independence between X_1, X_2 and X_3,

$$\log(\eta_{ijk}) = \lambda + \lambda_i^1 + \lambda_j^2 + \lambda_k^3, \quad \forall\, i, j, k;$$

the conditional independence of X_1 and X_2 given X_3,

$$\log(\eta_{ijk}) = \lambda + \lambda_i^1 + \lambda_j^2 + \lambda_k^3 + \lambda_{ik}^{13} + \lambda_{jk}^{23}, \quad \forall\, i, j, k;$$

and the independence of X_1 from (X_2, X_3),

$$\log(\eta_{ijk}) = \lambda + \lambda_i^1 + \lambda_j^2 + \lambda_k^3 + +\lambda_{jk}^{23}, \quad \forall\, i, j, k.$$

When all the interaction terms are nonnull, the loglinear model is called *saturated*.

B.1 Maximum Likelihood Estimation of the Parameters

Let n_{ijk} be the cell counts of a sample of size n, and assume that the objective is to find maximum likelihood estimates of the interaction terms. The loglikelihood function is

$$\ell(\boldsymbol{\theta}|n_{ijk}) = c + \sum_{ijk} n_{ijk} \log(\theta_{ijk}) = c + \sum_{ijk} n_{ijk} \log(\eta_{ijk}). \tag{B.1}$$

According to the null interaction terms of the chosen loglinear model, equation (B.1) shows the minimal sufficient statistics for the parameters of interest. For instance, under the independence model only the marginal tables for respectively X_1, X_2 and X_3 are the minimal sufficient statistics, while for the saturated model the overall table n_{ijk} constitutes the minimal sufficient statistics. The ML estimates of the interaction terms (and consequently of θ_{ijk}) are sometimes directly computable from the tables representing the minimal sufficient statistics. When this is not the case, the ML estimates of θ_{ijk} can be found with iterative methods. For instance, this is the case for the loglinear model

$$\log(\eta_{ijk}) = \lambda + \lambda_i^1 + \lambda_j^2 + \lambda_k^3 + \lambda_{ij}^{12} + \lambda_{ik}^{13} + \lambda_{jk}^{23}, \quad \forall\, i, j, k, \tag{B.2}$$

where only the interaction term λ_{ijk}^{123} is assumed null for all i, j, k. In this case, (B.1) shows that the minimal sufficient statistics are the marginal contingency tables of (X_1, X_2) (with cell counts $n_{ij.}$), (X_1, X_3) (with cell counts $n_{i.j}$) and (X_2, X_3) (with cell counts $n_{.jk}$). The iterative procedure known as iterative proportional fitting (IPF) (Deming and Stephan, 1940) is usually applied. In the case of absence of the three-way interaction term, the IPF algorithm is applied as follows.

(i) Fix arbitrary values $\eta_{ijk}^{(0)}$ with an interaction structure no more complex than the fixed loglinear model. For instance, for model (B.2) without the three-way interaction term, set $\eta_{ijk}^{(0)} = 1$ for all i, j, k.

(ii) Adapt the fixed expected counts to minimal sufficient statistics, i.e.,

$$\eta_{ijk}^{(1)} = \eta_{ijk}^{(0)} \frac{n_{ij.}}{\eta_{ij.}^{(0)}}, \qquad \forall\, i, j, k,$$

$$\eta_{ijk}^{(2)} = \eta_{ijk}^{(1)} \frac{n_{i.k}}{\eta_{i.k}^{(1)}}, \qquad \forall\, i, j, k,$$

$$\eta_{ijk}^{(3)} = \eta_{ijk}^{(2)} \frac{n_{.jk}}{\eta_{.jk}^{(2)}}, \qquad \forall\, i, j, k.$$

(iii) Iterate step (ii) until the maximum difference between the counts of the minimal sufficient statistics tables and the corresponding fitted values is below a fixed positive threshold.

It can be proved (see Agresti, 1990) that the IPF leads to the ML estimates of $\boldsymbol{\theta}$. Furthermore, it can also be used for those loglinear models whose parameters can be directly estimated from the minimal sufficient tables.

Appendix C

Distance Functions

Let $\mathbf{x}_a$, $a = 1, \ldots, n$, be a set of n P-dimensional records. The distance between two records can be computed in various ways. Formally, a generic real-valued function d is said to be a *distance function* if, for any two records $\mathbf{x}_a$ and $\mathbf{x}_b$, the following properties hold (see Mardia *et al.*, 1979, pp. 375–381):

- symmetry, $d_{ab} = d_{ba}$;
- nonnegativity, $d_{ab} \geq 0$;
- identity, $d_{aa} = 0$.

Moreover, d is said to be a *metric* if the following properties also hold:

- identity of indiscernibles, $d_{ab} = 0$ if and only if $a = b$;
- triangle inequality, $d_{ab} \leq d_{ac} + d_{cb}$.

When the P-dimensional records are observations of P quantitative variables $\mathbf{X}$ measured over n sample units, a class of distance functions is described by the Minkowsky metric:

$$d_{ab} = \left[\sum_{p=1}^{P} c_p^{\lambda} \left| x_{ap} - x_{bp} \right|^{\lambda} \right]^{1/\lambda}, \qquad \lambda \geq 1,$$

where c_p is a scaling factor for the pth variable. The best-known metrics based on the Minkowsky are:

(i) for $\lambda = 1$, the *Manhattan metric* (*city-block*),

$$d_{ab} = \sum_{p=1}^{P} c_p \left| x_{ap} - x_{bp} \right|;$$

(ii) for $\lambda = 2$, the *Euclidean metric*,

$$d_{ab} = \sqrt{\sum_{p=1}^{P} c_p^2 \left(x_{ap} - x_{bp}\right)^2};$$

(iii) letting $\lambda \to +\infty$, the *Chebyshev metric*,

$$d_{ab} = \max_p \left\{c_p \left|x_{ap} - x_{bp}\right|\right\}.$$

Variables can be standardized by means of the standard deviation, setting $c_p = 1/\sigma_p$, or using the range,

$$c_p = 1/R_p, \text{ with } R_p = \max_a \left\{x_{ap}\right\} - \min_a \left\{x_{ap}\right\},$$

or other suitable functions of these measures of variability (in order to attenuate the impact of possible outliers, a trimmed version of the range can be used). The variables are not standardized setting $c_p = 1$, $p = 1, \ldots, P$. Variable standardization can be independent of the entire range of the variables as in the *Canberra metric*:

$$d_{ab} = \sum_{p=1}^{P} \frac{\left|x_{ap} - x_{bp}\right|}{x_{ap} + x_{bp}}, \qquad x_{ap} > 0,\ x_{bp} > 0.$$

Finally, it is worth mentioning the *Mahalanobis distance*, defined as

$$d_{ab} = \left(\mathbf{x}_a - \mathbf{x}_b\right)' \Sigma_{\mathbf{XX}}^{-1} \left(\mathbf{x}_a - \mathbf{x}_b\right), \tag{C.1}$$

where $\Sigma_{\mathbf{XX}}$ is the covariance matrix of $\mathbf{X}$. When $\Sigma_{\mathbf{XX}}$ is not known, it is possible to use one of its estimates. This distance takes into account the statistical relationship among the $\mathbf{X}$ variables.

The computation of distance in the presence of a set of mixed mode variables is slightly more difficult. Two alternatives are available: (i) transform all categorical variables into continuous variables and apply one of the previous distance functions; (ii) apply a distance function that takes into account the different nature of the variables. In this last case, Gower's dissimilarity coefficient is particularly appropriate. The idea is essentially that of obtaining a general distance measure by computing an average of suitable distances obtained for each variable:

$$d_{ab} = \frac{1}{P} \sum_{p=1}^{P} c_p d_{abp},$$

where $c_p = 1$ for binary variables (a nominal variable should be split into as many dummy variables as there are categories of the variable) and $c_p = 1/R_p$ for continuous and categorical ordinal variables; see for instance, Mardia *et al.*

(1979, p. 383). Usually, the city-block metric is recommended for the *Gower-type measure of distance*:

$$d_{abp} = \left| x_{ap} - x_{bp} \right|.$$

Particular attention must be paid in the computation of distance when binary variables are involved. If the two categories of a binary variable exhaust all the possibilities, as in the variable 'Gender', distances can be easily computed so that $d_{ab} = 0$ if the categories of the variable coincide, and 1 if they do not. Sometimes binary variables represent the absence or presence of an event (e.g. the variable 'educational status' may consist of two categories 'graduate' and 'nongraduate'). In this case, a distance should be considered equal to zero only when the event is present for both records a and b, while in all other cases the distance is positive. Note, in particular, that it is also positive when the event is absent for both a and b.

A variation of Gower's dissimilarity measure it that of weighting differently the distances on the single variables according to the relative importance of the variables:

$$d_{ab} = \frac{\sum_{p=1}^{P} g_p c_p d_{abp}}{\sum_{p=1}^{P} g_p},$$

where g_p is the weight assigned to the pth variable. Obviously, setting $g_p = 0$ means discarding the variable from the computation of distance. The definition of weights does not follow a particular statistical rule. Most of the time, they are chosen by subjective evaluation. When the number of variables P is large, it could be better to try to identify which variables should be included in distance computations ($g_p = 1$) and which should be excluded ($g_p = 0$). In presence of mixed mode variables, some categorical variables may be used to define donation classes. This simplifies the computations given that distances have to be computed only among units belonging to the same class.

Appendix D

Finite Population Sampling

In finite population sampling the objective of the inference is that of estimating one or more characteristics of the finite population (mean, sum, ratios, etc.) by means of a sample selected from it. Several books are available on this topic, among them Cochran (1977) and Särndal *et al.* (1992). In the following a brief review of finite population sampling is given.

Let $\mathcal{P}$ be a finite population of size N. The elements (units) in $\mathcal{P}$ are usually labelled according to their position in the list (*sampling frame*):

$$\mathcal{P} = \{1, \ldots, a, \ldots, N\}.$$

Let Y denote the variable of interest and let y_a be the value of Y associated with the ath population element. Usually, the unknown population characteristics to be estimated are the total amount of Y in the population, $T_Y = \sum_{a \in \mathcal{P}} y_a$, or the population mean $\overline{y}_{\mathcal{P}} = T_Y / N$, or a function of one of these parameters.

In practice a sample, say $\mathcal{S}$, is selected from $\mathcal{P}$, $\mathcal{S} \subseteq \mathcal{P}$, and the estimates of unknown population characteristics are derived by processing the values of Y observed on the sample elements.

There are different approaches to drawing conclusions on finite population parameters from a sample. The approach commonly used assumes that the unknown values $y_1, \ldots, y_N$ are fixed quantities. Several possible samples can be selected from $\mathcal{P}$ with the given random selection mechanism, defined by the *sampling design*. The sample design has a central role in determining the statistical properties of the estimators computed from the sample. It is a function $p(\cdot)$ that assigns a probability $p(\mathcal{S})$ of selecting a sample to any sample $\mathcal{S}$. This kind of inference, based solely on the sampling design, is referred to as *design based*. In this framework, an estimator $\widehat{\theta}$ of the finite population parameter θ is evaluated in terms of its sampling distribution, usually summarized through the mean square error,

$$\mathrm{MSE}(\hat{\theta}) = \sum_{\mathcal{S}} p\,(\mathcal{S}) \left[\hat{\theta}\,(\mathcal{S}) - \theta\right]^2 = V_p(\hat{\theta}) + \left[B_p(\hat{\theta})\right]^2,$$

Statistical Matching: Theory and Practice M. D'Orazio, M. Di Zio and M. Scanu

where V_p denotes the *design variance* or *sampling variance*, while B_p represents the *design bias* of the estimator. The estimator is unbiased with respect to the design if $B_p(\hat{\theta}) = 0$, i.e. if its expectation with respect to the design is equal to θ:

$$E_p(\hat{\theta}) = \sum_{\mathcal{S}} p\,(\mathcal{S})\,\hat{\theta}\,(\mathcal{S}) = \theta.$$

A fundamental design unbiased estimator for the population total T_Y is the Horvitz–Thompson estimator (Horvitz and Thompson, 1952).

$$\hat{T}_{\mathrm{HT}} = \sum_{a\in\mathcal{S}} \frac{y_a}{\pi_a} = \sum_{a\in\mathcal{S}} \omega_a y_a,$$

where π_a represents the probability of the ath unit being included in the sample (*first-order inclusion probability*):

$$\pi_a = P\,(\mathcal{S} \ni a) = \sum_{\mathcal{S}\ni a} p\,(\mathcal{S})\,,$$

and the inverse of the inclusion probabilities, $\omega_a = 1/\pi_a$, gives the *design weights* of the population elements. The inclusion probabilities follow immediately from the chosen sampling design $p(\cdot)$. They must be strictly positive ($0 < \pi_a \leq 1$, for all $a \in \mathcal{P}$), ensuring that each population element has a chance of being selected in the sample. If this condition is not satisfied, the sampling design cannot be said probabilistic.

The Horvitz–Thompson estimator provides unreliable estimates when the inclusion probabilities π_a are not positively correlated with the unknown values y_a, $a \in \mathcal{P}$. In order to improve the precision of the estimators, the use of auxiliary variables (i.e. variables whose values are known in advance for all the elements in the population) related to Y is preferable. In finite population sampling the auxiliary variables, say $\mathbf{X}$, can be adopted at the design stage or at the estimation stage.

In the first case, one continuous auxiliary variable can be used in the definition of the inclusion probabilities by setting $\pi_a \propto x_a$. When the auxiliary variable is categorical, another strategy consists in partitioning the population $\mathcal{P}$ into homogeneous subpopulations (called *strata*), and drawing independent samples within each stratum (*stratified sampling*).

In the second case, the use of auxiliary information at the estimation stage implies an adjustment of the Horvitz–Thompson estimator. The generalized regression estimator (GREG) (see Särndal *et al.*, 1992, Chapter 6) is one of the most commonly used:

$$\hat{T}_{\mathrm{GREG},Y} = \hat{T}_{\mathrm{HT},Y} + \left(\sum_{a\in\mathcal{P}} \mathbf{x}_a - \sum_{a\in\mathcal{S}} \mathbf{x}_a\omega_a\right)' \hat{\mathbf{B}},$$

where

$$\hat{\mathbf{B}} = \left(\sum_{a\in\mathcal{S}} \omega_a c_a \mathbf{x}_a \mathbf{x}_a'\right)^{-1} \left(\sum_{a\in\mathcal{S}} \omega_a c_a \mathbf{x}_a y_a\right)$$

is the estimator of the vector of the regression coefficients of Y on $\mathbf{X}$. The factor c_a depends on the assumed regression model ξ of Y on $\mathbf{X}$ at finite population level, or more precisely on the structure of the conditional variances implied by ξ. As noted in Särndal *et al.* (1992, p. 227), in this context 'the role of the model ξ is to describe the finite population scatter ... The finite population looks as if it might have been generated in accordance with the model ξ. However, the assumption is never made that the population was really generated by the model ξ. The basic properties (approximate unbiasedness, ...) are not dependent on whether the model ξ holds or not. Our procedures are thus *model assisted*, but they are not model dependent.' In these terms, the goodness of fit of the model affects only the efficiency of the GREG estimator. Inference on population parameters is still design based and the introduction of the regression model aims only to improve the efficiency of survey estimators. This is the reason for the term *model assisted design-based inference.*

Note that the GREG estimator corresponds to a set of estimators, one for each specification of the factor c_a and of the type of auxiliary variables $\mathbf{X}$. For instance, when there is one categorical variable X, it is still possible to use the GREG estimator by using dummy variables for each X category, and the GREG estimator corresponds to a poststratified estimator.

An interesting feature of the GREG estimator is that the weighting system resulting from its application is *calibrated* to the known population totals of the auxiliary variables. Writing the GREG estimator as a linearly weighted sum of the observed values y_a,

$$\hat{T}_{\text{GREG},Y} = \sum_{a \in \mathcal{S}} \omega_a g_a y_a,$$

where

$$g_a = 1 + c_a \left(\sum_{a \in \mathcal{P}} \mathbf{x}_a - \sum_{a \in \mathcal{S}} \omega_a \mathbf{x}_a \right)' \left(\sum_{a \in \mathcal{S}} \omega_a c_a \mathbf{x}_a \mathbf{x}_a' \right)^{-1} \mathbf{x}_a,$$

the following result holds:

$$\sum_{a \in \mathcal{S}} \omega_a g_a \mathbf{x}_a = \sum_{a \in \mathcal{P}} \mathbf{x}_a.$$

This is an appealing feature in practical applications.

The GREG estimator belongs to the family of *calibration estimators*, i.e. those estimators that use *calibrated weights*. The term 'calibrated weights' refers to the final weights obtained by modifying the initial design weights ω_a as little as possible in order to satisfy a set of constraints called *calibration equations*. Formally, a calibration estimator can be written as

$$\hat{T}_{W,Y} = \sum_{a \in \mathcal{S}} \omega_a^{(1)} y_a,$$

where the new weights $\omega_a^{(1)}$ are computed by minimizing their overall distance

$$\sum_{a \in \mathcal{S}} d(\omega_a^{(1)}, \omega_a)$$

with respect to the corresponding initial design weights, ω_a, under the constraint

$$\sum_{a \in \mathcal{S}} \omega_a^{(1)} \mathbf{x}_a = \sum_{a \in \mathcal{P}} \mathbf{x}_a.$$

Various distance functions $d(\omega_a^{(1)}, \omega_a)$ are available; see Deville and Särndal (1992) for details. Note that when the Euclidean distance is adopted (see Appendix C) the calibration estimator reduces to the GREG estimator.

Appendix E

R Code

E.1 The R Environment

R is a language and environment for statistical computing and graphics. It furnishes a wide variety of statistical and graphical techniques, and is highly extensible. R is available as free software under the terms of the Free Software Foundation's GNU General Public License in source code form. For more information, visit `http://www.r-project.org/`.

The base R distribution and the additional packages are available on the CRAN family of Internet sites, `http://cran.r-project.org/`.

E.2 R Code for Nonparametric Methods

These R functions implement methods described in Sections 2.4.1 and 2.4.3.

```
RANDhd.mtc
function(rec, don, str=NULL)
{
# This function performs Random Hot Deck Matching
# Arguments:
# rec: can be a number, indicating the number of
#      recipients; or a vector, indicating the id of
#      recipients; must be a data.frame containing
#      variable str when donation classes have to be
#      used
#
# don: can be a number, indicating the number of
```

Statistical Matching: Theory and Practice M. D'Orazio, M. Di Zio and M. Scanu

```
#       donors; or a vector, indicating the id of
#       donors; must be a data.frame containing variable
#       str when donation classes have to be used
#
# str: when NULL, no donation classes are used;
#       otherwise it is the name of the variable,
#       contained in both rec and don, to be used to
#       build the donation classes
#
# Value:
# mtc.id: data.frame with two columns: the id of
#          recipient and the corresponding donor's id

# initial checks
 if(is.null(str)){
  if(length(rec)==1 && length(don)==1){
   nr <- rec
   rec.id <- 1:nr
   nd <- don
   lab.don<- 1:nd
  }
  else{
   nr <- length(rec)
   rec.id <- rec
   nd <- length(don)
   lab.don <- don
  }
  donors.id <- sample(lab.don, nr, replace=T)
 }


# random hot deck matching in classes
 else{
  nr <- nrow(rec)
  nd <- nrow(don)
  if(is.null(rownames(rec))) rec.lab <- 1:nr
  else rec.lab <- rownames(rec)

  if(is.null(rownames(don))) don.lab <- 1:nd
  else don.lab <- rownames(don)

  aa <- factor(rec[,str])
  lab.str <- levels(aa)
  ns <- length(lab.str)
```

```
  str.rec <- as.character(rec[ ,str])
  str.don <- as.character(don[ ,str])

  rec.id <- numeric(0)
  donors.id <- numeric(0)

  for(i in 1:ns){
   id.rec.i <- rec.lab[str.rec==lab.str[i]]
   rec.id <- c(rec.id, id.rec.i)
   nr.i <- length(id.rec.i)
   id.don.i <- don.lab[str.don==lab.str[i]]
   if(length(id.don.i)==0){
    cat("ERROR: no available donors in stratum ",
     lab.str[i], fill=T)
    stop()
   }
   aa <- sample(id.don.i, nr.i, replace=T)
   donors.id <- c(donors.id, aa)
  }
 }
 mtc.id <- cbind(rec.id=rec.id, don.id=donors.id)
 mtc.id
}


========================================================

strNNDhd.mtc
function (rec, don, str, x.vv, xvv.type, xvv.rng,
constr=FALSE)
{
#
# Nearest neighbour distance hot deck matching
# within classes identified by variable str in input
#
# Note that str is the name of the variable in both rec
# and don used to build the donation classes.

 if(is.null(rownames(rec))) rownames(rec) <- 1:nrow(rec)
 if(is.null(rownames(don))) rownames(don) <- 1:nrow(don)
 aa <- factor(rec[,str])
 lab.str <- levels(aa)
 ns <- length(lab.str)
 rec.str <- as.character(rec[,str])
```

```
 don.str <- as.character(don[,str])
 mtc.id <- matrix(NA,ncol=2)
 out.d <- numeric(0)
 out.nad <- as.list(numeric(ns))

 for(i in 1:ns){
  s.don <- don[don.str==lab.str[i],]
  tst <- sum(rec[,str]==lab.str[i])
  if(tst==0) {
   cat("No available donors in stratum", lab.str[i])
   stop()
  }
  s.rec <- rec[rec.str==lab.str[i],]
  appo <- NNDhd.mtc(s.rec, s.don, x.vv, xvv.type,
   xvv.rng, constr)
  mtc.id <- rbind(mtc.id, appo$mtc.id)
  out.d <- c(out.d, appo$dist.rd)
  out.nad[[i]] <- appo$nad
 }
 mtc.id <- mtc.id[-1, ]
 colnames(mtc.id) <- colnames(appo$mtc.id)
 if(constr) fine <- list(mtc.id=mtc.id,
  dist.rd=unlist(out.d), call=match.call() )
 else fine <- list(mtc.id=mtc.id, dist.rd=unlist(out.d),
  nad=unlist(out.nad), call=match.call() )
 fine
}
=========================================================

NNDhd.mtc
function (rec, don, x.vv, xvv.type, xvv.rng=NULL,
constr=FALSE)
{
#
#-Arguments
# rec: recipient data set (an R matrix or data.frame)
# don: recipient data set (an R matrix or data.frame)
# x.vv: variables common to both data sets to be used to
#       compute distances among units.
#     Distances are computed with Gower distance formula
#     function gower.dist
#
# xvv.type: type of matching variables:
#        "da": dichotomous, 0=absence, 1=presence,
```

```
#               but such that d(0,0)=1;
#         "dd": dichotomous such that d(0,0)=0;
#         "cn": categorical nominal, d(c1,c2)=0 if c1=c2,
#               1 otherwise;
#         "co": categorical ordered;
#         "cc": continuous.
# xvv.rng: ranges for the variables, if NULL ranges are
#          computed on rec and don concatenated
# constr: logical, if TRUE constrained matching is
#         carried out by means of the interface
#         lpAssign.fcn to functions in the R library
#         lpSolve
#
#-Value
# mtc.id: an R matrix with labels of recipient and
#         corresponding donors;
# dist.rd: donor-recipient matching distance
# nad: number of available donors at minimum distance
#      (only constr=F)
#

 x.rec <- data.matrix(rec[ ,x.vv])
 x.don <- data.matrix(don[ ,x.vv])
 if(is.null(xvv.rng)){
  xx <- rbind(x.rec[ ,x.vv],x.don[ ,x.vv])
  xvv.rng <- apply(xx, 2, max) - apply(xx, 2, min)
  xvv.rng[xvv.type=="cn"]<-1
 }

 nr <- nrow(x.rec)
 nd <- nrow(x.don)

 r.lab <- rownames(rec)
 if(is.null(r.lab)) r.lab <- 1:nr

 d.lab <- rownames(don)
 if(is.null(d.lab)) d.lab <- 1:nd

 dist.rd <- numeric(nr)
 nad <- numeric(nr)
 don.lab <- numeric(nr)
 mdist <- matrix(0, nr, nd)

 for(i in 1:nr){
```

```
  mdist[i,] <- gower.dist(x.rec[i,], x.don,
                 v.type=xvv.type, v.rng=xvv.rng)

# unconstrained nearest neighbour matching
  appo <- mdist[i,]
  dist.rd[i] <- min(appo) # distance recipient-donor
  appo <- d.lab[appo==min(appo)]
  nad[i] <- length(appo) # number of available donors
  if(length(appo)==1) don.lab[i] <- appo
  else don.lab[i] <- sample(appo, 1)
 }
 rec.lab <- r.lab
# constrained nearest neighbour matching:
 if(constr){
  if(nr > nd){
   cat("WARNING: There more recipients than donors;",
        fill=T)
   cat("some donors will be used more than once",
        fill=T)
  }
  appo <- lpAssign.fcn(mdist)
  don.lab <- d.lab[appo$solution[,2]]
  rec.lab <- r.lab[appo$solution[,1]]
  dist.rd <- appo$dist.rd
 }
# output
if(is.na(as.numeric(don.lab[1])))
  mtc.id <- cbind(rec.id=rec.lab, don.id=don.lab)
 else
  mtc.id <- cbind(rec.id=as.numeric(rec.lab),
                  don.id=as.numeric(don.lab))
 if(constr)
  fine <- list(mtc.id=mtc.id, dist.rd=dist.rd,
                             call=match.call())
 else
  fine <- list(mtc.id=mtc.id, dist.rd=dist.rd,
                   nad=nad, call=match.call())

 fine
}
=========================================================

gower.dist
function (x, y, v.type, v.rng)
```

```
{
#
# Computes the Gower distance among the values in vector
# x (values of p variables on a given unit) and all n
# rows of y (the same p variables observed on n units).
#
#-Arguments
# x: an R vector with p values
# y: an R matrix with dim=c(n, p)
# v.type: type of p variables:
#   "da": dichotomous, 0=absence, 1=presence,
#         with d(0,0)=1
#   "dd": dichotomous, 0=absence, 1=presence,
#         with d(0,0)=0
#   "cn": categorical nominal, d(c1,c2)=0 if c1=c2, 1
#         otherwise
#   "co": categorical ordered
#   "cc": continuous.
#
# v.rng: ranges for the variables; NA and 0 are not
#        admissible.
#  Please note that the range of variables of type "da",
#  "dd" and "cn" is set equal to 1!
#
#-Value
# a vector with the n distances among x and
#   all the units in y

 if(is.null(dim(y))) y <- matrix(y,ncol=1)

 p <- ncol(y)
 n <- nrow(y)
 if(length(x)!=p) {stop("Error:
     length of x must equal the no. of columns of y")}

 out.dd <- matrix(0, n, p)
 for(k in 1:p){
  if(v.type[k]=="da"){
   appo <- abs(x[k]-y[,k])
   appo[(x[k]==0 & y[,k]==0)] <- 1
   out.dd[,k] <- appo
  }
  else if(v.type[k]=="cn"){
   appo <- rep(0, n)
```

```
   appo[(x[k]==y[,k]) ] <- 0
   appo[(x[k]!=y[,k]) ] <- 1
   out.dd[,k] <- appo
  }
  else{
   rng.k <- v.rng[k]
   appo <- abs(x[k] - y[, k])/rng.k
   out.dd[,k] <- appo
  }
 }
 rowMeans(out.dd)
}
```

===

```
lpAssign.fcn
function (x)
{
#
# This is an interface to functions in library
# lpSolve  (maintained by Sam Buttrey)
# to solve the assignment problem.
#
# Library lpSolve must be loaded before running this
# function!
#
# Arguments:
# x: matrix of distances among the nrow(x)
#    units and the ncol(x) units
#
# Value:
# min.dist: the overall minimum distance
# dist.rd: vector of distances among the recipients
#    and the donors in the final solution
# solution: a two-column matrix with the rows
#     and the associated column units.
#
 nr <- nrow(x)
 nc <- ncol(x)
 if(nr==nc) {
  appo <- lp.assign(x)
 }
 else if(nr<nc){
  r.sig <- rep("==", nr)
```

```
  r.rhs <- rep(1, nr)
  c.sig <- rep("<=", nc)
  c.rhs <- rep(1, nc)
  appo <- lp.transport(x, r.sig, r.rhs, c.sig, c.rhs)
 }
 else{
  r.sig <- rep("==", nr)
  r.rhs <- rep(1, nr)
  c.sig <- rep(">=", nc)
  c.rhs <- rep(1, nc)
  appo <- lp.transport(x, r.sig, r.rhs, c.sig, c.rhs)
 }
 sol <- c(t(appo$solution) )
 sol <- sol * c(t(col(x)))
 sol <- sol[sol!=0]
 msol <- cbind(row=1:nr, col=sol)
 dist.rd <- x[msol]
 list(min.dist=appo$objval, solution=msol,
   dist.rd=dist.rd)

}
```

E.3 R Code for Parametric and Mixed Methods

These R functions implement methods described in Sections 2.1.2, 2.5.1, 3.2.4 and 3.6.1.

```
MoriSche.mtc
function (data.A, data.B, x.vv, rho.yz, macro=TRUE)
{
#
# This function implements a part of the theory
# presented in Moriarity and Scheuren (2001).
#
#-Arguments
# data.A: data set A={X,Y} (an R matrix or data.frame)
#       X must be continuous univariate or multivariate
#       while Y must be continuous univariate
# data.B: data set B={X,Z} (an R matrix or data.frame)
#       X must be continuous, univariate or multivariate
#       while Z must be continuous univariate
#
# xvv: names of the variables X common to both data sets
#       to be used as matching variables.
```

```
# rho.yz: a guess for Cor(Y,Z); if it is not admissible
#         the closest admissible value is considered
# macro: logical; if TRUE (default) the constrained
#        matching step is skipped and only "starting"
#        parameter estimates are given in output.
#
# NB the constrained matching is carried out using
#    the interface lpAssign.fcn to functions in library
#    lpSolve.
#    Library lpSolve must be loaded before running this
#    function!
#
#-Value:
# an R list with the following components:
#  rho.yz: initial guess for the Cor(Y,Z) given in the
#          arguments or the closest admissible
#          value for it.
#  oth.vc: estimates for error terms of variances of Y
#          (first row) and Z (second row)
#  call: the call to this function.
#
#  Moreover if macro=T
#  mu: an R vector (macro=T) with estimated means of the
#      variables;
#  vc: matrix with estimated Var-Cov
#  cor: matrix with estimated correlations.
#
# otherwise (macro=F)
#  fill.A: filled micro-data file obtained using A as
#          recipient
#  mtc.id: matrix with two columns: rows of A and
#          corresponding B donors
#  mtc.dist: overall matching distance.
#

 if(is.list(data.A))  data.A <- data.matrix(data.A)
 if(is.list(data.B))  data.B <- data.matrix(data.B)

 nA <- nrow(data.A)
 nB <- nrow(data.B)
#
# extracting matching variables X (assumed to be
#  continuous)
 p.x <- length(x.vv)
 x.A <- matrix(c(data.A[,x.vv]), ncol=p.x)
```

```
 pos.x.A <- match(x.vv, colnames(data.A))

 x.B <- matrix(c(data.B[,x.vv]), ncol=p.x)
 pos.x.B <- match(x.vv, colnames(data.B))
#
# preparing dependent variables Y and Z (assumed to be
#  continuous)
#
 y.A <- data.A[ ,-pos.x.A]
 y.lab <- colnames(data.A)[-pos.x.A]
 z.B <- data.B[ ,-pos.x.B]
 z.lab <- colnames(data.B)[-pos.x.B]
#
# summary statistics
#
 mu.x <- colMeans(rbind(x.B,x.A) )
 S.x <- var(rbind(x.B, x.A))

 mu.y <- mean(y.A)
 S.y <- var(y.A)
 S.xy <- var(cbind(x.A,y.A))
 v.x.y <- S.xy[-(p.x+1),(p.x+1)]

 mu.z <- mean(z.B)
 S.z <- var(z.B)
 S.xz <- var(cbind(x.B,z.B))
 v.x.z <- S.xz[-(p.x+1),(p.x+1)]

 if(p.x==1) vc <- rbind(c(c(S.x), v.x.y, v.x.z))
 else vc <- cbind(S.x, v.x.y, v.x.z)
 vc <- rbind(vc, c(c(v.x.y), S.y, NA))
 vc <- rbind(vc, c(c(v.x.z), NA, S.z))


 v.names <- c(x.vv, y.lab, z.lab)
 dimnames(vc) <- list(v.names, v.names)
 p <- length(v.names)
 y.pos <- p.x + 1
 z.pos <- p.x + 2
#
# checks if the input value for Cor(Y,Z), rho.yz, is
# admissible or randomly generates an admissible value
# for it
# (1) computes bounds for Cor(Y,Z)
```

```
 c.xy <- c(cor(x.A, y.A))
 c.xz <- c(cor(x.B, z.B))

 if(p.x==1){
  low.c <- c.xy*c.xz - sqrt( (1-c.xy^2)*(1-c.xz^2) )
  up.c <-  c.xy*c.xz + sqrt( (1-c.xy^2)*(1-c.xz^2) )
 }
 else{
  ic.x <- solve(cov2cor(S.x))
  mc1 <- matrix( rep(c.xy, p.x), ncol=p.x )
  mc2 <- matrix( rep(c.xz, p.x), ncol=p.x, byrow=T)
  cc <- mc1 * ic.x * mc2
  dd1 <- 1 - sum( mc1 * ic.x * t(mc1) )
  dd2 <- 1 - sum( t(mc2) * ic.x * mc2 )
  low.c <- sum(cc) - sqrt(dd1*dd2)
  up.c <- sum(cc) + sqrt(dd1*dd2)
 }
#
# (2) checks the assigned value for Cor(Y,Z) or takes
# the closest admissible value

 sum.rho.yz <- c(IN.rho.yz=rho.yz)
 if( (rho.yz<=up.c) && (rho.yz>=low.c) ) {
  cov.yz <- rho.yz * sqrt(S.y*S.z)
 }
 else{
  cat("Warning: value for rho.yz is not admissible:
           a new value has been substituted for it",
           fill=T)
  if(rho.yz > up.c) rho.yz <- up.c - 0.001
  if(rho.yz < low.c) rho.yz <- low.c + 0.001
  cov.yz <- rho.yz * sqrt(S.y*S.z)
 }
 sum.rho.yz <- c(sum.rho.yz, OUT.rho.yz=rho.yz)
 vc.0 <- vc
 vc.0[is.na(vc)] <- cov.yz
#
# Predicting values for Z in file A

 B.z.xy <- t(rbind(c(v.x.z, cov.yz)) %*% solve(S.xy))
 z.pred <- mu.z + cbind( t(t(x.A)- mu.x), y.A-mu.y ) %*%
  B.z.xy

 fi.3 <- rbind(c(v.x.z, cov.yz)) %*% solve(S.xy) %*%
```

```
  cbind(c(v.x.z, cov.yz))
 serr.z <- S.z - fi.3
 if(serr.z<0) serr.z <- 0
 z.ep <- z.pred + rnorm(nA, 0, sqrt(serr.z))
#
# Predicting values for Y in file B

 B.y.xz <- t(rbind(c(v.x.y, cov.yz)) %*% solve(S.xz))
 y.pred <- mu.y + cbind( t(t(x.B)-mu.x), z.B-mu.z ) %*%
  B.y.xz

 fi.6 <- rbind(c(v.x.y, cov.yz)) %*% solve(S.xz) %*%
  cbind(c(v.x.y, cov.yz))
 serr.y <- S.y - fi.6
 if(serr.y<0) serr.y <- 0
 y.ep <- y.pred + rnorm(nB, 0, sqrt(serr.y))
#
# if macro=T only the parameter estimates are given in
# output
#
 if(macro){
  vec.mu <- c(mu.x, mu.y, mu.z)
  names(vec.mu) <- v.names

  o.vc <- matrix(c(S.y, S.z, fi.6, fi.3, serr.y,
   serr.z), nrow=2)
  dimnames(o.vc) <- list(c(y.lab,z.lab),
   c("S","fi","S.err"))
# output
  final <- list(rho.yz=sum.rho.yz, mu=vec.mu, vc=vc.0,
    cor=cov2cor(vc.0), oth.vc=o.vc, call=match.call())
 }
#
# with macro=F the constrained matching step is carried
# out


 else{
  if(nA>nB) stop("The number of donors is less than the
                                         number of recipients")
  S.1 <- S.2 <- vc.0
  SS <- S.1 + S.2
  rid.SS <- SS[(p.x+1):p, (p.x+1):p]
  irSS <- solve(rid.SS)
```

```
# last alternative in Table 1 in Moriarity and Scheuren
# distance recipient-donor on A=(y.A, z.ep) and
# B=(y.ep,z.B)

  new.B <- cbind(y.ep, z.B)
  madist <- matrix(0, nA, nB)
  for(i in 1:nA){
   new.A <- c(y.A[i], z.ep[i])
   madist[i,] <- mahalanobis(new.B, new.A, irSS,
     inverted=T)
  }
# to perform constrained matching the function lp.assign
# in library lpSolve can be used (only small problems)
  appo <- lpAssign.fcn(madist)
  mor.mind <- appo$min.dist
  mor.lab <- appo$solution[,2]

# fills A data file
  fill.A <- cbind(x.A, y.A, z.B[mor.lab])
  colnames(fill.A) <- c(x.vv, y.lab, z.lab)

# computes parameter estimates
  o.vc <- matrix(c(S.y, S.z, fi.6, fi.3, serr.y,
  serr.z), nrow=2)
  dimnames(o.vc) <- list(c(y.lab,z.lab),
   c("S","fi","S.err"))

# output
  final <- list(rho.yz=sum.rho.yz, oth.vc=o.vc,
   fill.A=fill.A, mtc.id=appo$solution,
   mtc.dist=mor.mind, call=match.call())
 }
 final
}

=========================================================

MLmixed.mtc
function (data.A, data.B, x.vv, prho.yz=0, macro=TRUE)
{
#
# ML estimates of the parameters of the
# multivariate normal distribution for X, Y, and Z
# when matching is based on the external information
```

```
# on prho.yz. If Macro=FALSE, the two filled data sets
# are given in output. The filled data sets are obtained
# following Moriarity and Scheuren (2001)
#
# ==> Arguments:
# data.A: data set A={X,Y} (an R matrix or data.frame)
# data.B: data set B={X,Z} (an R matrix or data.frame)
# xvv: names of the variables X common to both data sets
#      (AKA the matching variables)
# prho.yz: the initial value guessed for Cor[(Y,Z)|X]
# macro: logical; if TRUE (default) the constrained
#        matching step is skipped and only "starting"
#        parameter estimates are given in output.
#
# NB the constrained matching is carried out using
#    the interface lpAssign.fcn to functions in library
#    lpSolve.
# Library lpSolve must be loaded before running this
# function!
#
#-Value:
# an R list with the following components:
#  prho.yz: initial guess for the Cor(Y,Z|X) given in
#           the arguments.
#  res.var: estimates for error terms of variances of
#           Y|(X,Z) and Z|(X,Y)
#  call: the call to this function.
#
#  Moreover if macro=T
#  mu: an R vector (macro=T) with estimated means of the
#      variables;
#  vc: matrix with estimated Var-Cov
#  cor: matrix with estimated correlations.
#
# otherwise (macro=F)
#  fill.A: filled micro-data file obtained using A as
#          recipient;
#  mtc.id: matrix with two columns: rows of A and
#          corresponding B donors;
#  mtc.dist: overall matching distance.

 if(is.list(data.A))  data.A <- data.matrix(data.A)
 if(is.list(data.B))  data.B <- data.matrix(data.B)
 nA <- nrow(data.A)
```

```
 nB <- nrow(data.B)
#
# extracting matching variables X (assumed to be
# continuous)
 p.x <- length(x.vv)
 x.A <- matrix(c(data.A[,x.vv]), ncol=p.x)
 pos.x.A <- match(x.vv, colnames(data.A))
 x.B <- matrix(c(data.B[,x.vv]), ncol=p.x)
 pos.x.B <- match(x.vv, colnames(data.B))
#
# preparing dependent variables Y and Z (assumed to be
# continuous)
#
 y.A <- data.A[ ,-pos.x.A]
 y.lab <- colnames(data.A)[-pos.x.A]
 z.B <- data.B[ ,-pos.x.B]
 z.lab <- colnames(data.B)[-pos.x.B]
#
# regression in file B: Z vs. X
#
 lm.B <- lm(z.B ~ x.B)
 res.B <- residuals(lm.B)
 se.B <- sqrt(mean(res.B2))  # ML estimate of V(Z|X)
#
# regression in file A: Y vs. X
#
 lm.A <- lm(y.A ~ x.A)
 res.A <- residuals(lm.A)
 se.A <- sqrt(mean(res.A2))  # ML estimate of V(Y|X)
#
# ML estimates for X variables
 mu.x <- colMeans(rbind(x.A, x.B))
 S.x <- var(rbind(x.A, x.B))*((nA+nB-1)/(nA+nB))
#
# ML estimates for Y
 mu.y <- sum(coefficients(lm.A)*c(1,mu.x))
 S.yx <- coefficients(lm.A)[-1] %*% S.x
 S.y <- se.A^2 + S.yx %*% solve(S.x) %*%  t(S.yx)
#
# ML estimates for Z
 mu.z <- sum(coefficients(lm.B)*c(1,mu.x))
 S.zx <- coefficients(lm.B)[-1] %*% S.x
 S.z <- se.B^2 + S.zx %*% solve(S.x) %*%  t(S.zx)
#
```

```
# ML estimates for Y,Z, given the starting value for
# Cor[(Y,Z)|X]

 S.YZgX <- prho.yz * (se.A * se.B) # partial
# Cov[(Y,Z)|X]
 S.YZ <- S.YZgX + S.yx %*% solve(S.x) %*% t(S.zx)
 S.ZgYX <- se.B^2 - S.YZgX %*% solve(se.A^2) %*%
    t(S.YZgX)
 S.YgZX <- se.A^2 - S.YZgX %*% solve(se.B^2) %*%
    t(S.YZgX)

 mu <- c(mu.x, mu.y, mu.z)
 v.names <- c(x.vv, y.lab, z.lab)
 names(mu) <- v.names

 res.var <- c(YgZX=c(S.YgZX), ZgYX=c(S.ZgYX))

 vc.1 <- rbind(S.yx, S.zx)
 vc.2 <- matrix(c(S.y, S.YZ, S.YZ, S.z), 2, 2)
 vc <- rbind(S.x, vc.1)
 vc <- cbind(vc, rbind(t(vc.1),vc.2) )
 dimnames(vc) <- list(v.names, v.names)
#
# if macro=TRUE only ML estimates of parameters are
# given in output


 if(macro){
  final <- list(prho.yz=prho.yz, res.var=res.var,
   mu=mu, vc=vc, cor=cov2cor(vc), call=match.call())
 }
 else{
  if(nA>nB) stop("The number of donors is less than the
                                    number of recipients")
#
# Predicting values for Z in file A

  z.pred <- mu.z + rbind(vc[z.lab,c(x.vv,y.lab)]) %*%
   solve(vc[c(x.vv,y.lab),c(x.vv,y.lab)]) %*%
   t(cbind(sweep(x.A, 2, mu.x), y.A-mu.y))
  z.ep <- c(z.pred) + rnorm(nA, 0, sqrt(S.ZgYX))
#
# Predicting values for Y in file B
```

```
  y.pred <- mu.y + rbind(vc[y.lab,c(x.vv,z.lab)]) %*%
   solve(vc[c(x.vv,z.lab),c(x.vv,z.lab)]) %*%
   t(cbind(sweep(x.B, 2, mu.x), z.B-mu.z))
  y.ep <- c(y.pred) + rnorm(nB, 0, sqrt(S.YgZX))
  irSS <- solve(vc[c(y.lab,z.lab), c(y.lab,z.lab)])

# distance recipient-donor on A=(y.A, z.ep) and
# B=(y.ep, z.B)

  new.B <- cbind(y.ep, z.B)
  madist <- matrix(0, nA, nB)
  for(i in 1:nA){
   new.A <- c(y.A[i], z.ep[i])
   madist[i,] <- mahalanobis(new.B, new.A, irSS,
    inverted=T)
  }

# to perform constrained matching the function lp.assign
# in library lpSolve can be used (only small problems)
  appo <- lpAssign.fcn(madist)
  mor.mind <- appo$min.dist
  mor.lab <- appo$solution[,2]

# fills A data files
  fill.A <- cbind(x.A, y.A, z.B[mor.lab])
  colnames(fill.A) <- c(x.vv, y.lab, z.lab)

# output
  final <- list(prho.yz=prho.yz, res.var=res.var,
   fill.A=fill.A, mtc.id=appo$solution,
   mtc.dist=mor.mind, call=match.call())
 }
 final
}
```

E.4 R Code for the Study of Uncertainty

This R function implements the method described in Section 4.5.1.

```
emcat.c
function(s, start, cc, prior = 1, showits = T,
maxits = 1000, eps = 0.0001)
```

```
{
# This is a modified version of function em.cat
# in library cat, original version by Joseph L. Schafer
# ported to R by Ted Harding and Fernando Tusell.
# Library cat needs to be loaded before running
# this function!
#
# An argument is introduced:
# cc: a vector of length 2 containing the positions
#     of the cells in the table involved in the
#     following constraint
#     Pr(cc[1]) >= Pr(cc[2])
#

 if(length(prior) == 1) prior <- rep(prior, s$ncells)
 w <- !is.na(prior)
 if(missing(start)) {
  start <- rep(1, s$ncells)
  start[!w] <- 0
 }
 else {
#  if(any(start[w] == 0)) {
#   warning("Starting value on the boundary")
#  }
  if(any(!w)) {
   if(any(start[!w] != 0)) {
 stop("Starting value has nonzero
   elements for structural zeros")
   }
  }
 }
 prior <- as.double(prior)
 start <- as.double(start)
 start.0 <- as.double(start)
 it <- 0
 convgd <- FALSE
 if(showits)
  cat(paste("Iterations of EM:", "\n"))
 cond <- numeric(0)
 j <- 1
 while((!convgd) & (it < maxits)) {
  old <- start
  start <- .Fortran("estepc",
   s$ncells,
```

```
   start,
   start,
   s$npatt,
   s$p,
   s$r,
   s$mdpgrp,
   s$ngrp,
   s$mobs,
   s$nmobs,
   s$d,
   s$jmp,
   as.integer(0),
   integer(s$p),
   integer(s$p),
   PACKAGE="cat") #
#
#----------------------------------------------------
# checks output of each EM iteration
#----------------------------------------------------
#
  if(start[[13]] == 1)
   stop("Bad parameter value: assigns zero prob.
        to an observed event") #
  else start <- start[[3]]
  start[w] <- (start[w] + prior[w] - 1) #


  start[w] <- start[w]/sum(start[w])
  if(any(start < 0))
   stop("Estimate outside the parameter space. Check
        prior.")
  start[w][start[w] < (1e-007/sum(w))] <- 0 #
#
#----------------------------------------------------
# modified version of Schafer's code starts here
#----------------------------------------------------
#
  new <- start
  c1 <- cc[1]
  c2 <- cc[2]
  if(new[c2] > new[c1]) {
   alfa <- (new[c1] - new[c2])/(old[c2] - new[c2] -
    old[c1] + new[c1])
   new <- alfa * old + (1 - alfa) * new #
```

```
    if(sum(start - new) < 1e-005) {
     cond <- c(cond, 2)
     j <- j + 1
     aa <- start.0[c2] + start.0[c1]
     start <- start.0
     start[c2] <- 0.5^j * aa
     start[c1] <- aa - start[c2]
    }
    else {
     cond <- c(cond, 1)
     start <- new
    }
   }
   else {
    cond <- c(cond, 0)
    start <- new
   }
   it <- it + 1
   if(showits)
    cat(paste(format(it), "...", sep = ""))
    convgd <- all(abs(old - start) <= (eps * abs(old)))
  }
  if(showits)
   cat("\n")
  names(cond) <- 1:it
  list(iter = it, hist = cond, j.val = j, out.em = start)
 }
```

E.5 Other R Functions

This R function is used in the application described in Section 7.2 and computes the absolute and partial η^2 described in Section 7.2.3.

```
eta.fcn
function (glm.yx)
{
# function to compute the table of eta and
# partial eta coefficients:
#
# eta= SS_{effect}/SS_{total}
# partial.eta = SS_{effect}/(SS_{effect}+SS_{error})
#
# starting from the ANOVA table.
```

```
# glm.xy is an lm or glm object.
#
 anova.yx <- anova(glm.yx)
 dev.tot <- anova.yx[1,4]
 dev.res <- deviance(glm.yx)
 anova.yx <- anova.yx[-1,-4]
 eta <- anova.yx[,2]/dev.tot
 eta.p <- anova.yx[,2]/(dev.res+anova.yx[,2])
 out <- cbind(anova.yx[,-3], eta=eta, eta.p=eta.p)
 out <- out[order(out[,"eta"], decreasing = T),]
 list( dev.tot=dev.tot, dev.res=dev.res, anova=out)
}
```

References

Abello, R. and Phillips, B. (2004) Statistical matching of the HES and NHS: an exploration of issues in the use of unconstrained and constrained approaches in creating a basefile for a microsimulation model of the pharmaceutical benefits scheme. Technical report, Australian Bureau of Statistics. Methodology Advisory Committee Paper, June.

Adamek, J.C. (1994) Fusion: Combining data from separate sources. *Marketing Research: A Magazine of Management and Applications* **6**, 48–50.

Agresti, A. (1990) *Categorical Data Analysis*. New York: Wiley.

Anagnoson, J.T. (2000) Microsimulation of public policy. In G.D. Garson (ed.), *The Handbook of Public Information Systems*. New York: Marcel Dekker.

Anderson, T.W. (1957) Maximum likelihood estimates for a multivariate normal distribution when some observations are missing. *Journal of the American Statistical Association* **52**, 200–203.

Anderson, T.W. (1984) *An Introduction to Multivariate Statistical Analysis*, 2nd edn. New York: Wiley.

Antoine, J. (1987) A case study illustrating the objectives and perspectives of fusion techniques. In H. Henry (ed.), *Readership Research: Theory and Practice*, pp. 336–351. Amsterdam: Elsevier Science.

Antoine, J. and Santini, G. (1987) Fusion techniques: alternative to single source methods. *European Research* **15**, 178–187.

Baker, K. (1990) The BARB/TGI fusion. Technical report, Ken Baker Associates, Ickenham, UK.

Baker, K., Harris, P. and O'Brien, J. (1989) Data fusion: an appraisal and experimental evaluation. *Journal of the Market Research Society* **31**, 152–212.

Bakker, B.F.M. and Winkels, J.W. (1998) Why integration of household surveys?–Why POLS? *Netherlands Official Statistics* **13**, 5–7.

Banca d'Italia, (2002) Italian Household Budget in 2000. Supplement to the *Statistical Bulletin, Methodological Notes and Statistical Information*, **12**(6).

Barr, R.S. and Turner, J.T. (1990) Quality issues and evidence in statistical file merging. In G.E. Liepins and V.R.R. Uppuluri (eds), *Data Quality Control: Theory and Pragmatics*, pp. 245–313. New York: Marcel Dekker.

Barry, J.T. (1988) An investigation of statistical matching. *Journal of Applied Statistics* **15**, 275–283.

Statistical Matching: Theory and Practice M. D'Orazio, M. Di Zio and M. Scanu

Battistin, E., Miniaci, R. and Weber, G. (2003) What do we learn from recall consumption data? Technical Report Temi di Discussione del Servizio Studi no. 466, Banca d'Italia.

Bergsma, W.P. and Rudas, T. (2002) Marginal models for categorical data. *Annals of Statistics* **30**, 140–159.

Bernardo, J.M. and Smith, A.F.M. (2000) *Bayesian Theory*. Chichester: Wiley.

Bordt, M., Cameron, G.J., Gibble, S.F., Murphy, B.B., Rowe, G.T. and Wolfson, M.C. (1990) The social policy simulation database and model: an integrated tool for tax/transfer policy analysis. *Canadian Tax Journal* **38**, 48–65.

Box, G.E.P. and Tiao, G.C. (1992) *Bayesian Inference in Statistical Analysis*. New York: Wiley.

Breiman, L., Friedman, J.H., Olshen, R.A. and Stone, C.J. (1984) *Classification and Regression Trees*. Belmont, CA: Wadsworth.

Buck, S. (1960) A method of estimation of missing values in multivariate data suitable for use with electronic computer. *Journal of the Royal Statistical Society, B* **22**, 302–306.

Buck, S.F. (1989) Single source data–the theory and the practice. *Journal of the Market Research Society* **31**, 489–500.

Budd, E.C. (1971) The creation of a microdata file for estimating the distribution of income. *Review of Income and Wealth* **17**, 317–333.

Burkard, R.E. and Derigs, U. (1980) *Assignment and Matching Problems: Solution Methods with FORTRAN-Programs*. Berlin: Springer-Verlag.

Capotorti, A. and Vantaggi, B. (2002) Locally strong coherence in inferential processes. *Annals of Mathematics and Artificial Intelligence* **35**, 125–149.

Cassel, C.M. (1983) Statistical matching–statistical prediction. What is the difference? An evaluation of statistical matching and a special type of prediction using data from a survey on living conditions. *Statistisk Tidskrift* **5**, 55–63.

Cheng, P.E. and Chu, C.K. (1996) Kernel estimation of distribution functions and quantiles with missing data. *Statistica Sinica* **6**, 63–78.

Citoni, G., Di Nicola, F., Lugaresi, S. and Proto, G. (1991) Statistical matching for tax-benefit microsimulation modelling: a project for Italy. Technical Report 'Gruppo di Lavoro sulle Analisi delle Politiche Redistributive', Istituto di Studi per la Programmazione Economica, November.

Cochran, W.G. (1977) *Sampling Techniques*, 3rd edn. New York: Wiley.

Cohen, M.L. (1991) Statistical matching and microsimulation models. In C.F. Citro and E.A. Hanushek (eds), *Improving Information for Social Policy Decisions: The Uses of Microsimulation Modeling. Vol. II: Technical Papers*. Washington, DC: National Academy Press.

Coletti, G. and Scozzafava, R. (2002) *Probabilistic Logic in a Coherent Setting*. Dordrecht: Kluwer.

Coli, A. and Tartamella, F. (2000a) The link between national accounts and households micro data. Paper presented to the Meeting of the Siena Group on Social Statistics, Maastricht (Netherlands), 22–24 May.

Coli, A. and Tartamella, F. (2000b) A pilot social accounting matrix for Italy with a focus on households. Paper presented to the 26th General Conference of the International Association for Research in Income and Wealth, Cracow (Poland), 27 August–2 September. Available at http://www.iariw.org/c2000.asp (accessed December 2005).

Coli, A., Tartamella, F., Sacco, G., Faiella, I., Scanu, M., D'Orazio, M., Di Zio, M., Siciliani, I., Colombini, S. and Masi, A. (2005) La costruzione di un archivio di microdati sulle famiglie italiane ottenuto integrando l'indagine ISTAT sui consumi delle famiglie italiane e l'indagine Banca d'Italia sui bilanci delle famiglie italiane. Technical report, *Documenti 12/2006*, Istituto Nazionale di Statistica, Rome (in Italian).

Conti, P.L. and Scanu, M. (2005) On the evaluation of matching noise produced by nonparametric imputation techniques. Technical Report 2005/7, Dipartimento di Statistica, Probabilità e Statistiche Applicate, Università di Roma 'La Sapienza'.

Cox, D.R. and Wermuth, N. (1996) *Multivariate Dependencies*. London: Chapman & Hall.

Dawid, A.P. (1979) Conditional independence in statistical theory. *Journal of the Royal Statistical Society, B* **41**, 1–31.

de Finetti, B. (1974) *Theory of Probability*. London: Wiley.

De Waal, T. (2003) Solving the error localization problem by means of vertex generation. *Survey Methodology* **29**, 71–79.

Decoster, A. and Van Camp, G. (1998) The unit of analysis in microsimulation models for personal income taxes: fiscal unit or household? Technical Report DPS 98.33, Centrum voor Economische Studiën, Department Economie, Katholieke Universiteit Leuven.

DeGroot, M.H. (1987) Record linkage and matching systems. In S. Kotz and N.L. Johnson (eds) *Encyclopedia of Statistical Sciences*, Vol. 7, pp. 649–654. Wiley.

DeGroot, M.H. and Goel, P.K. (1976) The matching problem for multivariate normal data. *Sankhyā, B* **38**, 14–29.

DeGroot, M.H., Feder, P.I. and Goel, P.K. (1971) Matchmaking. *Annals of Mathematical Statistics* **42**, 578–593.

Deming, W.E. and Stephan, F.F. (1940) On a least squares adjustment of a sampled frequency table when the expected marginal totals are known. *Annals of Mathematical Statistics* **11**, 427–444.

Dempster, A.P., Laird, N.M. and Rubin, D.B. (1977) Maximum likelihood from incomplete data via the EM algorithm. *Journal of the Royal Statistical Society, B* **39**, 1–38.

Denk, M. and Hackl, P. (2003) Data integration and record matching: an Austrian contribution to research in official statistics. *Austrian Journal of Statistics* **32**, 305–321.

Deville, J.C. and Särndal, C.E. (1992) Calibration estimators in survey sampling. *Journal of the American Statistical Association* **87**, 376–382.

D'Orazio, M., Di Zio, M. and Scanu, M. (2002) Statistical matching and official statistics. *Rivista di Statistica Ufficiale* **2002/1**, 5–24.

D'Orazio, M., Di Zio, M. and Scanu, M. (2005a) A comparison among different estimators of regression parameters on statistically matched files through an extensive simulation study. Technical Report, *Contributi* 2005/10, Istituto Nazionale di Statistica, Rome.

D'Orazio M., Di Zio, M. and Scanu, M. (2005b) Statistical matching for categorical data: displaying uncertainty and using logical constraints. *Journal of Official Statistics*. To appear.

Dubins, L.E. (1975) Finitely additive conditional probabilities, conglomerability and disintegration. *Annals of Probability* **3**, 89–99.

Edgett, G.L. (1956) Multiple regression with missing observations among the independent variables. *Journal of the American Statistical Association* **51**, 122–131.

Ettlinger, M.P., O'Hare, J.F., McIntyre, R.S., King, J., Fray, E.A. and Miransky, N. (1996) *Who Pays? A Distributional Analysis of the Tax Systems in all 50 States*. Citizens for Tax Justice and The Institute of Taxation and Economic Policy, Washington, DC.

Eubank, R. (1988) *Spline Smoothing and Nonparametric Regression*. New York: Marcel Dekker.

Everitt, B.S. (1984) *An Introduction to Latent Variable Models*. London: Chapman & Hall.

Ezzati-Rice, T.M., Fahimi, M., Judkins, D. and Khare, M. (1993) Serial imputation of NHANES III with mixed regression and hot–deck techniques. *Proceedings of the Section on Survey Research Methods of the American Statistical Association*, pp. 292–296.

Fellegi, I.P. and Holt, D. (1976) A systematic approach to automatic edit and imputation. *Journal of the American Statistical Association* **71**, 17–35.

Filippello, R., Guarnera, U. and Jona Lasinio, G. (2004) Use of auxiliary information in statistical matching. In *Proceedings of the XLII Conference of the Italian Statistical Society*, pp. 37–40. Bari (Italy), 9–11 June 2004. Padua: CLEUP

Gelman, A., Carlin, J.B., Stern, H.S. and Rubin, D.B. (2004) *Bayesian Data Analysis*, 2nd edn. Boca Raton, FL: Chapman & Hall/CRC.

Goel, P.K. and Ramalingam, T. (1989) *The Matching Methodology: Some Statistical Properties*. New York: Springer–Verlag.

Haberman, S.J. (1977) Product models for frequency tables involving indirect observation. *Annals of Statistics* **5**, 1124–1147.

Hansen, P. and Jaumard, B. (1990) Algorithms for the maximum satisfiability problem. *Computing* **44**, 279–303.

Härdle, W. (1992) *Applied Nonparametric Regression*. Cambridge: Cambridge University Press.

Horowitz, J.L. and Manski, C.F. (2000) Nonparametric analysis of randomized experiments with missing covariate and outcome data. *Journal of the American Statistical Association* **95**, 77–84.

Horvitz, D.G. and Thompson, D.J. (1952) A generalization of sampling without replacement from a finite universe. *Journal of the American Statistical Association* **47**, 663–685.

Istat, (2002) I consumi delle famiglie, anno 2000. *Annuario Settore Famiglia e Società*, no. 7 (in Italian).

Jephcott, J. and Bock, T. (1991) The application and validation of data fusion. *Journal of the Market Research Society* **40**, 185–205.

Judge, G.G., Griffiths, W.E., Hill, R.C. and Lee, T.C. (1980) *The Theory and Practice of Econometrics*. New York: Wiley.

Kadane, J.B. (1978) Some statistical problems in merging data files. In Department of Treasury, *Compendium of Tax Research*, pp. 159–179. Washington, DC: US Government Printing Office. Reprinted in 2001: *Journal of Official Statistics*, **17**, 423–433.

Kamakura, W.A. and Wedel, M. (1997) Statistical data fusion. *Journal of Marketing Research* **34**, 485–498.

Kenward, M.G., Goetghebeur, E.J.T. and Molenberghs, G. (2001) Sensitivity analysis for incomplete categorical data. *Statistical Modelling* **1**, 31–48.

Keribin, C. (2000) Consistent estimation of the order of mixture models. *Sankhyā, A* **62**, 49–66.

Klevmarken, N.A. (1986) Comment on Paass (1986) In G.H. Orcutt, J. Merz and H. Quinke (eds), *Microanalytic Simulation Models to Support Social and Financial Policy*, pp. 421–422. Amsterdam: Elsevier Science.

Lazzeroni, L.C., Schenker, N. and Taylor, J.M.G. (1990) Robustness of multiple-imputation techniques to model misspecification. *Proceedings of the Section on Survey Research Methods, American Statistical Association*, pp. 260–265.

Little, R.J.A. (1988) Missing-data adjustments in large surveys. *Journal of Business and Economic Statistics* **6**, 287–296.

Little, R.J.A. (1993) Pattern-mixture models for multivariate incomplete data. *Journal of the American Statistical Association* **88**, 125–134.

Little, R.J.A. and Rubin, D.B. (2002) *Statistical Analysis with Missing Data*, 2nd edn. Hoboken, NJ: Wiley.

Liu, T.P. and Kovacevic, M.S. (1994) Statistical matching of survey datafiles: a simulation study. *Proceedings of the Section on Survey Research Methods of the American Statistical Association*, pp. 479–484.

Lord, F.M. (1955) Estimation of parameters from incomplete data. *Journal of the American Statistical Association* **50**, 870–876.

Manski, C.F. (1995) *Identification Problems in the Social Sciences*. Cambridge, MA: Harvard University Press.

Mardia, K.V., Kent, J.T. and Bibby, J.M. (1979) *Multivariate Analysis*. London: Academic Press.

Martini, A. and Trivellato, U. (1997) The role of survey data in microsimulation models for social policy analysis. *Labour* **11**, 83–112.

Matthai, A. (1951) Estimation of parameters from incomplete data with application to design of sample surveys. *Sankhyā* **11**, 145–152.

McLachlan, G.J. and Basford, K.E. (1988) *Mixture Models: Inference and Applications to Clustering*. New York: Marcel Dekker.

McLachlan, G.J. and Peel, D. (2000) *Finite Mixture Models*. New York: Wiley.

Meng, X.L. and Rubin, D.B. (1993) Maximum likelihood via the ECM algorithm: a general framework. *Biometrika* **80**, 267–278.

Moore, J.H. (2004) Measuring defined benefit plan replacement rates with PenSync. *Monthly Labor Review* **127**(11), 57–68.

Moriarity, C. and Scheuren, F. (2001) Statistical matching: a paradigm for assessing the uncertainty in the procedure. *Journal of Official Statistics* **17**, 407–422.

Moriarity, C. and Scheuren, F. (2003) A note on Rubin's statistical matching using file concatenation with adjusted weights and multiple imputation. *Journal of Business and Economic Statistics* **21**, 65–73.

Moriarity, C. and Scheuren, F. (2004) Regression-based statistical matching: recent developments. *Proceedings of the Section on Survey Research Methods, American Statistical Association*.

Neapolitan, R.A. (2004) *Learning Bayesian Networks*. Upper Saddle River, NJ: Prentice Hall.

Nielsen, S.F. (2001) Nonparametric conditional mean imputation. *Journal of Statistical Planning and Inference* **99**, 129–150.

O'Brien, S. (1991) The role of data fusion in actionable media targeting in the 1990's. *Marketing and Research Today* **19**, 15–22.

Okner, B.A. (1972) Constructing a new data base from existing microdata sets: the 1966 merge file. *Annals of Economic and Social Measurement* **1**(3), 325–342.

Okner, B.A. (1974) Data matching and merging: an overview. *Annals of Economic and Social Measurement* **3**(2), 347–352.

Paass, G. (1985) Statistical record linkage methodology: state of the art and future prospects. In *Bulletin of the International Statistical Institute, Proceedings of the 45th Session*, Vol. LI, Book 2. Voorburg, Netherlands: ISI.

Paass, G. (1986) Statistical match: evaluation of existing procedures and improvements by using additional information. In G.H. Orcutt, J. Merz and H. Quinke (eds) *Microanalytic Simulation Models to Support Social and Financial Policy*, pp. 401–422. Amsterdam: Elsevier Science.

R Development Core Team 2004 *R: A Language and Environment for Statistical Computing*. Vienna: R Foundation for Statistical Computing.

Radner, D.B., Allen, R., Gonzalez, M.E., Jabine, T.B. and Muller, H.J. (1980) *Report on Exact and Statistical Matching Techniques.* Statistical Policy Working Paper 5, Federal Committee on Statistical Methodology.

Rässler, S. (2002) *Statistical Matching: A Frequentist Theory, Practical Applications and Alternative Bayesian Approaches*. New York: Springer-Verlag.

Rässler, S. (2003) A non-iterative Bayesian approach to statistical matching. *Statistica Neerlandica* **57**(1), 58–74.

Rässler, S. (2004) Data fusion: identification problems, validity, and multiple imputation. *Austrian Journal of Statistics* **33**(1–2), 153–171.

Rässler, S. and Fleischer, K. (1999) An evaluation of data fusion techniques. In *Proceedings of the XVI International Methodology Symposium–Statistics Canada*, 2–5 May 1999, pp. 129–136. Ottawa: Statistics Canada.

Redmond, G., Sutherland, H. and Wilson, M. (1998) *The Arithmetic of Tax and Social Security Reform: a User's Guide to Microsimulation Methods and Analysis.* Cambridge: Cambridge University Press.

Renssen, R.H. (1998) Use of statistical matching techniques in calibration estimation. *Survey Methodology* **24**, 171–183.

Roberts, A. (1994) Media exposure and consumer purchasing: an improved data fusion technique. *Marketing and Research Today* **22**, 159–172.

Rodgers, W.L. (1984) An evaluation of statistical matching. *Journal of Business and Economic Statistics* **2**, 91–102.

Rubin, D.B. (1974) Characterizing the estimation of parameters in incomplete-data problems. *Journal of the American Statistical Association* **69**, 467–474.

Rubin, D.B. (1976) Inference and missing data. *Biometrika* **63**, 581–592.

Rubin, D.B. (1986) Statistical matching using file concatenation with adjusted weights and multiple imputations. *Journal of Business and Economic Statistics* **4**, 87–94.

Rubin, D.B. (1987) *Multiple Imputation for Nonresponse in Surveys*. New York: Wiley.

Ruggles, N. (1999) The development of integrated data bases for social, economic and demographic statistics In N. Ruggles and R. Ruggles (eds) *Macro- and Microdata Analyses and Their Integration*, pp. 410–478. Cheltenham: Edward Elgar.

Ruggles, N. and Ruggles, R. (1974) A strategy for merging and matching microdata sets. *Annals of Economic and Social Measurement* **1**(3), 353–371.

Särndal, C.E., Swensson, B. and Wretman, J. (1992) *Model-Assisted Survey Sampling*. New York: Springer-Verlag.

Schafer, J.L. (1997) *Analysis of Incomplete Multivariate Data*. London: Chapman & Hall.

Seber, G.A.F. (1977) *Linear Regression Analysis*. New York: Wiley.

Silverman, B.W. (1986) *Density Estimation for Statistics and Data Analysis*. London: Chapman & Hall.

Sims, C.A. (1972) Comments on Okner (1972). *Annals of Economic and Social Measurement* **1**(3), 343–345.

Singh, A.C., Armstrong, J.B. and Lemaitre, G.E. (1988) Statistical matching using log linear imputation. *Proceedings of the Section on Survey Research Methods, American Statistical Association*, pp. 672–677.

Singh, A.C., Mantel, H., Kinack, M. and Rowe, G. (1990) On methods of statistical matching with and without auxiliary information. Technical Report SSMD-90-016E, Methodology Branch, Statistics Canada.

Singh, A.C., Mantel, H., Kinack, M. and Rowe, G. (1993) Statistical matching: Use of auxiliary information as an alternative to the conditional independence assumption. *Survey Methodology* **19**, 59–79.

Sutherland, H., Taylor, R. and Gomulka, J. (2001) Combining household income and expenditure data in policy simulations. Technical Report MU0101, Department of Applied Economics, University of Cambridge.

Szivós, P, Rudas, T. and Tóth, I.G. (1998) A tax-benefit microsimulation model for Hungary. Technical report, TÁRKI Social Research Institute. Available at `http://www.tarki.hu/research/microsim/micro2.html`.

Tanner, M.A. and Wong, W.H. (1987) The calculation of posterior distributions by data augmentation. *Journal of the American Statistical Association* **82**, 528–550.

United Nations (1993) Social accounting matrices. In United Nations (ed.), *System of National Accounts,* Chapter XX. New York: UN.

Van Buuren, S. and Oudshoorn, C.G.M. (1999) Flexible multivariate imputation by MICE. Technical Report PG/VGZ/99.054, TNO, Leiden.

Van Buuren, S. and Oudshoorn, C.G.M. (2000) Multivariate imputation by chained equations. Technical Report PG/VGZ/00.038, TNO, Leiden.

van der Laan, P. (2000) Integrating administrative registers and household surveys. *Netherlands Official Statistics* **15**, 7–15.

Vantaggi, B. (2005) The role of coherence for the integration of different sources. In F.G. Cozman, R. Nau and T. Seidenfeld (eds), *Proceedings: 4th International Symposium on Imprecise Probabilities and Their Applications*, pp. 269–378. Pittsburgh: Brightdocs.

Wand, M. and Jones, C. (1995) *Kernel Smoothing*. London: Chapman & Hall.

Wiegand, J. (1986) Combining different media surveys: The German partnership model and fusion experiments. *Journal of the Market Research Society* **28**, 189–208.

Wilks, S.S. (1932) Moments and distributions of estimates of population parameters from fragmentary samples. *Annals of Mathematical Statistics* **3**, 163–194.

Williamson, J. (2004) Philosophies of probability: objective Bayesianism and its challenges. In A. Irvine (ed.) *Philosophy of Mathematics*, vol. 4 of the Handbook of the Philosophy of Science. Amsterdam: Elsevier. To appear.

Winkels, J.W. and Everaers, P.C.J. (1998) Design of an integrated survey in the Netherlands. The case of POLS. *Netherlands Official Statistics* **13**, 8–11.

Winkler, W.E. (1993) Improved decision rules in the Fellegi–Sunter model of record linkage. *Proceedings of the Section on Survey Research Methods, American Statistical Association*, pp. 274–279.

Winkler, W.E. (1995) Matching and record linkage. In G.P. Cox, D.A. Binder, B.N. Chinnappa, A. Christianson, M. Colledge and P.S. Kott (eds), *Business Survey Methods*, pp. 355–384. New York: Wiley.

Wolfson, M., Gribble, S., Bordt, M., Murphy, B. and Rowe, G. (1987) The social policy simulation database: an example of survey and administrative data integration. In J.W. Coombs and M.P. Singh (eds), *Proceedings: Symposium on Statistical Uses of Administrative Data*, pp. 201–229. Ottawa: Statistics Canada.

Wolfson, M., Gribble, S., Bordt, M., Murphy, B.B. and Rowe, G.T. (1989) The social policy simulation database and model: an example of survey and administrative data integration. *Survey of Current Business* **69**, 36–41.

Yoshizoe, Y. and Araki, M. (1999) Statistical matching of household survey files. Technical Report 10, ITME (Information Technology and the Market Economy) Project of the Japan Society for the Promotion of Science.

Index

Statistical Matching: Theory and Practice M. D'Orazio, M. Di Zio and M. Scanu

WILEY SERIES IN SURVEY METHODOLOGY

Established in part by WALTER A. SHEWHART AND SAMUEL S. WILKS

Editors: *Robert M. Groves, Graham Kalton, J. N. K. Rao, Norbert Schwarz, Christopher Skinner*

The ***Wiley Series in Survey Methodology*** covers topics of current research and practical interests in survey methodology and sampling. While the emphasis is on application, theoretical discussion is encouraged when it supports a broader understanding of the subject matter.

The authors are leading academics and researchers in survey methodology and sampling. The readership includes professionals in, and students of, the fields of applied statistics, biostatistics, public policy, and government and corporate enterprises.

*BIEMER, GROVES, LYBERG, MATHIOWETZ, and SUDMAN · Measurement Errors in Surveys
BIEMER and LYBERG · Introduction to Survey Quality
BRADBURN, SUDMAN, and WANSINK ·Asking Questions: The Definitive Guide to Questionnaire Design—For Market Research, Political Polls, and Social Health Questionnaires, *Revised Edition*
BRAVERMAN and SLATER · Advances in Survey Research: New Directions for Evaluation, No. 70
CHAMBERS and SKINNER (editors) Analysis of Survey Data
COCHRAN · Sampling Techniques, *Third Edition*
COUPER, BAKER, BETHLEHEM, CLARK, MARTIN, NICHOLLS, and O'REILLY (editors) · Computer Assisted Survey Information Collection
COX, BINDER, CHINNAPPA, CHRISTIANSON, COLLEDGE, and KOTT (editors) · Business Survey Methods
*DEMING · Sample Design in Business Research
DILLMAN · Mail and Internet Surveys: The Tailored Design Method
GROVES and COUPER · Nonresponse in Household Interview Surveys
GROVES · Survey Errors and Survey Costs
GROVES, DILLMAN, ELTINGE, and LITTLE · Survey Nonresponse
GROVES, BIEMER, LYBERG, MASSEY, NICHOLLS, and WAKSBERG · Telephone Survey Methodology
GROVES, FOWLER, COUPER, LEPKOWSKI, SINGER, and TOURANGEAU · Survey Methodology
*HANSEN, HURWITZ, and MADOW · Sample Survey Methods and Theory, Volume 1: Methods and Applications
*HANSEN, HURWITZ, and MADOW · Sample Survey Methods and Theory, Volume II: Theory
HARKNESS, VAN DE VIJVER, and MOHLER · Cross-Cultural Survey Methods
KALTON and HEERINGA · Leslie Kish Selected Papers
KISH · Statistical Design for Research
*KISH · Survey Sampling
KORN and GRAUBARD · Analysis of Health Surveys
LESSLER and KALSBEEK · Nonsampling Error in Surveys
LEVY and LEMESHOW · Sampling of Populations: Methods and Applications, *Third Edition*
LYBERG, BIEMER, COLLINS, de LEEUW, DIPPO, SCHWARZ, TREWIN (editors) · Survey Measurement and Process Quality
MAYNARD, HOUTKOOP-STEENSTRA, SCHAEFFER, VAN DER ZOUWEN · Standardization and Tacit Knowledge: Interaction and Practice in the Survey Interview

*Now available in a lower priced paperback edition in the Wiley Classics Library.

PORTER (editor) · Overcoming Survey Research Problems: New Directions for Institutional Research, No. 121
PRESSER, ROTHGEB, COUPER, LESSLER, MARTIN, MARTIN, and SINGER (editors) · Methods for Testing and Evaluating Survey Questionnaires
RAO · Small Area Estimation
REA and PARKER · Designing and Conducting Survey Research: A Comprehensive Guide, *Third Edition*
SÄRNDAL and LUNDSTRÖM · Estimation in Surveys with Nonresponse
SCHWARZ and SUDMAN (editors) · Answering Questions: Methodology for Determining Cognitive and Communicative Processes in Survey Research
SIRKEN, HERRMANN, SCHECHTER, SCHWARZ, TANUR, and TOURANGEAU (editors) · Cognition and Survey Research
SUDMAN, BRADBURN, and SCHWARZ · Thinking about Answers: The Application of Cognitive Processes to Survey Methodology
UMBACH (editor) · Survey Research Emerging Issues: New Directions for Institutional Research No. 127
VALLIANT, DORFMAN, and ROYALL · Finite Population Sampling and Inference: A Prediction Approach